Nuclear Reactor Physics and Operation

This book serves as a thorough reference for students, researchers, and professionals in nuclear engineering and reactor physics, offering a detailed exploration of the core principles behind nuclear reactor theory, neutron transport, neutronic analysis, and reactor core design and calculations. Each chapter includes at least one example to illustrate the topics covered, and the latter half focuses on key areas relevant to operating reactors – reactor kinetics/dynamics and in-core fuel management. Building on the foundational physics presented in the first half, it develops reactivity models using realistic reactor cross-section data and advanced analytic tools. This book is a valuable resource for engineers and scientists in the nuclear industry, as well as senior and graduate students in Nuclear Engineering, Mechanical Engineering, and Physics.

Key Features

- Offers an in-depth examination of reactor physics, encompassing neutron interactions, reactor kinetics, reactor dynamics, fuel cycles, and safety factors, to provide a comprehensive understanding of nuclear reactor operation and design
- Contains clear explanations of complex theories and mathematical formulations, accompanied by illustrative diagrams, figures, and examples to facilitate comprehension
- Features structured chapters with learning objectives, summaries, review questions, and problem sets at varying levels of difficulty to reinforce understanding and encourage active engagement with the material

Bahman Zohuri is currently an Adjunct Professor in Artificial Intelligence Science at Golden Gate University, San Francisco, California, USA, who runs his own consulting company and was previously a consultant at Sandia National Laboratory.

Seyed Kamal Mousavi Balgehshiri is currently working as a Researcher on the design and development of a Test Blanket Module for burning Minor Actinides and nuclear wastes and also producing Tritium in the Reversed Field Pinch and TOKAMAK configurations at University of Genova, Italy.

Guglielmo Lomonaco is an Associate Professor in the Department of Mechanical, Energy, Management, and Transportation Engineering (DIME) at University of Genova, Italy. He specializes in advanced nuclear systems and is a pioneer in developing hybrid fusion–fission systems, making significant contributions to the sustainable energy transition.

Nuclear Reactor Physics and Operation

Neutronic Analysis for Nuclear Reactor Design

Bahman Zohuri, Seyed Kamal Mousavi Balgehshiri,
and Guglielmo Lomonaco

CRC Press
Taylor & Francis Group
Boca Raton London New York

CRC Press is an imprint of the
Taylor & Francis Group, an **informa** business

Designed cover image: Shutterstock

First edition published 2026
by CRC Press
2385 NW Executive Center Drive, Suite 320, Boca Raton FL 33431

and by CRC Press
4 Park Square, Milton Park, Abingdon, Oxon, OX14 4RN

CRC Press is an imprint of Taylor & Francis Group, LLC

Library of Congress Cataloging-in-Publication Data
Names: Zohuri, Bahman, author. | Mousavi Balgehshiri, Seyed Kamal, author. | Lomonaco, Guglielmo, author.
Title: Nuclear reactor physics and operations : neutronic analysis for nuclear reactor design /
Bahman Zohuri, Seyed Kamal Mousavi Balgehshiri and Guglielmo Lomonaco.
Description: First edition. | Boca Raton, FL : CRC Press, 2025. | Includes bibliographical references and index. |
Identifiers: LCCN 2024054973 (print) | LCCN 2024054974 (ebook) | ISBN 9781032966229 (hardback) |
ISBN 9781032966243 (paperback) | ISBN 9781003590194 (ebook)
Subjects: LCSH: Nuclear reactors. | Nuclear physics.
Classification: LCC QC786.5.Z64 2025 (print) | LCC QC786.5 (ebook) | DDC 621.48/31–dc23/eng/20250403
LC record available at https://lccn.loc.gov/2024054973
LC ebook record available at https://lccn.loc.gov/2024054974

ISBN: 9781032966229 (hbk)
ISBN: 9781032966243 (pbk)
ISBN: 9781003590194 (ebk)

DOI: 10.1201/9781003590194

Typeset in Times
by Deanta Global Publishing Services, Chennai, India

This book is dedicated to my son, Sasha Zohuri, and grandchildren, Dariush, Donya, and Kian

Bahman Zohuri

This book is dedicated to my niece, Sogol, and my sister, Maryam

Seyed Kamal Mousavi Balgehshiri

This book is dedicated to my wife Felicita (the light of my life) and to my children Amalia and Ettore (the reason of my life)

Guglielmo Lomonaco

Contents

About the Authors

Bahman Zohuri currently works for Galaxy Advanced Engineering, Inc., a consulting firm that he started in 1991 when he left both the semiconductor and defense industries after many years working as a chief scientist. After graduating from the University of Illinois, USA, in the field of physics and applied mathematics, he went to the University of New Mexico, USA, where he studied nuclear engineering and mechanical engineering. He joined Westinghouse Electric Corporation, where he performed thermal hydraulic analysis and studied natural circulation in an inherent shutdown, heat removal system (ISHRS) in the core of a liquid metal fast breeder reactor (LMFBR) as a secondary fully inherent shutdown system for secondary loop heat exchange. All these designs were used in nuclear safety and reliability engineering for a self-actuated shutdown system. He designed a mercury heat pipe and electromagnetic pumps for large pool concepts of an LMFBR for heat rejection purposes for this reactor around 1978, when he received a patent for it. He was subsequently transferred to the defense division of Westinghouse, where he oversaw dynamic analysis and methods of launching and controlling MX missiles from canisters. The results were applied to MX launch seal performance and muzzle blast phenomena analysis (i.e., missile vibration and hydrodynamic shock formation). He was also involved in analytical calculations and computations in the study of non-linear ion waves in rarefying plasma. The results were applied to the propagation of so-called soliton waves and the resulting charge collector traces in the rarefaction characterization of the corona of laser-irradiated target pellets. As part of his graduate research work at Argonne National Laboratory, he performed computations and programming of multi-exchange integrals in surface physics and solid state physics. He earned various patents in areas such as diffusion processes and diffusion furnace design while working as a senior process engineer at various semiconductor companies, such as Intel Corp., Varian Medical Systems, and National Semiconductor Corporation. He later joined Lockheed Martin Missile and Aerospace Corporation as Senior Chief Scientist and oversaw the research and development (R&D) and the study of the vulnerability, survivability, and both radiation and laser hardening of different components of the Strategic Defense Initiative, known as Star Wars.

This included payloads (i.e., IR sensor) for the Defense Support Program, the Boost Surveillance and Tracking System, and Space Surveillance and Tracking Satellite against laser and nuclear threats. While at Lockheed Martin, he also performed analyses of laser beam characteristics and nuclear radiation interactions with materials, transient radiation effects in electronics, electromagnetic pulses, system-generated electromagnetic pulses, single-event upset, blast, thermomechanical, hardness assurance, maintenance, and device technology.

He spent several years as a consultant at Galaxy Advanced Engineering serving Sandia National Laboratories, where he supported the development of operational hazard assessments for the Air Force Safety Center in collaboration with other researchers and third parties. Ultimately, the results were included in Air Force Instructions issued specifically for directed energy weapons operational safety. He completed the first version of a comprehensive library of detailed laser tools for airborne lasers, advanced tactical lasers, tactical high-energy lasers, and mobile/tactical high-energy lasers, for example.

He also oversaw SDI computer programs, in connection with Battle Management C^3I and artificial intelligence, and autonomous systems. He is the author of several publications and holds several patents, such as for a laser-activated radioactive decay and results of a through-bulkhead initiator. He has published the following works: *Heat Pipe Design and Technology: A Practical Approach*; *Dimensional Analysis and Self-Similarity Methods for Engineering and Scientists*; *High Energy Laser (HEL): Tomorrow's Weapon in Directed Energy Weapons Volume I*; and recently a book on the subject of *Directed Energy Weapons and Physics of High Energy Laser.* He has published other books, including *Thermodynamics in Nuclear Power Plant Systems* and *Thermal-Hydraulic Analysis of Nuclear Reactors.*

Seyed Kamal Mousavi Balgehshiri is a PhD researcher focused on the design and development of a Test Blanket Module (TBM) for burning Minor Actinides (MA) and nuclear waste while producing Tritium in the RFP-TOKAMAK at the University of Genova. He also conducts strategic studies on nuclear energy programs, reviewing global progress in advanced nuclear reactor development, energy planning, and modeling. With different countries pursuing diverse strategies to ensure energy security, meet climate goals, and transition to net-zero emissions, he notes that today's short-term decisions will have long-term effects. Despite the changing landscape, energy security and effective strategy adoption for diversifying the energy portfolio remain key priorities for policymakers.

One of his main interests is strategic energy planning, particularly SWOT (Strengths, Weaknesses, Opportunities, and Threats) analysis in nuclear energy macroplanning. He believes that accurate situational analysis is critical for companies and organizations to develop strategies that achieve their goals.

He holds a BSc in chemical engineering from the University of Tabriz, Iran, where he focused on chemical analysis in laboratory settings. He later earned an MSc nuclear engineering, specializing in the nuclear fuel cycle and materials at Amirkabir University of Technology. During his MSc studies, he began researching advanced nuclear fuels, nuclear fuel cycles, and reactors.

He is also interested in research on nuclear fission reactors for space applications, particularly for powering equipment (kilowatt-class) in space. He recognizes the importance of nuclear energy for both spacecraft propulsion and providing power for equipment needed in space exploration, especially for missions beyond the solar system.

Another key area of his research focuses on the strategies of advanced nuclear countries in diversifying and securing nuclear fuel supplies based on the "reactor–fuel cycle" network. He emphasizes that choosing the best long-term strategy for a reliable nuclear fuel supply is essential for the development of nuclear energy. Additionally, his work includes Technology Readiness Assessment (TRA) for advanced nuclear fuels and Advanced Small Modular Reactors (ASMR). He highlights the importance of evaluating the Technology Readiness Level (TRL) and identifying Critical Technology Elements (CTEs) when introducing new technologies to determine a maturity plan.

Guglielmo Lomonaco is a distinguished academic and researcher with extensive expertise in nuclear engineering. Currently, he holds the position of Associate Professor at the Department of Mechanical, Energy, and Management Engineering (DIME) at the University of Genova, where he has been a faculty member since 2011. His academic achievements are marked by a PhD in Electrical and Thermal Energetics from the University of Pisa, Italy, complemented by an MSc in Nuclear Engineering and a BSc in Nuclear and Industrial Safety, both awarded with the highest honors. His postdoctoral research at the University of Pisa focused on Neutronic and Thermal-Fluid-Dynamic GCFR Analyses, during which he contributed to significant advancements in the field. His expertise has been further recognized through the prestigious National Scientific Qualification (ASN) for full professorship (prima fascia), a testament to his contributions and leadership in nuclear engineering.

He has played a crucial role in several high-profile European Union (EU) research projects over his career, progressing from Junior Expert to Senior Expert and Scientific Officer. His work spans multiple frameworks, including the High-Temperature Reactor Physics and Fuel Cycle Studies (HTR-N and HTR-N1), ReActor for Process Heat And Electricity (RaPHaEl), Gas Cooled Fast Reactor (GCFR), Plutonium and Minor Actinide Management in Thermal High-Temperature Gas-Cooled Reactors (PuMA), European Gas Cooled Fast Reactor (GoFastR), solving CHAllenges in Nuclear DAta (CHANDA), and Graduate and Executive Nuclear Training and Lifelong Education (GENTLE) Euratom projects, where he has made substantial contributions to the advancement of nuclear safety, reactor technology, and innovative nuclear systems. As the University of Genoa's

Unit Manager for the OCAPIE project, funded by Fondazione Compagnia San Paolo, and the PRIN UNIGE Research Unit Responsible for the Fusion–Fission Hybrid Pilot Reactor for Sustainable Energy Transition project, funded by the Italian Ministry of University. He has demonstrated his leadership in driving critical research initiatives.

In addition to his research, he is deeply committed to education. He has been co-responsible for the course Nuclear System for Energy Production at the University of Genova since 2010 and has led several other key courses, including Nuclear Plants and Advanced Nuclear Plants. His teaching extends to prominent institutions such as the University of Pisa, University of Florence, and Politecnico di Milano, where he has lectured on a wide range of topics, including nuclear power plants, complex plant design, and energy and sustainable development. His role in academia is further underscored by his involvement in doctoral programs, serving as a member of the Collegio dei Docenti for the PhD in Nuclear Engineering at the University of Pisa and for the PhD in Mechanical, Energy, and Management Engineering at the University of Genoa.

He is an influential figure in the scientific community, with his editorial contributions including guest editor roles for special issues in international journals such as *Science and Technology of Nuclear Installations*, *Sustainability*, "Journal of Nuclear Engineering", "Frontiers in Nuclear Engineering", and *Energies*. As the Editor-in-Chief of the *Global Journal of Energy Technology Research Updates*, he continued to shape the discourse in nuclear and energy technology. He is also a member of various international scientific associations and serves as a reviewer and editorial board member for leading journals in nuclear engineering.

His research interests are vast, encompassing nuclear energy and technology, innovative nuclear fuel cycles, neutronics, thermal hydraulics, and computational fluid dynamics (CFD). He has a particular focus on advanced nuclear systems, including High-Temperature Reactors (HTR), Lead-cooled Fast Reactors (LFR), Gas-cooled Fast Reactors (GFR), Molten Salt Reactors (MSR), Accelerator-Driven Systems (ADS), Small Modular Reactors (SMR), and nuclear space reactors. He is also a pioneer in the development of hybrid fusion–fission systems, contributing significantly to the sustainable energy transition.

With over 100 publications in international books, journals, and conference proceedings, his research is widely recognized and cited, reflecting his impact on the field. As of March 2024, his h-index stands at 14, according to the SCOPUS database, highlighting the breadth and influence of his work. His contributions to nuclear engineering, both as a researcher and educator, make him a leading figure in the pursuit of innovative and sustainable energy solutions, positioning him as a key contributor to the future of global energy technology.

Preface

These notes are presented for the understanding of reactor physics and neutronic analysis and how to design a nuclear power plant reactor for internal educational purposes. In this book, we provides a foundational overview of the key concepts involved in designing a nuclear reactor and the essential nuclear physics associated with it. While we don't go into every detail, it outlines the fundamental knowledge required for understanding reactor design based on fission principles.

As we know, nuclear energy is a source of sustainable energy that creates energy free from carbon monoxide and dioxide, and, for that matter, any other greenhouse gases and pollution, to meet the demand for growth in electricity driven by the global growth of population.

Nuclear energy is one of the most efficient and reliable forms of energy available to meet the global demand for sustainable energy. The new generation of reactors (i.e., Generation IV) guarantee the reliability, safety, and proliferation of nuclear power plants for energy production. More and more plants are being commissioned or retrofitted worldwide due to the demand for clean sources of energy to meet the increasing electricity demand.

It is impossible to overstate the importance of energy. Just thinking about where humanity would be without it is enough to demonstrate this point. As in the past, energy will play a vital role in shaping future industries, cities, nations, and the world. That is why we believe that energy is a critical factor in shaping future paradigms in any target entity or world, and that we must have a better understanding of the role that energy plays in the world today and in the future. In this book, we briefly look at the definition of energy and its different forms and review some data related to energy consumption in the world.

Furthermore, as a source of clean energy, we believe the future of nuclear power technology, despite the challenges it faces, is an important option for all the world to meet future energy needs without emitting carbon monoxide (CO) and dioxide (CO_2), or other greenhouse gases and other atmospheric pollutants. We also believe it is more efficient than other comparable sources of renewable energies, such as solar, wind, etc. Globally, renewables made up 29% of electricity generation in 2020, with much of this from hydropower (16.8%). A record amount of over 256 GW of renewable power capacity was added globally during 2020 and continues to be the focal point for climate and energy solutions. Electricity demand is a direct function of population growth globally and is also driven by the present century's extraordinary technological developments.

In this book, we provide nine chapters featuring high-level technical approaches to nuclear reactors with an introduction to neutron and neutronic analysis.

Chapter 1 provides a comprehensive introduction to neutrons and the principles of neutronics analysis. The neutron is a subatomic particle, designated by the symbol "n" or chemically presented as "n_0". This means it has a neutral charge (neither positive nor negative electrically), with a mass slightly greater than that of a proton. Protons and neutrons constitute the nuclei of atoms, and since protons and neutrons behave similarly within the nucleus, and each has a mass of approximately 1 atomic mass unit, they are both referred to as nucleons. Their properties and interactions are described by nuclear physics. Neutronics analysis focuses on the behavior of neutrons within all types of reactors, as discussed throughout the entire book.

In Chapter 2, Fick's laws of diffusion—formulated by Adolf Fick in 1855—are introduced to describe the process of diffusion. Fick's First Law is used to derive the Second Law, which is mathematically equivalent to the diffusion equation and can be used to determine the diffusion coefficient.

Chapter 3 covers nuclear reactor physics, a field that focuses on the applied study and engineering of chain reactions to achieve a controlled rate of fission in a nuclear reactor for energy production. Most nuclear reactors use a chain reaction to induce a controlled rate of nuclear fission in fissile material, releasing both energy and free neutrons. A reactor consists of an assembly of nuclear fuel (a reactor core), usually surrounded by a neutron moderator such as regular water, heavy water, graphite, or zirconium hydride, and fitted with mechanisms such as control rods that control the rate of the reaction. The physics of nuclear fission has several quirks that affect the design and behavior of nuclear reactors. This chapter presents a general overview of the physics of nuclear reactors and their behavior.

Chapter 4 discusses nuclear reactor kinetics, which addresses how a reactor responds to transient changes in neutron flux caused by deviations from the critical state due to reactivity insertions. Such situations arise during operational changes such as control rod motion, environmental changes such as a change in boron concentration, or due to accidental disturbances in the reactor steady state operation. In this chapter, we study the task of investigating the time-dependent behavior of nuclear reactors.

Furthermore, Chapter 5 introduces nuclear energy concepts, covering topics from neutron behavior to fast neutron spectrum. Global interest in fast reactors has been growing since their inception in 1960 because they can provide efficient, safe, and sustainable energy. Their closed fuel cycle can support long-term nuclear power development as part of the world's future energy mix and decrease the burden of nuclear waste. It is generally recognized that the long-term development of nuclear power as part of the world's future energy mix will require fast reactor technology with a closed fuel cycle. The fast neutron spectrum allows fast reactors to largely increase the energy yield from natural uranium as compared to thermal reactors. This high utilization of fuel can extend nuclear power programs for thousands of years and provide significant improvements in nuclear waste management. It is for these reasons that fast reactors have been under development for decades in several countries, primarily as breeders and, in recent years, also as high-level waste burners.

As part of reactor physics and reactor design, Chapter 6 covers the theory of two-group diffusion along with the analysis of bare and reflected reactors. In the two-group theory treatment, we consider a thermal energy group and combine all neutrons of higher energy into a fast energy group. The chapter covers one-region bare reactors, and the neutron moderation driven by Fermi Age Theory, which is also dealt with in previous chapters. However, most reactors are bare type ones and almost all of them have reflectors to reduce neutron leakage, as well as having additional regions of fertile materials. Here in this chapter, we introduce multiregional reactors through the group diffusion method, since Fermi Age Theory cannot be used to describe the slowing down of neutrons in this type of reactor system, due to mathematical computational difficulty, including the two-group method.

Additionally, this book touches on nuclear instrumentation and control, such as the gamma thermometer, which is discussed in Chapter 7 in the context of its use within a nuclear reactor core. Gamma thermometers are used in nuclear reactors for power monitoring and, potentially, for water coolant level monitoring—particularly from the perspectives of safety and probabilistic risk analysis.

Chapter 8 covers the topic of nuclear reactor statics, where we outline the typical approach to the solution of complex multidimensional problems, such as neutron balance equations. This involves developing suitable approximations along individual dimensions that finally allow an approximate solution to the full problem. The chapter specifically discusses dependencies along individual dimensions, treatment of pure space dependence, energy dependencies with the neutron spectrum, resonance cross-sections, approaches of angular dependencies, and multi-group diffusion theory.

Finally, Chapter 9 covers the topic of dynamics and control of nuclear reactors, in general. Nuclear reactor dynamics includes the theories and predictive methods that are at the heart of nuclear reactor safety. Time-dependent reactor behavior is explained in both mathematical and physical terms. This chapter also holistically explains the logic behind the working formulas and calculational methods for reactor transients and illustrates typical dynamic responses. The classical concept of point kinetics is developed in three steps, with a discussion of various possible solutions to the kinetics problem. In the previous chapter, we discussed statics and instability of reactors; however, in reality, nuclear reactors are extremely unstable, and this issue in itself evokes the matter of the nuclear reactor's dynamics. Transient analyses, in association with the integrity of the first and the second circuits of nuclear reactors, are a matter of reactor dynamics, which represents control and instrumentation, important factors in ensuring the safe and economic operation of nuclear power plants. System modeling, simulation, dynamic analysis, control system design, and controller implementation are the basic issues of dynamics and control of nuclear energy facilities. Especially, for the emerging advanced nuclear energy systems, dynamic system modeling and simulation are a prerequisite for the control system design.

As the authors of this book, we aim to shed light on nuclear technology as a promising and clear source of energy, particularly with the emergence of Generation IV reactors. Our goal is to help readers gain a solid understanding of nuclear reactors through straightforward technical language.

Bahman Zohuri
San Mateo, California
United States of America

Seyed Kamal Mousavi Balgehshiri
Genova, Liguria
Italy

Guglielmo Lomonaco
Genova, Liguria
Italy

Acknowledgments

I am indebted to the many people who aided me, encouraged me, and supported me beyond my expectations. Some are not around to see the results of their encouragement in the production of this book, yet I hope they know of my deepest appreciation. I especially want to thank many of my friends to whom I am deeply indebted. They have continuously supported me without hesitation and have always kept me going in the right direction.

Above all, I offer very special thanks to my late mother and father, and to my children, in particular, my son Sasha and my grandchildren. They have provided constant interest and encouragement, without which this book would not have been written. Their patience with my many absences from home and long hours in front of the computer to prepare the manuscript are especially appreciated.

Bahman Zohuri

Here, I appreciate and thank the esteemed and wise man Dr. Zohuri's compassionate guidance which showed me the right way to progress and gain knowledge. I dedicate this book to my late father, who was a symbol of effort and attempt and who was a great advisor during my education, and also to my mother who always encouraged me with compassion. I am overwhelmed in all humbleness and gratefulness to fully acknowledge all those who have helped me to put these ideas well above the level of simplicity and into something concrete.

Seyed Kamal Mousavi Balgehshiri

First and foremost, I want to express my deepest gratitude to my beloved wife, whose unwavering support and love have been my guiding light. To my wonderful children, my daughter and my son: your joy and curiosity inspire me every day. I am profoundly grateful to those who believed in me, including my past mentor Prof. Nicola Cerullo: your faith has been a cornerstone of my journey. Lastly, to my esteemed colleagues and collaborators: your support and hard work have made this project a reality. Thank you all for being part of this incredible journey.

Guglielmo Lomonaco

1 Introduction of Neutron and Neutronics

1.1 INTRODUCTION

The arrangement of electrons around an atom's hefty nucleus largely determines its chemical characteristics. The number of protons, or atomic number, determines the charge of the nucleus, which in turn determines the electron configuration. The number of neutrons is the neutron number. Neutrons do not affect the electron configuration, but the sum of atomic and neutron numbers is the mass of the nucleus.

Isotopes are atoms of a chemical element that differ solely in their neutron numbers. For instance, carbon, which has an atomic number of 6, has two isotopes: the less common carbon-13, with seven neutrons, and the more prevalent carbon-12, with six neutrons. Fluorine is one of the elements that only has one stable isotope in nature. Tin (Sn) has 10 stable isotopes, although other elements have several stable isotopes.

Both the atomic and neutron numbers affect an atomic nucleus's characteristics. The long-range electromagnetic force repels the protons in the nucleus because of their positive charge, while the considerably stronger yet short-range nuclear force holds the nucleons tightly together. Except for the single-proton hydrogen nucleus, all nuclei require neutrons to remain stable. Nuclear fission and fusion generate large amounts of neutrons. Through processes including fission, fusion, and neutron capture, the neutrons play a major role in the nucleosynthesis of chemical elements inside stars [1].

Nuclear power cannot be produced without neutrons. Neutrons were utilized to cause a wide variety of nuclear transmutations in the 10 years following James Chadwick's 1932 discovery of the particle. The fact that if a fission event generated neutrons, each of these neutrons may trigger more fission events – a cascade known as a nuclear chain reaction – was immediately apparent once nuclear fission was discovered in 1938. The first nuclear weapon (Trinity, 1945) and the first self-sustaining nuclear reactor (Chicago Pile-1, 1942, Figure 1.1) were the results of these discoveries and events.

Free neutrons are produced for use in irradiation and neutron scattering investigations by specialized neutron sources such as neutron generators (Figure 1.2), research reactors, and spallation sources. Additionally, emitted neutrons are employed in nuclear power to divide heavier atoms and create surplus heat, which is then used to generate electricity.

It should be noted that any device that emits neutrons, regardless of the method by which they are produced, is considered a neutron source. Physics, engineering, medicine, nuclear weapons, petroleum exploration, biology, chemistry, and nuclear power – all make use of neutron sources.

In a similar vein, nuclear reactors used solely as neutron sources are known as research reactors. In contrast to power reactors used for heat generation, electricity production, or maritime propulsion, they are also known as non-power reactors.

With a mean lifespan of around 15 minutes (Figure 1.3), a free neutron decays spontaneously to produce a proton, an electron, and an antineutrino. Depending on the dose, free neutrons can be a health danger since they indirectly produce ionizing radiations rather than directly ionizing atoms. Due to cosmic ray showers and the inherent radioactivity of spontaneously fissionable materials in the Earth's crust, there is a little natural "neutron background" flux of free neutrons on Earth. Refer to Figure 1.4 [1].

DOI: 10.1201/9781003590194-1

FIGURE 1.1 Chicago Pile-1 reactor core.

(Source: https://en.wikipedia.org/)

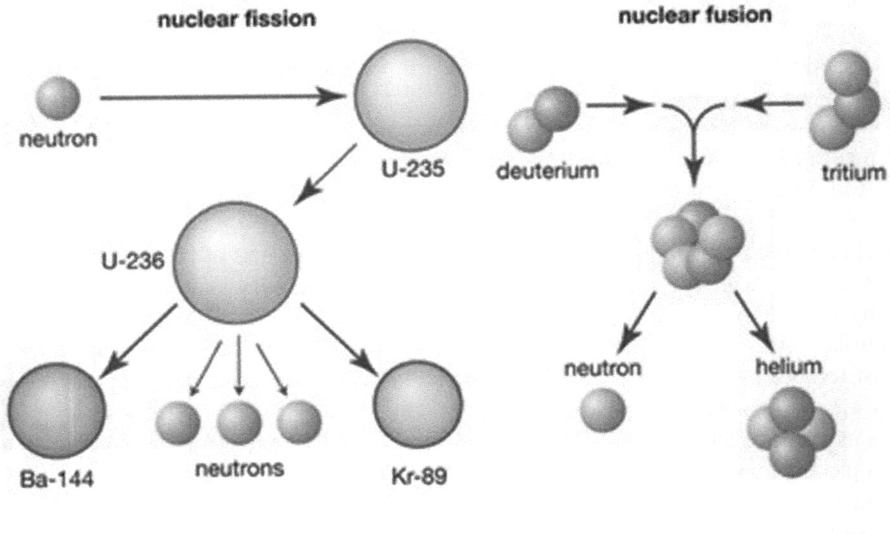

FIGURE 1.2 Neutron generation process.

(Source: Merriam-Webster Inc.)

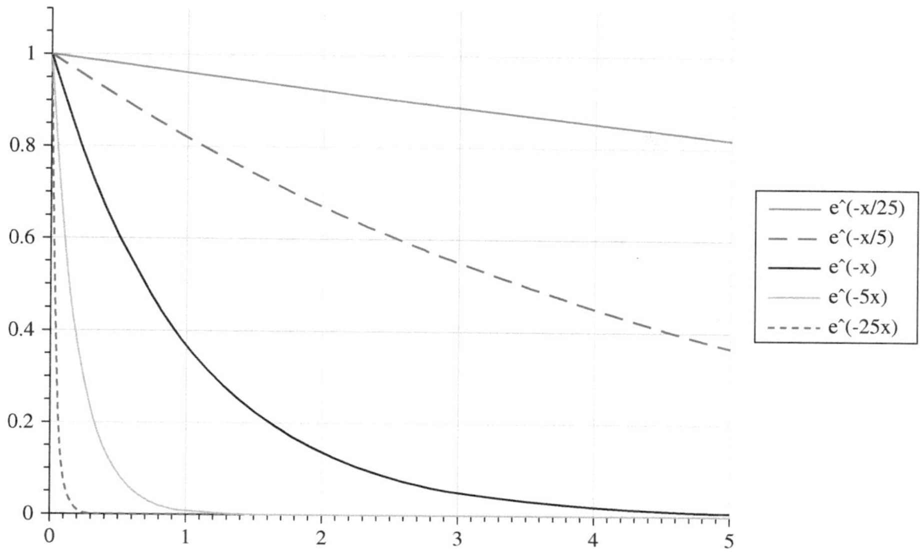

FIGURE 1.3 Illustration of antineutrino mean lifetime plots.

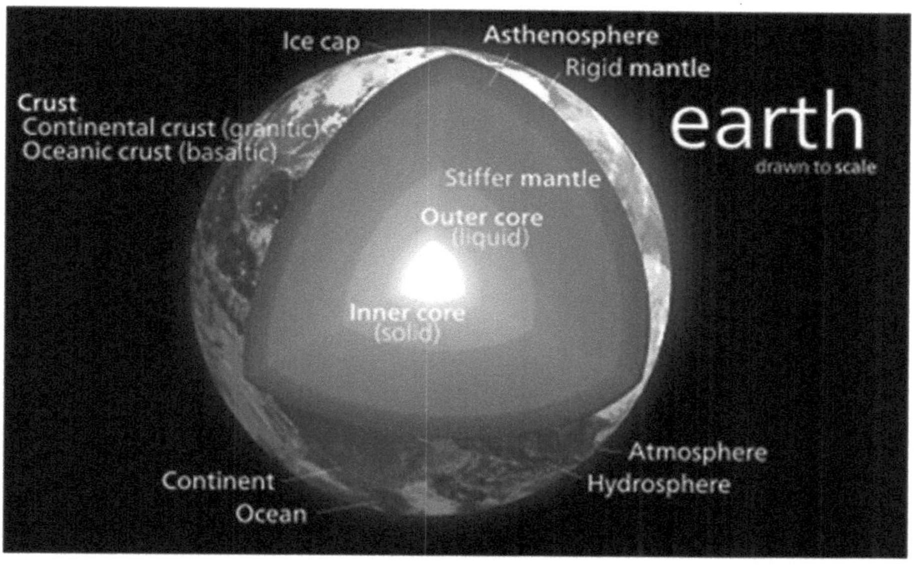

FIGURE 1.4 Earth's imaginary crust.

Bear in mind that Fission Neutron Energy is considered "fast neutron" type as illustrated in Figure 1.5, and these free neutrons have a kinetic energy level close to 1 MeV (1.6×10^{-13} J), hence a speed of ≈14,000 km/s (≈5% of the speed of light, i.e., 300×10^6 m/s). They are named fission energy or fast neutrons to distinguish them from lower-energy thermal neutrons and high-energy neutrons produced in cosmic showers or accelerators.

Nuclear fission is a reaction in which the nucleus of an atom splits into two or more smaller nuclei. The fission process often produces gamma photons and releases a very large amount of energy, even by the energetic standards of radioactive decay.

Fast neutrons are produced by nuclear processes such as nuclear fission, as previously explained. As noted earlier, fission-produced neutrons have a mean energy of 2 MeV (for ^{235}U fission neutrons),

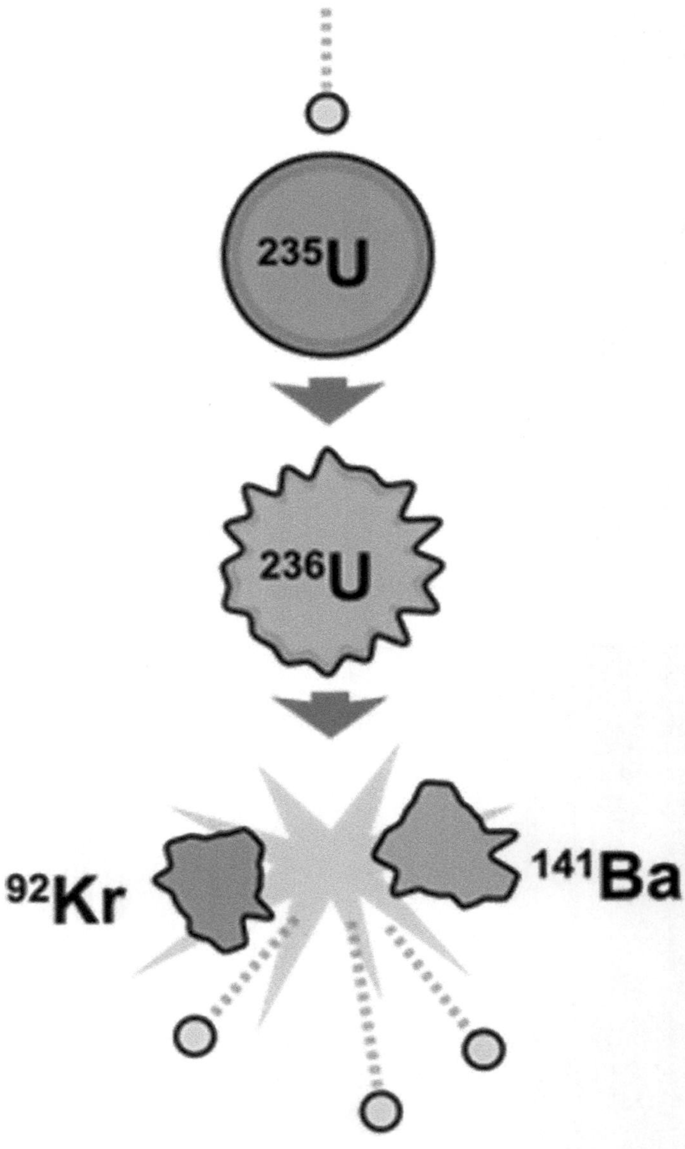

FIGURE 1.5 Nuclear fission reaction activity illustration.

a mode of just 0.75 MeV, and a Maxwell–Boltzmann distribution of kinetic energies (Figure 1.6) that spans from 0 to ≈14 MeV. Therefore, more than half of them do not meet the criteria for being fast, as shown in Figure 1.7 [1] (and therefore, have a distant possibility of commencing fission in fertile materials, such as ^{238}U and ^{232}Th).

Moderation is the process by which fast neutrons can be converted to thermal neutrons. Neutron moderators are used for this purpose. Neutrons are usually moderated in reactors using graphite, heavy water, or light water.

Similarly, the most energetic neutrons, with a kinetic energy of 14.1 MeV and a velocity of 17% of the speed of light, are created when two isotopes of hydrogen, D (Deuterium) + T (Tritium) → He (Helium) + n (Neutron), fuse.

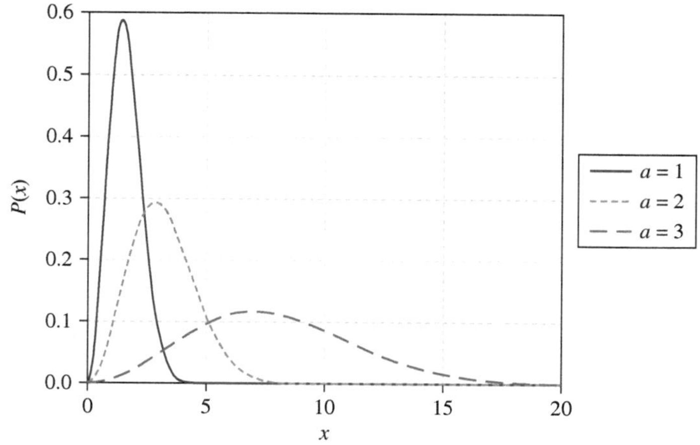

FIGURE 1.6 Maxwell–Boltzmann distribution of kinetic energies illustration.

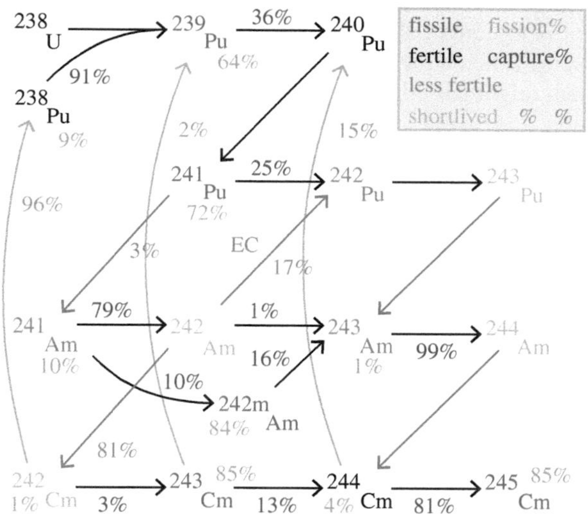

FIGURE 1.7 Fertile materials table.

(Source: https://en.wikipedia.org/wiki/Neutron)

Even when the kinetic energy of the deuterium and tritium nuclei is a thousandth of that of the 14.1 MeV that would be generated, D–T fusion is the simplest fusion process to start and may achieve near-peak rates.

Even non-fissile heavy nuclei can be effectively fissioned by 14.1 MeV neutrons, which have roughly 10 times the energy of fission neutrons. On average, these high-energy fissions generate more neutrons than fissions by lower-energy neutrons. For the transformation of transuranic waste, this makes D–T fusion neutron sources – like the planned tokamak power reactors – useful. By knocking neutrons loose from nuclei, 14.1 MeV neutrons can also create neutrons.

However, it is less likely that these extremely high-energy neutrons will only be collected without resulting in fission or spallation. For these reasons, D–T fusion's 14.1 MeV neutrons are widely used in nuclear weapon design to increase fission. Normally non-fissile materials, including depleted uranium (Uranium-238), can undergo fission when exposed to fusion neutrons; these materials have been utilized in thermonuclear weapon systems. Reactor-grade plutonium and other materials that are inappropriate or challenging to produce into primary fission bombs can also fission when exposed to fusion neutrons. Ordinary non-weapon grade materials become so problematic in several nuclear proliferation talks and treaties because of this physical feature.

Neutrons from other fusion processes are far less energetic. Half of the time, D–D fusion yields a 2.45-MeV neutron and helium-3, and the other half yields tritium and a proton but no neutron. No neutron is produced by D–^{3}He fusion [1].

1.2 NEUTRON INTERACTION WITH MATTER

Neutron assays can be carried out in different ways depending on how neutrons interact with materials. The amount of fissile material that the assay device can safely hold is limited by neutron interactions with the assay material, which also impacts how neutron results are interpreted. The foundation of a neutron detector is some neutron interaction with the detector's substance. Additionally, shielding materials must interact with neutrons in order to safeguard radiation.

This section provides fundamental information about neutron interactions that are important to nuclear material measurements.

The idea of cross-section is one attempt to explain the tiny interactions between neutrons and matter. Another attempt is to convey the likelihood that a specific interaction between a neutron and a nucleus would occur. When a lot of neutrons with the same energy are aimed at a thin layer of material, some of them might pass through without interaction, some might interact and change their energy and direction, and others might not come out of the sample at all.

For instance, the ratio of the number of neutrons that do not emerge to the number that were initially impacted on the layer represents the chance that a neutron will not emerge from a sample (i.e., that it will be absorbed or caught).

The chance of neutron absorption divided by the areal atom density, or the number of target atoms per unit area of the layer, is the cross-section for absorption: In light of the high number of atoms involved, the cross-section area must be a tiny fraction of a square centimeter. This kind of cross-section is known as microscopic cross-section and is denoted by the symbol σ because it represents the likelihood of neutron interaction with a single nucleus.

Examining the likelihood of a single neutron trying to penetrate a thin layer of material with area A and N target nuclei, each with cross-sectional area s, is another method for comprehending the idea of the microscopic cross-section. The chance of a single neutron striking one of these nuclei is approximately equal to the ratio of the total target area Ns to the area of layer A. The sum of all the nuclei's areas is then N_s. In other words, N_s/A or $(N/A)_s$, the areal target density times s, is the likelihood that a single neutron will collide with a nucleus.

However, cross-sections for neutron interactions are more complex than just the target's geometrical cross-sectional area at the atomic level. An effective cross-sectional area for the interaction might be created by substituting the s from the previous paragraph for this σ. The interaction's cross-section maintains the same area measurements as σ.

A heavy nucleus has a physical cross-sectional size of around 2×10^{-24} cm^2. Nonetheless, most nuclei have interaction cross-sections that fall between 10^{-27} and 10^{-21} cm^2. The barn, represented by the symbol (b), an alternative unit of area, is employed to prevent the discomfort of working with such small quantities. It is defined as 10^{-24} cm^2, meaning that a heavy nucleus's physical cross-sectional area is roughly equivalent to 2 barns. The interaction cross-sections of several neutrons vary from 0.001 to 1,000 barns.

As seen in Figure 1.8, which shows the many kinds of interactions and their cross-sections, a neutron can interact with the nucleus in a variety of ways. Different types of neutron interactions are depicted in this picture. The incoming and outgoing particles are indicated by the letters in parentheses, and the total cross-section σ_t indicates the likelihood that any interaction will occur.

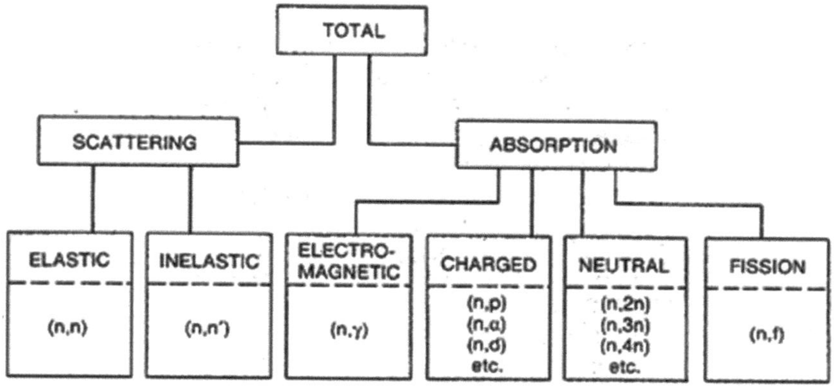

FIGURE 1.8 Neutronic reaction and notation type.

We can give a succinct indicator of an interaction of interest in collision using a straightforward conventional notation. For example, if a neutron n impinges on a target nucleus T, forming a resultant nucleus R and releasing an outgoing particle g, then this interaction is shown as $T(n,g)R$. The heavy nuclei are shown outside the parentheses. To denote a type of interaction without regard for the nuclei involved, only the portion in parentheses is shown, and an example of an (n,p) reaction is written in the form of ${}^{5}B(n,p){}^{5}$Be accordingly [15].

There are two main forms of interactions: scattering and absorption. The speed and direction of a neutron change when it is dispersed by a nucleus, but the nucleus retains the same amount of protons and neutrons as before the encounter. Radiation will eventually be released since the nucleus will have some recoil velocity and could be left in an excited state. Fission can be generated or a variety of radiations can be released when a neutron is absorbed by a nucleus [16].

In summary, the interaction of neutrons with matter can be presented as follows:

- Since neutrons are chargeless, they interact with nuclei or target nuclei through physical collisions.
- A neutron might scatter off or combine with the nucleus.
- When the neutron combines with a nucleus, some type of particle might be emitted (e.g., proton, alpha particles) and/or a "prompt" gamma rays.
- Neutrons, like other indirectly ionizing radiations (e.g., gamma rays), can travel substantial distances.
- The probability that a given type of reaction will occur depends on
  - The neutron energy
  - The identity of the target nuclide

In general, the kinds of reactions that can occur and their likelihood are determined by the neutron's kinetic energy, and they are categorized based on energy. On the specific categorization values, there is a disagreement.

Classification of neutrons according to their kinetic energies is as follows [17]:

- Cold (0–0.025 eV): At thermal equilibrium with extremely cold surroundings.
- Thermal (≈0.025 eV): Most applicable to thermal reactors.
- Epithermal (0.025–0.4 eV): More energy than thermal, utilized in some reactor designs.
- Cadmium (0.4–0.5 eV): It is substantially absorbed by ^{113}Cd.
- Epicadmium (0.5–1 eV): Above the cadmium cutoff, not absorbed by cadmium.
- Slow (1–10 eV): Just above the epicadmium range.

- Resonance (10–300 eV): High absorption peak, reactors should avoid prolonged exposure.
- Intermediate (300 eV–1 MeV): The transition between resonance and fast neutrons.
- Fast (1–20 MeV): Produced by fission, essential for fast reactors.
- Relativistic neutrons (>20 MeV): They have extremely high energy.

It is important to note that this detailed difference in neutron energy is unnecessary in reactor physics. In terms of reactor physics, neutrons may be loosely classified into three energy ranges.

- Thermal neutrons (0.025 eV–1 eV)
- Resonance neutrons (1 eV–1 keV)
- Fast neutrons (1 keV–10 MeV)

One way to describe neutrons is that they are born fast and will slow down as a result of scattering. This is known as moderation and is commonly used in the research of nuclear reactors. Ultimately, a target nucleus absorbs the neutrons, thus we may write:

$$\text{Fast neutron} \rightarrow \text{Thermal neutron} \rightarrow \text{Capture}$$

This can be illustrated in Figure 1.9, in general.

The primary categorization of nuclear reactors is based on the average energy of the neutrons that cause the majority of the fissions in the reactor core. From this perspective, nuclear reactors are separated into two categories [17].

- **Thermal reactors**: Almost all contemporary reactors rely on thermal neutrons to continue the chain reaction. Neutron moderators in reactors delay neutron fission until they reach thermal equilibrium with the atoms ($E < 1$ eV).
- **Fast neutron reactors**: Fast reactors have no neutron moderator and need less-moderating main coolants due to the utilization of fast neutrons ($E > 1$ keV) to generate fission in the fuel.

The basic distinction between these two reactor designs is their neutron cross-sections, which are very energy-reliant. In fast reactors, the capture-to-fission ratio is significantly reduced. Furthermore, the quantity of neutrons generated by each fission event differs, with fast reactors producing more neutrons than thermal reactors. These considerable disparities are mostly caused by changes in

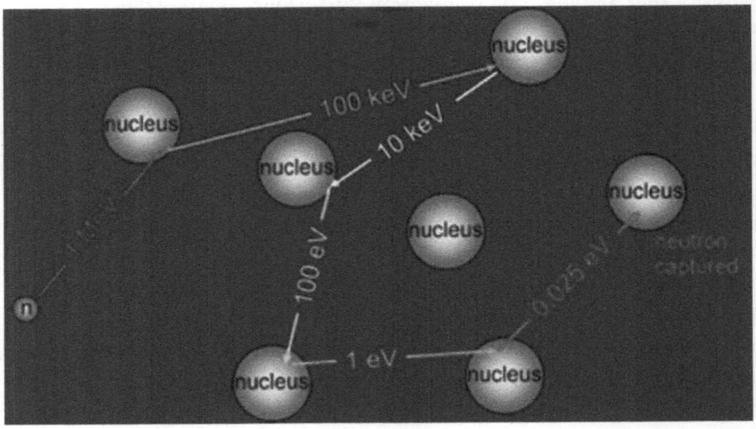

FIGURE 1.9 Typical fate of neutron illustration.

neutron fluxes. As a result, a thorough understanding of the neutron energy distribution within a reactor core is required.

Resonance absorption in ^{232}Th and ^{238}U is a key phenomenon in HTR (high temperature reactor) reactor physics. The quantity of absorption in these fertile materials influences the core reactivity and conversion factor, which affect the reactor's overall burnup behavior. Furthermore, the Doppler effect in these resonances is critical in defining the temperature coefficient of the core and hence the reactor's dynamic behavior and safety features. The transport theory approaches we have seen, so far, are theoretically ideal for treating resonance absorption in any reactor.

As previously stated, neutron cross-sections may be recognized as one means of measuring the way they interact with matter, and they are represented by a symbol that specifies the likelihood of the contact, which is dependent on both the nuclide and the neutron energy. The cross-sectional unit is the barn, which measures 10^{-24} cm^2.

Per the illustration in Figure 1.8, and the typical interaction of neutrons with matter either "Elastic" or "Inelastic", we can create these interactions as depicted in Figure 1.10.

Thus, from a microscopic cross-section point of view, the total cross-section σ_T for a given nuclide is the sum of the individual cross-sections for that nuclide, which is defined as

$$\sigma_T = \sigma_{el} + \sigma_{inel} + \sigma_p + \sigma_\alpha + \sigma_{2n} + \sigma_{np} + \sigma_\gamma + \sigma_f \qquad \text{Eq. 1.1}$$

This is what we have learned, so far, for per atom of a given nuclide. Now, we can pay attention to neutron macroscopic cross-section and its total is designated by Σ_T, that is the sum of all the atoms of a given nuclide in a cubic centimeter, written as cm^{-1} (i.e., cm^2/cm^3).

$$\Sigma_T = N\sigma_T = (6.02\times 10^{23})(\sigma_T)(\rho/A) \qquad \text{Eq. 1.2}$$

where

Σ_T = Total macroscopic cross-section (cm^{-1})
N = Number of atoms of the nuclide per cm^3
σ_T = Microscopic cross-section (cm^2)
ρ = Density (g/cm^3)
A = Isotopic mass (g/mole)

The microscopic removal cross-section (σ_R) and the macroscopic removal cross-section (Σ_R) are, sometimes, used in neutron-shielding calculations.

Scattering: Elastic	$^{A}M(n,n)^{A}M$	$\sigma_{el}, \sigma_s, \sigma_{n,n}$
Inelastic	$^{A}M(n,n')^{A}M$	$\sigma_{inl}, \sigma_i, \sigma_{n,n'}$
Proton	$^{A}M(n,p)^{A}N$	σ_p
Alpha	$^{A}M(n,\alpha)^{A-3}N$	σ_α
Neutron	$^{A}M(n,2n)^{A-1}M$	σ_{2n}
Neutron-proton	$^{A}M(n,np)^{A-1}N$	σ_{np}
Capture	$^{A}M(n,\gamma)^{A+1}M$	σ_γ
Fission	$^{A}M(n,fp)$	σ_f

FIGURE 1.10 Types of interactions' demonstrations.

The removal cross-section is approximately two-thirds to three-fourths of the total cross-section and in neutron-shielding calculations, we sometimes use the mass attenuation coefficient symbolized as Σ_R/ρ for this purpose.

Furthermore, according to Schaeffer (1973), the mass attenuation coefficient (Σ_R/ρ) for fast neutrons can be approximated with your choice with one of the following [21]:

$$\begin{aligned} \Sigma_R/\rho &= 0.19Z^{-0.743} \quad cm^2/g \quad (Z \le 8) \\ &= 0.125Z^{-0.565} \quad cm^2/g \quad (Z > 8) \\ \Sigma_R/\rho &= 0.206A^{-1/3}Z^{-0.294} \sim 0.206(AZ)^{-1/3} \\ \Sigma_R/\rho &= 0.206A^{-0.58} \end{aligned} \qquad \text{Eq. 1.3}$$

1.2.1 Elastic Scattering

Figure 1.11 illustrates elastic scattering, defined as a billiard ball-type collision in which the target nucleus kinetic energy is equal to 0.4 MeV and the total kinetic energy is same before and after the impact.

For nearly all nuclides and neutron energies, elastic scattering is the most common interaction. When a target nucleus has the same mass as the neutron, most energy may be transmitted from the neutron to the target nucleus. As a result, moderators including water, paraffin, plastic, and graphite delay neutrons by elastic scattering; the lower the target's atomic mass number, the more effective the moderator.

A typical elastic scattering curve as seen in Figure 1.12 shows the dependence of the ^{1}H target elastic scattering cross-section on neutron energy.

1.2.2 Inelastic Scattering

Inelastic scattering is a type of scattering collision where kinetic energy is not conserved.

The total kinetic energy was greater before the collision than after the scattering collision and is graphically depicted in Figure 1.13. The kinetic energy of excited target nucleus was 0.3 MeV.

When inelastic scattering occurs, the target nucleus receives the remaining energy as excitation energy; as a result, when the dispersed nucleus de-excites, it releases one or more gamma rays.

High nuclei and high-energy neutrons are most likely to be involved in inelastic scattering, which is uncommon. It is occasionally employed as a moderating device for neutrons with extremely high energies. As seen in Figure 1.14, the curve of such a collision illustrates how ^{56}Fe inelastic scattering

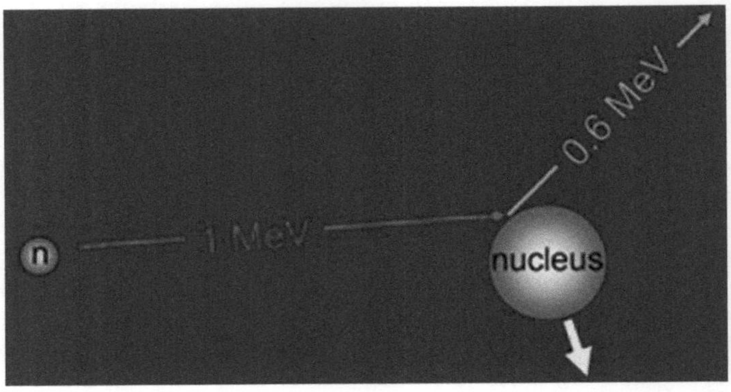

FIGURE 1.11 Neutron elastic scattering behavior.

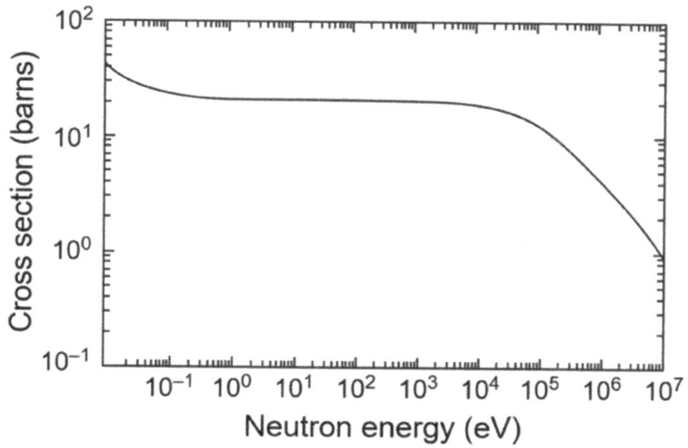

FIGURE 1.12 A typical neutron elastic scattering curve.

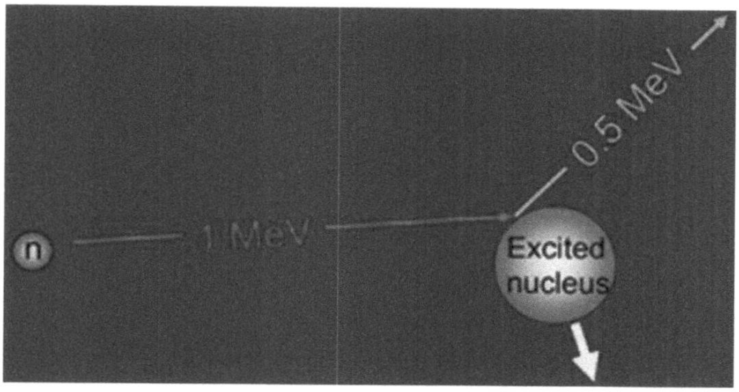

FIGURE 1.13 Neutron inelastic scattering behavior.

cross-section varies with neutron energy. Remember that the minimal energy to ignite the target nucleus is needed because inelastic scattering does not happen below a particular threshold.

There are a lot more details involved in neutron scattering and collision with matter that are beyond the scope of this book and the reader(s) should refer to books on the subject of particle physics or nuclear physics for further information.

1.3 NEUTRON SLOWING DOWN

So far, we have seen all we need to know about neutron characteristics in the above sections, it is finally time to move on to the slowing-down equation and the idea behind the neutron slowing-down scenario. We can recall the steady state transport equation for a homogeneous and infinite system (no time, space, or angle dependence, only energy).

$$\Sigma_t(E)Q(E) = \int_0^{\infty} dE'\Sigma_s(E' \rightarrow E)Q(E') + S(E) \quad \text{Eq. 1.4}$$

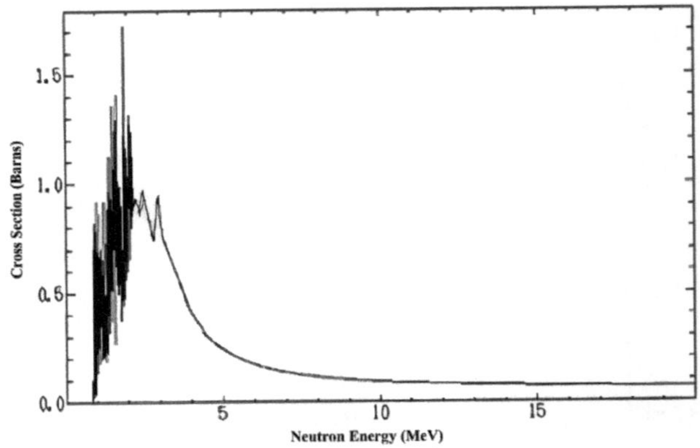

FIGURE 1.14 A typical neutron inelastic scattering curve.

For the energy, which is the only variable remaining, we obtain an integral equation, and we succeed! In fact, the slowing-down equation is just the Boltzmann equation reduced to the case of only one variable, the neutron velocity v – or a related variable, and the kinetic energy or lethargy u – considered in the slowing down domain.

We define two quantities, the arrival density $\rho(E)$ and the slowing down current $q(E)$.

1.3.1 The Arrival Density

ρdE is the number of neutrons, for unit time and volume, that arrive in the energy interval between E and $E + dE$.

$$\rho(E) = \int_E^\infty dE'\Sigma_s(E')\cdot P(E' \to E)\cdot \phi(E') \qquad \text{Eq. 1.5}$$

with $P(E' \to E)dE$ denoting the probability that a neutron scattered at energy E' will be transferred in the energy dE between and $E + dE$.

We often write

$$\Sigma_s(E' \to E) = \Sigma_s(u')\cdot P(u' \to u) \qquad \text{Eq. 1.6}$$

Thus, we will write

$$\rho(E) = \int_E^\infty dE'\Sigma_s(E' \to E)\cdot \phi(E') \qquad \text{Eq. 1.7}$$

We will often consider the lethargy as a variable, and consequently, write the arrival density as follows:

$$\rho(u) = \int_{-\infty}^u \Sigma_s(u' \to u)\cdot \phi(u')du' \qquad \text{Eq. 1.8}$$

1.3.2 The Slowing-Down Current

$q(E)$ is the number of neutrons, for unit time and volume, scattered from an energy $E' > E$ to an energy $E'' < E$ as illustrated in Figure 1.15.

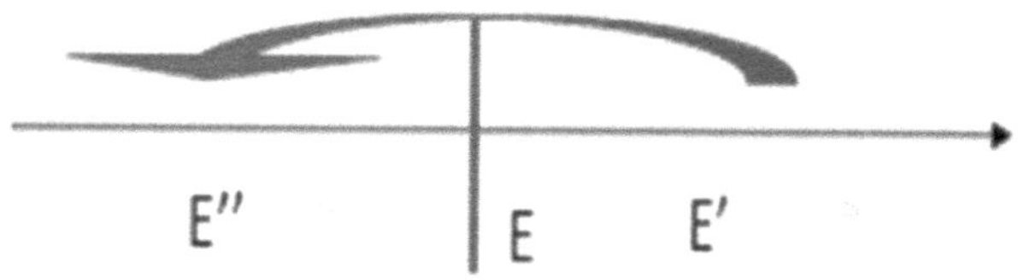

FIGURE 1.15 Current slowing-down illustration.

$$q(E) = \int_{-\infty}^{u} \int_{u}^{\infty} \Sigma_s(u' \rightarrow u'')\phi(u')du'du'' \qquad \text{Eq. 1.9}$$

1.4 THE FIRST FORM OF THE SLOWING-DOWN EQUATION

The first form of the slowing-down equation involves the "arrival density" described as

$$\rho(u) = \int_{-\infty}^{u} \Sigma_s(u' \rightarrow u)\phi(u')du' \qquad \text{Eq. 1.10}$$

The neutrons produced directly at the lethargy (inside) of interest by emission from the source must be added to the neutrons that arrive at lethargy u (within du) following a scattering event.

Let us write this second density as S.

There is just one potential consequence for the neutrons entering or produced at the point u (within du) in the current situation, and that is a collision (whether it results in absorption or scattering). Consequently, the total collision rate $\Sigma_t \phi du$ where t denotes total is equal to the sum of the two densities (arrival density and emission density).

$$Sdu + \rho du = \Sigma_t \phi du \qquad \text{Eq. 1.11}$$

Therefore, one can write

$$S(u) + \rho(u) = S(u) + \int_{-\infty}^{u} \Sigma_s(u' \rightarrow u)\phi(u')du' = \Sigma(u)\phi(u) \qquad \text{Eq. 1.12}$$

This kind of integral equation can only be resolved numerically unless you are really fortunate. The majority of calculation codes employ the first form.

1.5 THE SECOND FORM OF THE SLOWING-DOWN EQUATION

A second count can be undertaken to determine transfer characteristics. We can count the neutrons that leap over a lethargy u rather than the neutrons that sink to a lethargy u (inside du) [18].

This is known as the slowing-down current, written as $q(u)$ and defined above.

Let us imagine, the following slow-down current obeys the equation, then

$$q(u + du) - q(u) = \frac{dq(u)}{u} du = S(u)du - \Sigma_a(u)\phi(u)du \qquad \text{Eq. 1.13}$$

We can simplify by du that

$$\frac{dq(u)}{du} = S(u) - \Sigma_a(u)\phi(u) \qquad \text{Eq. 1.14}$$

or, with respect to the energy

$$\frac{dq}{dE} = S(E) - \Sigma_a \phi(E) p \qquad \text{Eq. 1.15}$$

When combined with the equation given in the definition, this equation becomes the second version of the slowing-down equation.

The slowing-down equation is theoretically equivalent in both variants. In fact, we will ultimately get the following equation if we derive the equation with regard to

$$\frac{dq(u)}{du} = \Sigma_s(u)\phi(u) - \rho(u) \qquad \text{Eq. 1.16}$$

By replacing this with the second form of the slowing-down equation, we re-obtain the first form. Now, let us consider elastic scattering only. In the lethargy variable, we have for $u - \varepsilon < u' < u$, therefore,

$$P(u' \to u) = \frac{e^{-(u-u')}}{1-\alpha}$$
$$\Sigma_s(u' \to u) = \frac{\Sigma_s(u')\, e^{-(u-u')}}{1-\alpha} \qquad \text{Eq. 1.17}$$

Otherwise

$$P(u' \to u) = 0$$
$$\Sigma_s(u' \to u) = 0 \qquad \text{Eq. 1.18}$$

The slowing-down equation was written in two forms, and the principle was explained. In summary, we put them in the following form:

1. Slowing-down in hydrogen without absorption
   a. $A = 1$
   b. $\Sigma_a = 0$
2. Slowing-down in hydrogen with absorption
   a. $A = 1$
   b. $\Sigma_a > 0$
3. Slowing-down in a heavier nucleus without absorption
   a. $A > 1$
   b. $\Sigma_a = 0$
4. Slowing-down in a heavier nucleus with absorption
   a. $A > 1$
   b. $\Sigma_a > 0$

Let us try a few examples for clarification on this matter, as follows:

Example 1
Case

$$\begin{cases} A = 1 \\ \Sigma_a = 0 \end{cases}$$

In this first example, only hydrogen is considered, and there is no absorption. For hydrogen, $\alpha = 0$. Consequently, the slowing-down equation with a monoenergetic source becomes

$$\Sigma_s(E)\phi(E) = \int_E^{\frac{E}{\alpha}} dE'\left(\frac{\Sigma_s(E')}{(1-\alpha)E'}\right)\phi(E') + S(E)$$

$$\Sigma_s(E)\phi(E) = \int_E^{\frac{E}{\alpha}} dE'\left(\frac{\Sigma_s(E')}{(1-\alpha)E'}\right)\phi(E') + S_0\delta(E-E_0)$$

The collision density $F(E)$ is defined as

$$F(E) = \Sigma_s(E)\phi(E)$$

So, we obtain the following equation.

$$F(E') = \int dE'\left(\frac{F(E')}{E'}\right) + S_0\delta(E-E_0)$$

The delta function term is eliminated by separating the source neutrons from the collided flux.

$$F(E) = F_c(E) + \underbrace{C\delta(E-E_0)}_{\text{neutron with 0 collision}}$$

Therefore, the equation becomes

$$F_c(E) + C\delta(E-E_0) = \int_E^{E_0} dE'\left(\frac{F_c(E')}{E'}\right) + \frac{C}{E_0} + S_0\delta(E-E_0)$$

This implies that $C = S_0$, then

$$F_c(E) = \underbrace{\int_E^{E_0} dE'\left(\frac{F_c(E')}{E'}\right)}_{\substack{\text{neutrons with}\\ \text{several collisions}}} + \underbrace{\frac{S_0}{E_0}}_{\substack{\text{neutrons with}\\ \text{1 collision}}}$$

To transform the integral equation into a differential equation, the derivative of the preceding equation is taken according to

$$\frac{d}{dx}\int_{a(x)}^{b(x)} f(x,x')dx' = f(x,b)\frac{db}{dx} - f(x,a)\left(\frac{da}{dx}\right) + \int_{a(x)}^{b(x)}\left(\frac{\partial f(x,x')}{\partial x}\right)dx'$$

The result is

$$\frac{dF_c(E)}{dE} = \frac{d}{dE}\left[\int_E^{E_0}\left(\frac{F_c(E')}{E'}\right)\right] = -\frac{F_c(E)}{E}$$

The solution is the below equation with a constant D of integration.

$$F_c(E) = \frac{D}{E}$$

At $E = E_0$, the initial condition gives

$$F_c(E_0) = \frac{S_0}{E_0}$$

Consequently,

$$F_c(E) = \frac{S_0}{E}$$

Eventually, we get

$$F(E) = \frac{S_0}{E} + S_0\delta(E - E_0)$$

The solution then becomes

$$\phi(E) = \frac{S_0}{\Sigma_s(E)E} + \left(\frac{S_0}{\Sigma_s(E)}\right)\delta(E - E_0)$$

One can notice that the flux is proportional to $1/E$. This implies that the flux diverges when $E \to 0$, which occurs due to the non-absorption and infinite-system assumptions.

In the lethargy variable, we can write

$$\phi(E)dE = -\phi(u)du$$
$$\phi(u) = E\phi(E)$$

Hence, we have

$$\phi(u) = \frac{S_0}{\Sigma_s(u)} \simeq \text{constant}$$

The solution is, thus, constant in lethargy. This shows that the lethargy is the "right" variable to use in the slowing-down problems.

From the previous solution, we can compute the slowing-down current.

$$q(E) = \int_0^E dE'' \int_E^\infty \frac{dE'}{E'} \Sigma_s(E' \to E'')\phi(E') = S_0$$

In an infinite, non-absorbing medium, and for a given energy E, all source neutrons will sooner or later be slowed-down past E. See Figure 1.15.

For a distributed (in energy) source $S(E)$, we can use the previous solution as a Green function.

$$\phi(E) = \int_E^\infty dE\left(\frac{S(E)}{\Sigma_s(E}\right)\left[\frac{1}{E} + \delta(E - E_0)\right] = \frac{S(E)}{\Sigma_s(E)} + \int_E^\infty \left(\frac{dE}{E}\right)\left(\frac{S(E)}{\Sigma_s(E)}\right)$$

Example 2

Case

$$\begin{cases} A = 1 \\ \Sigma_a > 0 \end{cases}$$

In this example, a medium of hydrogen is assumed with $\Sigma_a > 0$. Actually, hydrogen does not absorb neutrons much. Therefore, this model represents a combination of hydrogen and a "infinite" mass-absorbing element, which does not slow neutrons.

In this case, the slowing-down equation becomes

$$\left[\Sigma_a(E)+\Sigma_s(E)\right]\phi(E)=\int_E^{E_0}\frac{dE'}{E'}\Sigma_s(E')\phi(E')+S_0\delta(E-E_0)$$

We introduce the collision density that we decompose as the sum of an uncollided component and a collided component F_0. Therefore, we get

$$F(E)=\Sigma_t(E)\phi(E)=F_c(E)+C\delta(E-E_0)$$

where F_c satisfies the equation,

$$F_c(E)=\int_E^{E_0}\left(\frac{dE'}{E'}\right)\left(\frac{\Sigma_s(E')}{\Sigma_t(E')}\right)F_c(E')+\left(\frac{\Sigma_s(E_0)}{\Sigma_t(E_0)}\right)\left(\frac{S_0}{E_0}\right)$$

After differentiation, we obtain the following differential equation

$$\frac{dF_c(E)}{dE}=-\left[\frac{\Sigma_s(E)}{E\Sigma_t(E)}\right]F_c(E)$$

The solution is

$$F_c(E)=\underbrace{\left(\frac{\Sigma_s(E_0)}{\Sigma_t(E_0)}\right)\left(\frac{S_0}{E}\right)}\exp\left[-\int_E^{E_0}\left(\frac{dE'}{E'}\right)\left(\frac{\Sigma_a(E')}{\Sigma_t(E')}\right)\right]$$

↓
Probability that the first collision at energy E_0 is a diffusion

For the flux, the solution is

$$\phi(E)=\left(\frac{S_0}{E\Sigma_t(E)}\right)\left(\frac{\Sigma_s(E_0)}{\Sigma_t(E_0)}\right)\exp\left[-\int_E^{E_0}\left(\frac{dE'}{E'}\right)\left(\frac{\Sigma_a(E')}{\Sigma_t(E')}\right)\right]$$

The flux is proportional to $1/E$ out of the resonances. It could be identified as the resonance escape probability, that is, the probability of being slowed down from energy E_0 to energy E, without being captured by the resonances.

$$p(E)=\exp\left[-\int_E^{E_0}\left(\frac{dE'}{E'}\right)\left(\frac{\Sigma_a(E')}{\Sigma_t(E')}\right)\right]$$

Example 3

Case

$$\begin{cases}A>1\\ \Sigma_a=0\end{cases}$$

For $A > 1$, we have $\alpha > 0$, and the slowing-down equation becomes

$$\Sigma_s(E)\phi(E) = \int_E^{\frac{E}{\alpha}} \frac{dE'\Sigma_s(E')}{(1-\alpha)E'}\phi_c(E') + \frac{S_0}{(1-\alpha)E_0}$$

After differentiating with respect to variable E and using collision density F_c once more, we get

$$\frac{dF_c}{dE} = \frac{1}{(1-\alpha)E}\left[F_c\left(\frac{E}{a}\right) - F_c(E)\right]$$

This is not an ordinary differential equation, but rather a differential-difference equation. It is considered a monoenergetic source at E_0. The uncollided neutrons are at E_0 , the once-collided neutrons are between E_0 and αE_0, and the twice-collided are between E_0 and $\alpha^2 E_0$.

This means that, concerning lethargy, the following explanation or information applies (Figure 1.16).

We need to solve separately in each interval.

For $A > 1$, the solution exhibits the Placzek transients using Figure 1.16, therefore, we can write

$F(u)$ is discontinuous at $u = u_0 + \varepsilon$

$\frac{dF}{du}$ is discontinuous at $u = u_0 + 2\varepsilon$

$\frac{d^2F}{du^2}$ is discontinuous at $u = u_0 + 3\varepsilon$

The asymptotic solution (valid for $u >\sim u_0 + 3\varepsilon$) is given by

$$\phi(u) \xrightarrow{u \gg u_0} \frac{S_0}{\xi \Sigma_s(u)}$$

In terms of energy,

$$\phi(E) \xrightarrow{E \ll E_0} \frac{S_0}{\xi E \Sigma_s(E)}$$

When we have there defined ξ, the average slowing-down power is obtained from the following equation. This is on average weighted by the corresponding scattering cross-sections.

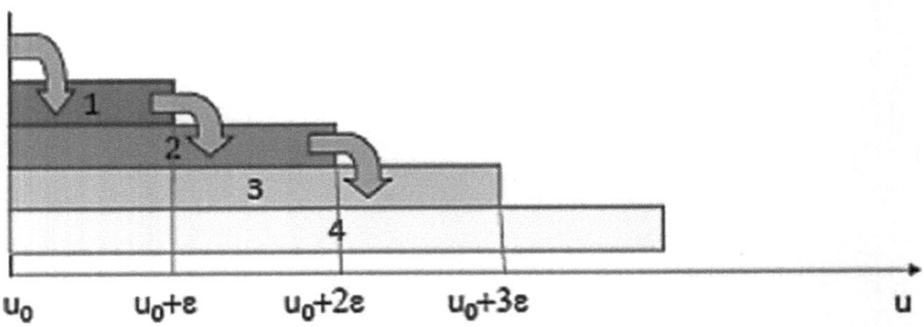

FIGURE 1.16 Placzek transients lethargy energy.

$$\xi(E) = \frac{\Sigma_i \left[\xi_i \sum_s^{(i)}(E) \right]}{\Sigma_i \left[\sum_s^{(i)}(E) \right]}$$

Example 4

Case

$$\begin{cases} A > 1 \\ \Sigma_a > 0 \end{cases}$$

The slowing-down equation with elastic and isotropic scattering is presented as

$$\left[\Sigma_a(E) + \Sigma_s(E)\right]\phi(E) = \int_E^{\frac{E}{\alpha}} dE' \left(\frac{\Sigma_s(E')}{(1-\alpha)E'} \right) \phi(E') + S(E)$$

Using the collision density $F(E)$ once again, we obtain

$$F(E) = \int_E^{\frac{E}{\alpha}} dE' \left(\frac{\Sigma_s(E')}{\Sigma_t(E')} \right) \left(\frac{F(E')}{(1-\alpha)E'} \right) + S(E)$$

The bad news, in this case, is that there is no simple solution. To solve that, we will need to wait until the next lecture on the resonant absorption.

1.6 FERMI AGE THEORY

The free neutrons exit the neutron source at various energy levels ranging from their neutron birth energy, which is the energy imparted to them in the reaction to birth energy plus acceleration energy (if the neutrons are traveling forward) or birth energy minus acceleration energy, if the neutrons bounce backward, as illustrated in Figure 1.17.

In addition, a free neutron has a mean lifespan of around 15 minutes and decays spontaneously into a proton, an electron, and an antineutrino. Depending on the dose, free neutrons can be a health hazard since they indirectly produce ionizing radiations rather than directly ionizing atoms.

1.6.1 The Slowing Density

The slowing density q is defined as the number of neutrons per unit volume per unit time slowing down beyond the energy E or its equivalent lethargy u as indicated for an infinite media as represented in Figure 1.18 [19].

Figure 1.9 shows E_0 as fission energy and E_{th} as thermal energy. All neutrons pass through each energy threshold, and the slowing down of density exceeding any energy E is a constant as follows:

$$q = q_0 \qquad \text{Eq. 1.19}$$

This is also the number of neutrons produced per unit volume per unit time at fission energy E_0 level. This is true if there is no absorption when slowing down, or if the absorption is minimal or inconsequential.

1.6.2 The Average Logarithmic Energy Decrement

For any scattering material with mass number A, the mean logarithmic reduction in energy between energies E_1 and E_2 per incident is independent of energy and is defined as

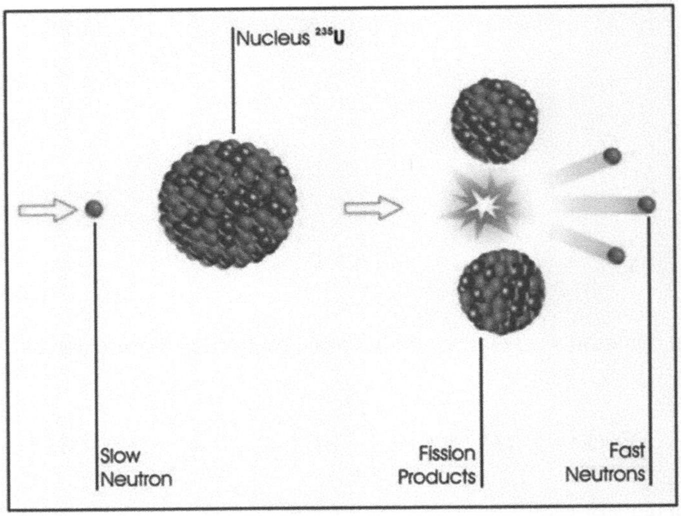

FIGURE 1.17 Slow neutron drives fission product.

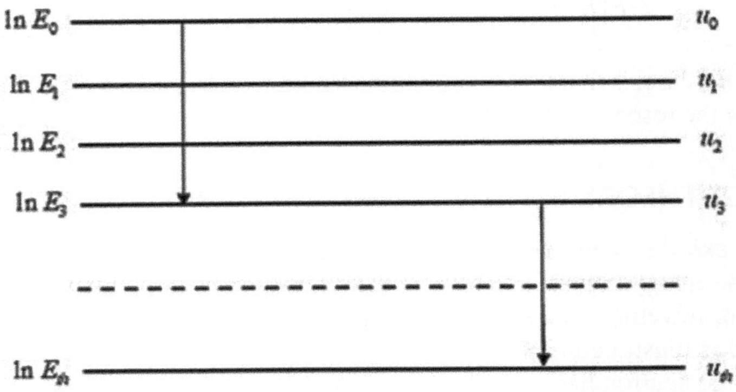

FIGURE 1.18 Slowing down of neutrons in an infinite medium.

$$\xi = \overline{\ln E_1 - \ln E_2} = \overline{\ln\left(\frac{E_1}{E_2}\right)} \quad \text{Eq. 1.20}$$

It also represents the increase in "lethargy" at every contact. The term lethargy refers to sloth. A decrease in energy is equivalent to an increase in lethargy. If $E_0$is the beginning energy, which is typically the average fission energy at 2 MeV, then the energy after slowing down to thermal energy is

$$E_{th} = E_0 e^{-u} \quad \text{Eq. 1.21}$$

Taking the natural logarithm of both sides' yields, we get

$$\ln\left(\frac{E_{th}}{E_0}\right) = -u$$

or

$$u = \ln\left(\frac{E_0}{E_{th}}\right) \quad \text{Eq. 1.22}$$

For any energy E, we can write

$$u = \ln\left(\frac{E_0}{E}\right) = \ln E_0 - \ln E \quad \text{Eq. 1.23}$$

and

$$du = -\frac{dE}{E} \quad \text{Eq. 1.24}$$

where the minus sign signifies a loss of energy.

If $E_{th} = 0.0323$ eV, then from Equation 1.23, the neutron's lethargy is

$$u = \ln\left(\frac{E_0}{E_{th}}\right) = \frac{2 \times 20^6}{0.0323} = 17.94 \approx 18 \quad \text{Eq. 1.25}$$

If we use the variable such as

$$\alpha = \left(\frac{A-1}{A+1}\right)^2 \quad \text{Eq. 1.26}$$

the average logarithmic energy decrease is

$$\xi = 1 + \left(\frac{\alpha}{1-\alpha}\right)\ln\alpha \approx \frac{2}{A + \frac{2}{3}} \quad \forall A \geq 10 \quad \text{Eq. 1.27}$$

The n number of interactions required to thermalize a neutron from fission energy E_0 to thermal energy E_{th} may be computed as follows:

$$n = \frac{\ln\left(\frac{E_0}{E_{th}}\right)}{\xi} \quad \text{Eq. 1.28}$$

For a hydrogen moderator

$$\xi = 1$$

From Equations 1.24 and 1.27, the number of collisions for a neutron to thermalize in a hydrogenous moderator is

$$n \approx \frac{18}{\xi} = \frac{18}{1} = 18 \quad \text{collisions}$$

In comparison, in graphite or a carbon moderator, from Equation 1.26, we get

$$\xi = \frac{2}{12 + \frac{2}{3}} = 0.1589$$

The number of collisions leading to thermalization is then calculated as

$$n = \frac{18}{0.1589} = 113 \quad \text{collisions}$$

It becomes clear that hydrogenous materials are superior neutron moderators than graphite.

If a reactor is made up of a combination of different materials, the average logarithmic energy decrease may be calculated from

$$\xi_{mixture} = \sum_{i=1}^{N} \frac{\xi_i \Sigma_{si}}{\Sigma_{si}}$$

where Σ_{si} is the macroscopic scattering cross-section for the $i-th$ isotope in the mixture and N is the number of isotopes in the mixture.

1.6.3 The Removal Cross-Section

We must determine the chance per unit length of a neutron losing all its energy above thermal. This is also known as the rapid neutron removal cross-section.

$$\Sigma_1 = \frac{\text{probability per unit length of a logarithmic energy loss}}{\text{total energy loss}}$$

Thus,

$$\Sigma_1 = \frac{\xi \Sigma_s}{\int_{E_{th}}^{E_0} \frac{dE}{E}} = \frac{\xi \Sigma_s}{\ln\left(\frac{E_0}{E_{th}}\right)} \qquad \text{Eq. 1.29}$$

1.6.4 The Inverse *E* Flux Dependence

For a lethargy range Δu, $\Delta u/\xi$ is the likelihood that a neutron will collide in the interval Δu, where ξ is the lethargy gain per collision. Since q is the number of neutrons per unit volume per unit time flowing through Δu, $q(\Delta u/\xi)$ represents the number of collisions per unit volume per unit time in the interval Δu. Δu replaced by du, and since

$$du = -\frac{dE}{E}$$

Then, it could be written

$$\frac{\text{Number of collisions in interval } dE}{\text{unit volume} \times \text{unit time}} = q\frac{du}{\xi} = -\frac{q}{\xi}\frac{dE}{E}$$

Since absorptions are insignificant, this is also equal to $-\Sigma_s \phi dE$ where ϕ is the neutron flux. Thus

$$-\Sigma_s \phi dE = \frac{q}{\xi}\frac{dE}{E}$$

or

$$\phi = \frac{q}{\xi \Sigma_s}\left(\frac{1}{E}\right) \qquad \text{Eq. 1.30}$$

For a weakly absorbing material, the slowing-down density $q = q_0$ is a constant. If the scattering cross-section is also constant, the flow will be inversely proportional to the neutron energy.

$$\phi \propto \left(\frac{1}{E}\right) \quad \text{Eq. 1.31}$$

1.6.5 The Fermi Age Equation

For a weakly absorbing material of finite size, a neutron balance equation for steady state and an interval of energy dE is as follows:

$$\begin{aligned}&[\textit{Neutrons produced in interval}\, dE]\\ &-[\textit{Neutrons absorbed in}\, dE]\\ &-[\textit{Neutrons leaking out of}\, dE] = 0\end{aligned} \quad \text{Eq. 1.32}$$

This can be expressed mathematically as

$$SdE - \Sigma_a \phi dE + D\nabla^2 \phi dE = 0 \quad \text{Eq. 1.33}$$

The source term for the interval dE is the number of neutrons slowing into dE minus the number of neutrons slowing out of dE.

$$SdE = q(E + dE) - q(E) = \left(\frac{\partial q}{\partial E}\right) dE \quad \text{Eq. 1.34}$$

If we can neglect the absorptions in the fast region, then $\Sigma_a \approx 0$, and we can write Equation 1.33 as

$$D(E)\nabla^2 \phi(E,r) dE = -\left(\frac{\partial q(E,r)}{\partial E}\right) dE \quad \text{Eq. 1.35}$$

where we are now show the dependence on the spatial position r and the energy E.

If we rewrite Equation 1.30 as

$$\phi(E,r) = \frac{1}{\xi \Sigma_s} \frac{q(E,r)}{E} \quad \text{Eq. 1.36}$$

then, Equation 1.18 can be written as

$$\left[\frac{D(E)}{\xi \Sigma_s E}\right] \nabla^2 q(E,r) = -\frac{\partial q(E,r)}{\partial E} \quad \text{Eq. 1.37}$$

Placing the dimension gradient on the left side and the energy derivative on the right side, we obtain

$$\nabla^2 q(E,r) = -\left[\frac{\xi \Sigma_s E}{D(E)}\right] \frac{\partial q(E,r)}{\partial E} \quad \text{Eq. 1.38}$$

The quantity can be defined as

$$d\tau = -\frac{D(E)}{\xi \Sigma_s} \frac{dE}{E} \quad \text{Eq. 1.39}$$

where τ is called the Fermi Age, then we can write the "Fermi Age Equation" as

$$\nabla^2 q = \frac{\partial q}{\partial \tau} \quad \text{Eq. 1.40}$$

The Fermi Age τ uses area units rather than time units. However, Equation 1.40 is similar to the parabolic diffusion equation found in mathematical physics, where the time is generally represented, thus the name.

The Fermi Age of thermal neutrons can be estimated from Equation 1.16 as

$$\tau = \int d\tau = -\int_{E_0}^{E_{th}} \frac{D}{\xi \Sigma_s} \frac{dE}{E} = +\left(\frac{\bar{D}_F}{\xi \bar{\Sigma}_s} \right) \ln \frac{E_0}{E_{th}} \quad \text{Eq. 1.41}$$

where Σ_s is the average of the macroscopic scattering cross-section Σ_s and the diffusion coefficient D throughout the energy range.

We may describe the Fermi Age in terms of the fast neutron removal cross-section using Equation 1.35.

$$\tau = \frac{\bar{D}_F}{\Sigma_1} \quad \text{Eq. 1.42}$$

It seems comparable to the thermal diffusion area expression.

$$L_{th}^2 = \frac{D_{th}}{\Sigma_{ath}} \quad \text{Eq. 1.43}$$

The diffusion coefficient may be calculated using the following formula if Σ_s is energy independent and rapid absorptions are minimal.

$$\bar{D}_F = \frac{1}{3\Sigma_s(1-\bar{\mu}_0)} \quad \text{Eq. 1.44}$$

On the other hand, Equation 1.19 may be used to determine the age $\bar{D}_F$ if experimental values are available.

1.6.6 Reactor Criticality Equation Based on Age Theory

One-group theory's thermal neutron diffusion equation may be written as follows:

$$D\nabla^2\phi - \Sigma_a \phi + k_\infty \ell_f \Sigma_a \phi = 0 \quad \text{Eq. 1.45}$$

where ℓ_f is the neutrons' fast non-leakage probability.

In the absence of rapid absorptions, $q(E) = q_0$ is stated. The resonance escape probability p is defined as follows, if there are absorptions.

$$p(E) = \frac{q(E)}{q_0} \quad \text{Eq. 1.46}$$

or

$$q(\tau) = pq(\tau = 0) \quad \text{Eq. 1.47}$$

Since the source term in the diffusion equation can be written as

$$\begin{aligned} S &= k_\infty \ell_f \Sigma_a \phi \\ &= \eta \varepsilon p f \ell_f \Sigma_a \phi \\ &= \left(\frac{\nu \Sigma_f}{\Sigma_{a,fuel}}\right) \varepsilon p \left(\frac{\Sigma_{a,fuel}}{\Sigma_a}\right) \ell_f \Sigma_a \phi \\ &= \nu \Sigma_f \varepsilon p \ell_f \phi \end{aligned} \quad \text{Eq. 1.48}$$

When the four-factor formula is used to define the infinite medium multiplication factor, we get

$$k_\infty = \eta \varepsilon p f \quad \text{Eq. 1.49}$$

The definition of the regeneration factor is

$$\eta = \frac{\nu \Sigma_f}{\Sigma_{a,fuel}} \quad \text{Eq. 1.50}$$

Thus, the fuel consumption factor is defined as follows:

$$f = \frac{\Sigma_{a,fuel}}{\Sigma_a} \quad \text{Eq. 1.51}$$

Equations 1.47 and 1.48 may be compared to determine that

$$\begin{aligned} S &= \nu \Sigma_f \varepsilon p \ell_f \phi \\ &= p \ell_f (\nu \Sigma_f \varepsilon \phi) \\ &= p \ell_f q(\tau = 0) \end{aligned} \quad \text{Eq. 1.52}$$

where

$$q(\tau = 0) = \nu \Sigma_f \varepsilon \phi \quad \text{Eq. 1.53}$$

Let us go back to the Age Equation and divide the variables as follows:

$$q(\tau, r) = F(\tau) R(r) \quad \text{Eq. 1.54}$$

and this is substituted in Equation 1.34 to get

$$\begin{aligned} \nabla^2 q(\tau, r) &= \frac{\partial q(\tau, r)}{\partial \tau} \\ \nabla^2 [F(\tau) R(r)] &= \frac{\partial [F(\tau) R(r)]}{\partial \tau} \\ F(\tau) \nabla^2 R(r) &= R9r) \frac{\partial F(\tau)}{\partial \tau} \\ \frac{\nabla^2 R(r)}{R(r)} &= \frac{1}{F(\tau)} \frac{\partial F(\tau)}{\partial \tau} \end{aligned} \quad \text{Eq. 1.55}$$

For a large unreflected or naked reactor, we may assume that the distribution of the slowing-down density and the heat flow is the same.

$$\frac{\nabla^2 R(r)}{R(r)} = \frac{\nabla^2 \phi(r)}{\phi(r)} = -B_g^2 \qquad \text{Eq. 1.56}$$

where B_g^2 is the geometric buckling.

Thus

$$\frac{1}{F(\tau)} \frac{dF(\tau)}{d\tau} = -B_g^2 \qquad \text{Eq. 1.57}$$

It is inevitably observed that a complete derivative has been used in place of the partial derivative.

Separating the variables and integrating, we get

$$\int \frac{dF(\tau)}{F(\tau)} = -B_g^2 \int d\tau$$

which has a solution

$$\ln F(\tau) - \ln C = -B_g^2 \tau$$

$$\ln \frac{F(\tau)}{C} = -B_g^2 \tau$$

Taking the exponential of both sides

$$F(\tau) = Ce^{-B_g^2 \tau} \qquad \text{Eq. 1.58}$$

Equation 1.54 gives us the value of the constant C, which we can then use in Equation 1.53 to estimate for the beginning condition at $\tau = 0$.

$$q(0, r) = CR(r) = \nu\varepsilon \Sigma_f \phi$$

From this, we get

$$C = \frac{\nu\varepsilon \Sigma_f \phi}{R(r)}$$

Substituting in Equation 1.58, we get

$$q(\tau) = \nu\varepsilon \Sigma_f \phi e^{-B_g^2 \tau} \qquad \text{Eq. 1.59}$$

The resonance escape chance is then multiplied by the number of neutrons slowing down to the thermal energy.

$$q_{th} = pq(\tau_{th}) = p\nu\varepsilon \Sigma_f \phi e^{-B_g^2 \tau_{th}} \qquad \text{Eq. 1.60}$$

But $q_{th} = S$, the diffusion equation source term for thermal neutrons indicates that the likelihood of rapid non-leakage is

$$\ell_f = e^{-B_{th}^2 \tau_{th}} \qquad \text{Eq. 1.61}$$

Equation 1.45 allows us to express the diffusion equation in terms of Fermi Age Theory as follows:

$$D\nabla^2\phi - \Sigma_a \phi + k_\infty e^{-B_g^2 \tau_{th}} \Sigma_a \phi = 0 \qquad \text{Eq. 1.62}$$

Dividing by D and replacing

$$\frac{D}{\Sigma_a} = L_{th}^2$$

$$\nabla^2\phi - \frac{\Sigma_a}{D}\phi + k_\infty e^{-B_g^2 \tau_{th}} \frac{\Sigma_a}{D}\phi = 0$$

$$\nabla^2\phi + \left[\frac{k_\infty e^{-B_g^2 \tau_{th}} - 1}{L_{th}^2}\right]\phi = 0$$

On dividing by ϕ and substituting, we obtain

$$\frac{\nabla^2\phi}{\phi} = -B_g^2$$

Then, we get

$$\frac{k_\infty e^{-B_g^2 \tau_{th}} - 1}{L_{th}^2} = B_g^2 \qquad \text{Eq. 1.63}$$

Rearranging, we get the Fermi Age criticality equation as

$$k_{eff} = 1 = \frac{k_\infty e^{-B_g^2 \tau_{th}}}{1 + L_{th}^2 B_g^2} \qquad \text{Eq. 1.64}$$

1.6.7 Special Cases of the Fermi Age Theory Criticality Equation

If the value of $B_g^2 \tau_{th}$ is small, we can expand the exponential as

$$e^{-B_g^2 \tau_{th}} \approx 1 - B_g^2 \tau_{th} \approx \frac{1}{1 + B_g^2 \tau_{th}} \qquad \text{Eq. 1.65}$$

It is comparable to the formula for the rapid non-leakage probability found in the one-group diffusion theory.

$$\ell_f = \frac{1}{1 + B_g^2 L_f^2} \qquad \text{Eq. 1.66}$$

We resolve the criticality if the value of L_f^2 is required in one- or two-group diffusion theory. The buckling equation (Equation 1.64) is

$$1 + L_{th}^2 B_g^2 = k_\infty e^{-B_g^2 \tau_{th}}$$

by trial and error or iteration. Equation 1.65 provides a solid first approximation and results in the following:

$$(1+L_{th}^2 B_g^2)(1+\tau B_g^2) \simeq k_\infty$$

$$1+(L_{th}^2+\tau)B_g^2 + L_{th}^2 B_g^4 \simeq k_\infty$$

In comparison to the second power of buckling, the fourth power of buckling for the large reactor might be regarded as modest, so

$$1+(L_{th}^2+\tau)B_g^2 \simeq k_\infty$$

$$1+M^2 B_g^2 \simeq k_\infty$$

Here the "migration area, M" (i.e., see Section 1.6) is defined as

$$M^2 = (L_{th}^2+\tau) \quad \text{Eq. 1.67}$$

The criticality equation for a large reactor derived from the modified one-group diffusion theory is as follows:

$$k_{eff} = 1 = \frac{k_\infty}{1+M^2 B_g^2} \quad \text{Eq. 1.68}$$

Knowing B_g^2, one can calculate L_f^2 from Equation 1.68 as

$$\ell_f = \frac{1}{1+B_g^2 L_f^2} = e^{-B_g^2 \tau_{th}}$$

or

$$L_f^2 = \frac{e^{-B_g^2 \tau_{th}} - 1}{B_g^2} \quad \text{Eq. 1.69}$$

1.6.8 Criticality Calculation of a Moderated Homogeneous Reactor on Age Theory

We demonstrate the application of Fermi Age theory to criticality calculations using a graphite-moderated reactor core with a migration area [2, 3].

$M^2 = 3,040$ [cm²]

The carbon moderator to uranium atomic ratio was

$$\frac{N_C}{N_U} = 10^4$$

Regeneration factor

$$\eta = 2.08$$

Fuel utilization factor

$$f = 0.988$$

Resonance escape probability

$$p \simeq 10$$

Fast fission factor

$$\varepsilon \simeq 1$$

Age in graphite

$\tau_{th}(graphite) = 350$ [cm^2]

From Equation 1.26, the infinite medium multiplication factor is

$$\begin{aligned} k_\infty &= \eta \varepsilon p f \\ &= 2.08 \times 1 \times 10 \times 0.988 \\ &= 2.06 \end{aligned}$$

Using Equation 1.41 for criticality in Fermi Age Theory, we get

$$\begin{aligned} 1 &= \frac{k_\infty e^{-B_g^2 \tau_{th}}}{1 + L_{th}^2 B_g^2} \\ &= \frac{k_\infty e^{-B_g^2 \tau_{th}}}{1 + (M^2 - \tau_{th}) B_g^2} \\ &= \frac{2.06 e^{-350 B_g^2}}{1 + (3040 - 350) B_g^2} \end{aligned}$$

Rearranging, we get

$$1 + 2690 B_g^2 = 2.06 e^{-350 B_g^2}$$

To get the value of the critical buckling, we can infer the following transcendental equation, which can be solved visually or iteratively.

$$B_{g,n+1}^2 = \frac{2.06 e^{-350 B_{g,n}^2} - 1}{2690} \qquad \text{Eq. 1.70}$$

The buckling's optimal initial value is 1 that can be calculated using the modified one-group diffusion theory as follows:

$$B_1^2 = 3.4900 \times 10^{-4}$$

The following outcomes are obtained by using the iteration formula in Equation 1.70.

$$B_2^2 = 3.0600 \times 10^{-4}$$
$$B_3^2 = 3.1627 \times 10^{-4}$$
$$B_4^2 = 3.1380 \times 10^{-4}$$
$$B_5^2 = 3.1440 \times 10^{-4}$$
$$B_6^2 = 3.1426 \times 10^{-4}$$
$$B_7^2 = 3.1429 \times 10^{-4}$$
$$B_8^2 = 3.1428 \times 10^{-4}$$
$$B_5^2 = 3.1428 \times 10^{-4}$$

At the ninth iteration, the result converged to four decimal places.

The critical radius for a spherical reactor core would be

$$R_c = \left(\frac{\pi^2}{3.148 \times 10^{-4}} \right)^{1/2} = 177.21\,[cm] \quad (168.17)$$

For a cube, the critical side length would be

$$a_c = \left(\frac{3\pi^2}{3.148 \times 10^{-4}} \right)^{1/2} = 306.94\,[cm] \quad (291.28)$$

For a cylinder with a height to diameter ratio of unity

$$H = 2R$$

the critical radius would be

$$R_c = \left(\frac{8.25}{3.148 \times 10^{-4}} \right)^{1/2} = 162.02\,[cm] \quad (153.75)$$

$$H_c = 2R_c = 324.04\,[cm] \quad (307.50)$$

Next to each calculated value, the values derived from the modified one-group theory are displayed in parentheses for comparison. It is possible to estimate critical masses and critical volumes from these critical dimensions.

The validity of the fundamental premise of Fick's law, as described in this chapter, determines whether Fermi Age Equation 1.34 may be applied for slowing density $q = q(\vec{r}, \tau_{th})$. For the lethargy-dependent current, this can be expressed as follows:

$$\vec{J}(\vec{r},u) = -D(u)\nabla\phi(\vec{r},u) \qquad \text{Eq. 1.71}$$

Additionally, assume that the asymptotic slowing-down density follows the derived equation for an infinite media, as shown in Equation 1.72, and this holds true for a finite medium as well.

$$q(\vec{r},u) \simeq \xi\,\Sigma_t(u)\phi(\vec{r},u) \qquad \text{Eq. 1.72}$$

The slowing-down density $q(\vec{r},u)$ with capture is a product of the slowing-down density without capture and the resonance escape probability $p(u) = p(0 \to u)$ as shown [4].

$$q(\vec{r},u) = q_0(\vec{r},u)p(u) \qquad \text{Eq. 1.73}$$

where $p(u)$ is given by the following equation:

$$q(u) = q(0)\exp\left[-\int_0^u du'\left(\frac{\Sigma_a(u')\phi(u')}{q(u')}\right)\right]$$

$$p(u_1 \to u_2) = \frac{q(u_2)}{q(u_1)} = \exp\left[-\int_{u_1}^{u_2} du\left(\frac{\Sigma_a(u)\phi(u)}{\Sigma_t(u)q(u)}\right)\right] \quad u_1 < u_2 \qquad \text{Eq. 1.74}$$

and again, it can be obtained for an infinite medium. Taking a derivative of Equation 1.73 with respect to lethargy u provides

$$\frac{\partial q(\vec{r},u)}{\partial u} = p(u)\frac{\partial q_0(\vec{r},u)}{\partial u} - \Sigma_a(u)\phi(\vec{r},u) \qquad \text{Eq. 1.75}$$

When neutrons slow down and lose energy, their lethargy rises. This may be expressed as [4] by substituting Equations 1.71 and 1.73 into the balancing Equation 1.76.

$$-\frac{\partial q(\vec{r},u)}{\partial u} + S(\vec{r},u) = \nabla \cdot \vec{J}(\vec{r},u) + \Sigma_a \phi(\vec{r},u) \qquad \text{Eq. 1.76}$$

With $S(\vec{r},u) \to 0$, we obtain

$$\frac{\partial q(\vec{r},u)}{\partial u} = \frac{D(u)}{\xi \Sigma_t(u)}\nabla^2 q(\vec{r},u) \qquad \text{Eq. 1.77}$$

This equation was recast into the following form and reflected in Equation 1.34.

$$\frac{\partial q(\vec{r},\tau)}{\partial \tau} = \nabla^2 q_0(\vec{r},\tau)$$

With a new variable of $\tau = \tau_{th}$ known as Fermi Age Time is shown as

$$\tau = \tau_{th}(u) = \int_0^u \frac{D(u)}{\xi \Sigma_t(u)} du \qquad \text{Eq. 1.78}$$

1.7 MIGRATION LENGTH – MIGRATION AREA

The previous section of this chapter provided a clear definition of Fermi Age Theory. The basic idea was to solve Equation 1.34, also known as the Fermi Age Equation or the Age Diffusion Equation, by introducing a new variable called τ_{th}, which is defined as X1 and, despite having area, is in units of area. This variable is called Fermi Age Time. The capture-free slowing-down density $q = q(\vec{r},\tau_{th})$ for a finite medium is solved using Equation 1.34, which roughly accounts for the neutron leakage effect.

The Fermi Age Time τ_{th} variable, as described in the previous section and all of its subsections, acts as a time variable or chronological age and causes a neutron slowdown or lethargy increase from the source energy at E_0, as seen in Figure 1.9 and is represented by Equation 1.78.

The fundamental presumptions presented in Equations 1.71 and 1.72 determine whether the Fermi Age Equation for $q_0(\vec{r},\tau)$ is applicable as neutron lethargy grows or as they slowdown from the source energy at E_0, as showed again in Figure 1.9. This implies that the Fermi Age Theory solution for the slowing-down density $q(\vec{r},u)$for a finite media should mainly hold for a medium with [4] and following conditions:

a. Small leakage
b. Weak or slowly varying captures, typically involving heavy nuclei such that atomic number is $A \gg 1.0$

One-sixth of the mean square distance a fast neutron travels from birth until it is removed from the fast group by absorption or slowing down into the thermal group is the age τ_{th}, which may be physically interpreted as such.

Now for a large reactor with small leakage, the quantity M^2 is defined as Neutron Migration Area or Square of the Neutron Migration Length and is written as

$$\begin{cases} M^2 = L_2^2 + L_1^2 \\ M^2 = L^2 + L_s^2 \\ M^2 = L^2 + \tau_{th} \end{cases} \qquad \text{Eq. 1.79}$$

Sometimes, for historical reasons, Equation 1.56 is written in lateral form than the first term.

As previously mentioned, the migration area's physical meaning that M^2 is equivalent to one-sixth of the square of the average distance in all dimensions between the neutron's birth point as a fast neutron and its absorption as a thermal neutron. For this reason, the Fermi Age of τ for thermal neutrons was designated τ_{th} in earlier sections, which corresponds to their lifetime in a reactor core.

Neutron dispersion in a limited or infinite media is often dictated by

- The source distribution, including whether it is a multiplying environment or an external neutron source.
- The geometry (in a finite medium).
- $L^2 = D/\Sigma_a$ is the diffusion area. It is proportional to the distance thermal neutrons travel before they are absorbed.
- A neutron's slowing-down length, or L_s is directly proportional to the distance that fast neutrons traverse from birth to thermalization. The "age", often known as the "Fermi Age", is a parameter τ that is frequently utilized since it can be calculated from Fermi Age Theory.

In conclusion, let us concentrate on the slowing and diffusion length. Here, down length and the diffusion length's physical meaning are as follows:

The average distance (in all dimensions) between the neutron's birth point (as a thermal neutron) and its absorption is equal to one-sixth of the square, or L^2.

Similar to the diffusion length for thermal neutrons, the distance traveled during moderation is correlated with the Fermi Age. The slowing-down length is the square root of the Fermi Age, $\tau_{th} = L_s^2$, whereas the Fermi Age is the same as the slowing-down length squared, L_s^2. The slowing-down length has the following physical meaning: Similar to the diffusion length for thermal neutrons, the distance traveled during moderation is correlated with the Fermi Age.

The average distance (in all dimensions) between the neutron's birth point (as a fast neutron) and its thermalization point is equal to one-sixth of the square, or L_s^2.

Again, let us consider the quantity M^2, where

$$\begin{cases} M^2 = L^2 + L_s^2 \\ M^2 = L^2 + \tau_{th} \end{cases} \quad \text{Eq. 1.80}$$

This quantity is called the migration area or square of the migration length. The physical meaning of the migration area is the average distance (in all dimensions) between the neutron's birth point (as a fast neutron) and its absorption (as a thermal neutron) is equal to one-sixth of the square, or M^2.:

Because they have an impact on the critical size and neutron leakage, the distances covered by fast neutrons during moderation and thermal neutrons during diffusion in a reactor are significant for reactor design.

1.7.1 Effect on the Neutron Leakage

It may be inferred that the migration area is the primary determinant of the total non-leakage probability (P_{NL}) of big reactors.

$$P_{NL} = \frac{1}{1 + M^2 B^2} \quad \text{Eq. 1.81}$$

where

$$M^2 = \tau + L_d^2$$
$$M = \sqrt{\tau + L_d^2}$$

1. **Fast Non-Leakage Probability**

The Fermi Age hypothesis may be used to calculate the rapid non-leakage probability. The likelihood that a neutron will stay in the core and transform into a thermal neutron without being lost due to rapid leakage is also as follows:

$$P_f = e^{-\tau B^2} \approx \frac{1}{1 + \tau B^2} \quad \text{Eq. 1.82}$$

where τ is the neutron's Fermi Age, and B is the geometrical buckling (in the critical state, $B_g = B_m$), which is solely determined by the core's form and size. The value of B for tiny cores is greater than that for large cores. As a result, it is evident that fast neutron leakage is larger for tiny cores and is determined by the macroscopic slowing down of the power of the neutron moderator.

2. **Thermal Non-Leakage Probability**

The thermal non-leakage probability may be calculated using neutron diffusion theory. The following equation represents the likelihood that a thermal neutron ($P_f = P_{NLT}$) will remain in the core.

$$P_t = \frac{1}{1 + L_d^2 B^2} \quad \text{Eq. 1.83}$$

The diffusion length, L_d, is the geometrical buckling (in the case of critical state $B_g = B_m$), which is determined solely by the form and size of the core. The value of B for tiny cores is greater than that for large cores.

3. **Total Non-Leakage Probability**

The fast non-leakage probability (P_f) and thermal non-leakage probability (P_t) may be combined into a single component to calculate the proportion of all neutrons that do not leak out of the reactor core.

This factor is termed the total non-leakage probability and is given by the symbol P_{NL}. It may be stated by the following equation:

$$\begin{cases} P_{NL} = \left(\dfrac{1}{1+\tau B^2}\right) \times \left(\dfrac{1}{1+L_d^2 B^2}\right) \\ P_{NL} = \dfrac{1}{1+(\tau + L_d^2)B^2 + \tau L_d^2 B^4} \end{cases} \quad \text{Eq. 1.84}$$

For large reactors, we may recast this equation without significantly reducing accuracy by substituting the diffusion length L_d and Fermi Age τ with the migration length M in the one-group equation. For big reactors, the term B^4 is negligible and may thus be ignored. We may then write Equation 1.81 as is where M is the migration area (m^2), and the migration length is defined as the square root of that area. As can be seen, big reactors' overall non-leakage probability is mostly determined by their migratory area.

1.8 MAIN OPERATIONAL CHANGES AFFECTING THE NEUTRON LEAKAGE

In a heterogeneous water-moderated reactor, both P_f and P_t are impacted by a change in the moderator temperature, and since the feedbacks' orientations are same, the overall non-leakage probability that results is likewise sensitive to the moderator temperature change. Consequently, the likelihood of leakage increases as the moderator temperature rises. The Moderator Temperature Coefficient (MTC) of the majority of Pressurized Water Reactors (PWRs) is negative due to two primary impacts, including this one.

The core temperature (or moderator temperature) affects the thermal neutron leakage. Because of the thermal expansion of water, the moderator temperature affects the macroscopic cross-sections for elastic scattering reactions, particularly the atomic number density, $N_{H_2O}(\Sigma_s = \sigma_s H_{H_2O})$.

Additionally, the core temperature affects the neutron absorption microscopic cross-section (σ_a). The direction of both operations is the same. The diffusion coefficient ($D = 1/3\Sigma_{tr}$) rises and the absorption cross-section falls with increasing core temperature, leading to an increase in thermal neutron leakage. A component of the MTC is this physical mechanism.

The core temperature (or moderator temperature) also affects the rapid neutron leakage. Because water expands thermally, the moderator temperature affects the macroscopic cross-sections for the elastic scattering reaction ($\Sigma_s = \sigma_s H_{H_2O}$). The fast neutron leakage rises with the core's temperature. A component of the MTC is this physical mechanism. Neutron detectors placed around the reactor vessel record an increase in neutron flux caused by it.

Based on above discussions, Table 1.1 is provided.

Fuel burnup also has a substantial impact on the overall non-leakage likelihood in power reactors. None of the characteristics, such as the diffusion coefficient or geometrical buckling, are connected to this relationship. The overall non-leakage probability in power reactors is highly dependent on

TABLE 1.1
Material and Reactor Type Non-Leakage Probability

Material	MSDP(1)	MR(1)	$\tau^{1/2} = L_f$[cm] (Moderation Length)(2)	L_d[cm] (Diffusion Length)(3)	M[cm] (Migration Length)
H_2O	1.35	71	5.1	2.54	5.7
D_2O	0.176	5670	11.4	160	160
C	0.060	192	19.2	53.5	56.8

1) The macroscopic slowing down power (MSDP) and the moderating ratio (MR) are key characteristics of moderators. They are not influenced by the presence of absorption. On the other hand, they differ slightly with temperature of the moderator.
2) Moderation lengths in power reactors are usually longer due to the higher moderator temperature and lower density of water molecules than in the cold water. This effect causes the resonance escape probability, and the total non-leakage probability decreases as the moderator temperature increases – two main effects causing the Moderator Temperature Coefficient of most Pressurized Water Reactors to be negative.
3) Diffusion lengths in power reactors are usually shorter due to the presence of high absorption cross-sections in comparison with diffusion lengths in pure moderators.

the reload method and a specific fuel loading pattern. One of the most important factors in fuel and neutron economy is neutron leakage.

Core designers create minimal leakage loading patterns, which avoid placing new fuel assemblies in the reactor core's periphery, to improve neutron and fuel efficiency. Gasoline with the greatest fuel burnup is placed into the periphery places. Because of their lower k_{inf} and sense of a non-multiplying environment, these "high" burnup assemblies have a lower relative power by nature than ordinary assemblies.

In summary, a particular loading pattern has a considerable impact on this parameter. Neutron leakage often rises with fuel burnup, particularly in low-leakage loading patterns. Reducing the k_{inf} disparities between peripheral high-burnup assemblies and fresh fuel assemblies is what causes this process. For more detailed information, the readers should consult Stacey (2001) through Lewis and Miller (1993; references [5–14]).

1.9 MICROSCOPIC NEUTRON ABSORPTION CROSS-SECTION σ_α

The microscopic neutron absorption cross-section will be discussed in this section. The probability of a neutron absorption is represented by the absorption cross-section as σ_a. By splitting the whole cross-section into scattering and absorption cross-sections, the relative chances of an absorption reaction or a neutron scattering are shown.

$$\sigma_t = \sigma_s + \sigma_a \qquad \text{Eq. 1.85}$$

Given a collision, σ_a/σ_t is the probability that the neutron will be absorbed and σ_s/σ_t is the probability that the neutron will be scattered.

Please see the following plots and resources in Figures 1.19 and 1.20 as well as Table 1.2.

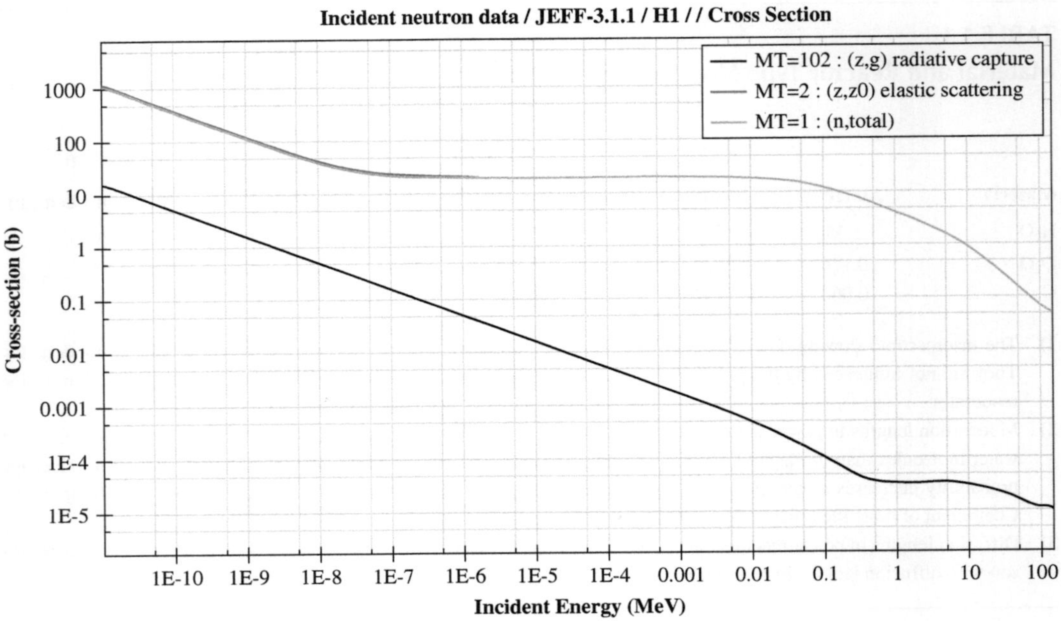

FIGURE 1.19 Hydrogen. Neutron absorption and scattering. Comparison of cross-sections.

(Source: JANIS (Java-based Nuclear Data Information Software); The JEFF-3.1.1 Nuclear Data Library)

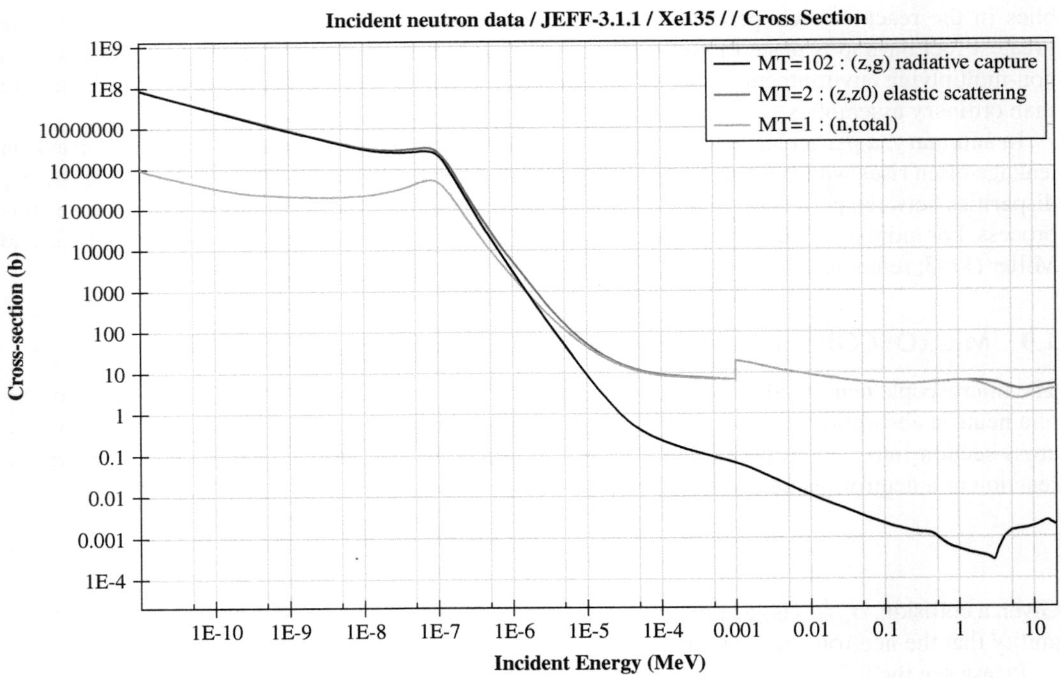

FIGURE 1.20 Hydrogen. Neutron absorption and scattering. Comparison of cross-sections.

(Source: JANIS (Java-based Nuclear Data Information Software); The JEFF-3.1.1 Nuclear Data Library)

TABLE 1.2
Table of Cross-Sections [20]

		Thermal neutron			Fast neutron		
		Scattering	**Capture**	**Fission**	**Scattering**	**Capture**	**Fission**
Moderator	H-1	20	0.2	–	4	0.00004	–
	H-2	4	0.0003	–	3	0.000007	–
	C-12	5	0.002	–	2	0.00001	–
Structural materials, others	Zr-90	5	0.006	–	5	0.006	–
	Fe-56	10	2	–	20	0.003	–
	Cr-52	3	0.5	–	3	0.002	–
	Ni-58	20	3	–	3	0.008	–
	O-16	4	0.0001	–	3	0.00000003	–
Absorber	B-10	2	200	–	2	0.4	–
	Cd-113	100	3	–	4	0.05	–
	Xe-135	400	2,000,000	–	5	0.0008	–
	In-115	2	100	–	4	0.02	–
Fuel	U-235	10	99	583	4	0.09	1
	U-238	9	2	0.00002	5	0.07	0.3
	Pu-239	8	269	748	5	0.05	2

REFERENCES

1. N. Feather. A history of neutrons and nuclei. Part 1. *Contemporary Physics*, 1(3), 191–203 (1960). https://doi.org/10.1080/00107516008202611.
2. J. R. Lamarsh, *Introduction to Nuclear Engineering*, Addison-Wesley Publishing Company, 1983.
3. M. Ragheb, *Lecture Notes on Fission Reactors Design Theory*, FSL-33, University of Illinois, 1982.
4. J. C. Lee, *Nuclear Reactor Physics and Engineering*, 1st edition, John Wiley and Sons Inc., 2020.
5. W. M. Stacey, *Nuclear Reactor Physics*, John Wiley & Sons, 2001.
6. S. Glasstone, *Nuclear Reactor Engineering: Reactor Systems Engineering*, 4th edition, Springer, 1994.
7. W. S. C. Williams, *Nuclear and Particle Physics*, 1st edition, Clarendon Press, 1991
8. G. R. Keepin, *Physics of Nuclear Kinetics*, 1st edition, Addison-Wesley Pub. Co, 1965.
9. R. R. Burn, *Introduction to Nuclear Reactor Operation*, McGraw-Hill, 1988.
10. U.S. Department of Energy, *Nuclear Physics and Reactor Theory: DOE Fundamentals Handbook*, Volume 1 and 2., U.S. Department of Energy, January 1993.
11. K. O. Ott, W. A. Bezella, *Introductory Nuclear Reactor Statics*, Revised edition, American Nuclear Society, 1989.
12. K. O. Ott, R. J. Neuhold, *Introductory Nuclear Reactor Dynamics*, American Nuclear Society, 1985.
13. D. L. Hetrick, *Dynamics of Nuclear Reactors*, American Nuclear Society, 1993.
14. E. E. Lewis, W. F. Miller, *Computational Methods of Neutron Transport*, American Nuclear Society, 1993.
15. NUREG/CR-5550, *Passive Nondestructive Assay of Nuclear Materials*, https://www.nrc.gov/docs/ML0914/ML091470585.pdf
16. M. T. Swinhoe, J. D. Hutchinson & P. M. Rinard. Neutron Interactions with Matter. In Springer eBooks (pp. 307–323) (2024b). https://doi.org/10.1007/978-3-031-58277-6_14
17. Z. Chai, A. Chatt, P. Bode, J. Kučera, R. Greenberg, & D. B. Hibbert. Vocabulary of radioanalytical methods (IUPAC Recommendations 2020). *Pure and Applied Chemistry*, 93(1), 69–111 (2020). https://doi.org/10.1515/pac-2019-0302.
18. Unknown, U. (n.d.). Chapter 7: Neutron Slowing Down - Part II. In *Nuclear Reactor Physics: Neutronics*. https://nuclearreactorphysics.weebly.com/uploads/1/6/5/0/16501436/neutronicsvii-neutronslowing-down.pdf
19. M. Ragheb, *Lecture Notes on Fission Reactors Design Theory*, FSL-33, University of Illinois, 1982; J. R. Lamarsh, *Introduction to Nuclear Engineering*, Addison-Wesley Publishing Company, 1983.
20. A. Z. Paydar, S. K. M. Balgehshiri, B. Zohuri, *Advanced Reactor Concept (ARC), A New Nuclear Power Plant Perspective Producing Energy*, 1st edition, Elsevier, July 15, 2023. Advanced Reactor Concepts (ARC) | ScienceDirect
21. H. A. Schaeffer, Fast neutron interaction cross sections and attenuation coefficients. In R. G. Jaeger (Ed.), *Engineering Compendium on Radiation Shielding* (Vol. II). Springer-Verlag, 1973.

2 Diffusion Equations and Fick's Laws

2.1 INTRODUCTION

The diffusion process – the movement of molecules from a region of higher concentration to one of lower concentration – is explained by Fick's Law of Diffusion. Adolf Fick established Fick's Law of Diffusion in 1855. Normal diffusion, often known as Fickian diffusion, is a diffusion process that complies with Fick's rules. Anomalous diffusion, also referred to as non-Fickian diffusion, is a diffusion process that does not follow Fick's Law.

The diffusion coefficient D is solved using Fick's Law of Diffusion.

These two laws are interconnected; for example, Fick's Second Law, which is comparable to the diffusing equation, is derived from Fick's First Law.

The following definition may be expressed using Fick's Law of Diffusion:

The concentration gradient is proportional to the molar flux resulting from diffusion.

The second derivative of concentration with space determines how quickly the concentration of the solution changes at a given position in space. Fick's First Law may be used to derive his Second Law, which is the same as the diffusion equation. These two laws are described as follows and behave in such a fashion.

2.2 FICK'S FIRST LAW

This law can be stated as, "the flow of a solute over a concentration gradient from a higher concentration to a lower concentration".

$$J_x = -D\frac{d\phi}{dx} \qquad \text{Eq. 2.1}$$

where

J_x = Diffusion flux in x-direction
D = Diffusivity
ϕ = Concentration
x = Position in x-direction

Here, more information is provided along with an expansion of the diffusion coefficient, $D = \frac{\delta_x^2}{2t}$, as seen in Figure 2.1 and Equation 2.2.

$$J_x = \frac{\delta_x^2}{2\tau}\frac{d\phi}{dx} \qquad \text{Eq. 2.2}$$

DOI: 10.1201/9781003590194-2

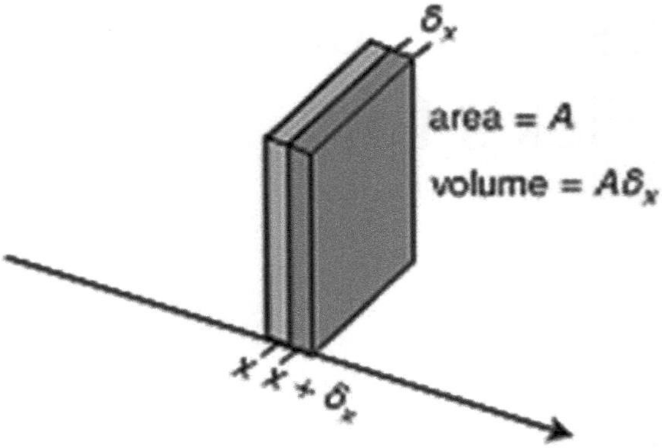

FIGURE 2.1 Fick's First Law.

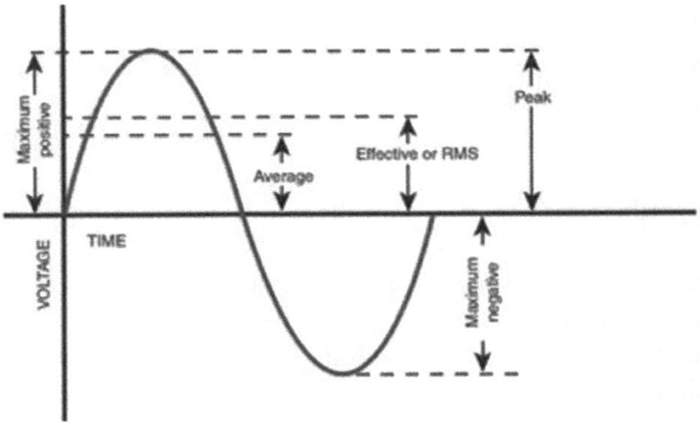

FIGURE 2.2 Root mean square.

where

J_x = Flux of molecules (or moles) per unit area unit time
δ_x = Root mean square displacement along the x-direction
τ = Average duration of random steps
$\frac{d\phi}{dx}$ = "Concentration gradient" is another name for the derivative of concentration with x.

Flux is in the negative x-direction if concentration rises with x.

Root mean square (RMS) in mathematics and its application is defined as the square root of the mean square (the arithmetic means of the squares of a set of numbers). The RMS is also known as the quadratic mean and is a particular case of the generalized mean with exponent 2. RMS can also be defined for a continuously varying function in terms of an integral of the squares of the instantaneous values during a cycle. See Figure 2.2.

The RMS is also known as the quadratic mean and is a particular case of the generalized mean with exponent 2. RMS can also be defined for a continuously varying function in terms of an integral of the squares of the instantaneous values during a cycle as illustrated in Figure 2.2.

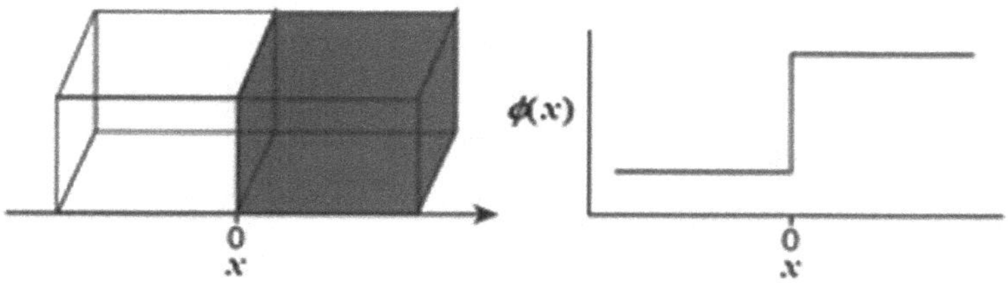

FIGURE 2.3 Experimental macroscopic diffusion.

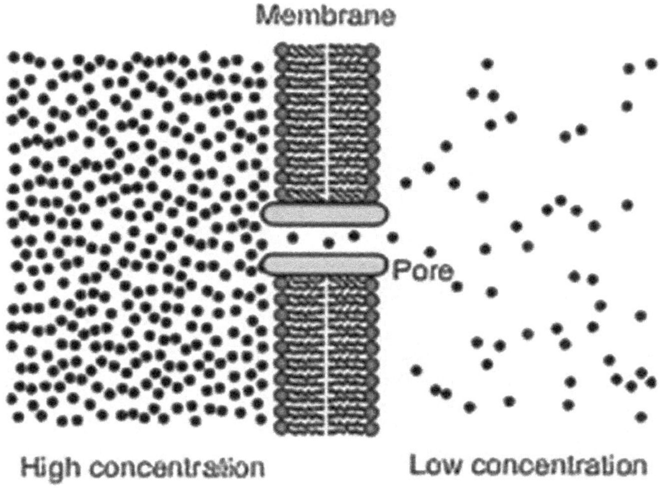

FIGURE 2.4 Molecular diffusion.

2.2.1 The Diffusion Coefficient, *D*

Consider the term from Fick's First Law shown in Equation 2.2 as $\delta_x^2/2\tau$, both δ_x and τ are parameters describing the Random Walk (RW) of molecules or large particles undergoing diffusion, and are constant for a given type of particle under defined solution conditions. In that case, we can define a new parameter known as diffusion coefficient D.

$$D = \frac{\delta_x^2}{2\tau} \quad \text{Eq. 2.3}$$

However, the most usual form of Fick's First Law is written as shown in Equation 2.1.

D can be experimentally determined, without knowing anything about the microscopic RW steps.

An idealized macroscopic diffusion experiment can be drawn as Figure 2.3, which shows $\phi(x)$ versus x change with time.

Diffusion of molecules across a cellular membrane with some plausible values, as shown in Figure 2.4.

- Diffusion coefficient for a small molecule: $\approx 10^{-10}$ m²/s to be considered later on.
- Thickness of a biological membrane: $\approx$ 3 nm
- Diameter of a molecular pore: $\approx$ 1 nm
- Concentrations: 50 mM and 5 mM

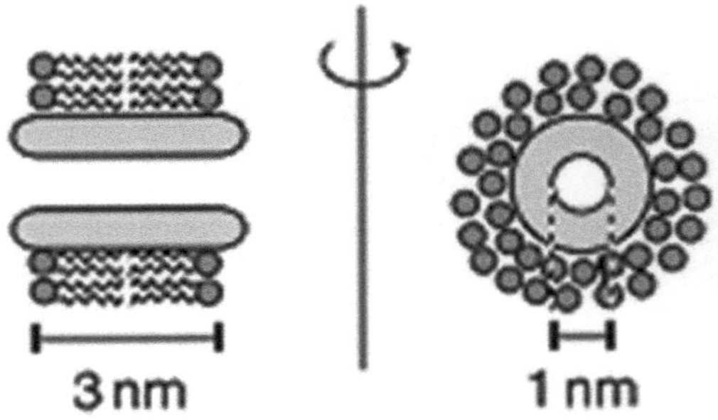

FIGURE 2.5 Membrane pore diffusion.

Applying Fick's First Law, we can write

$$\frac{d\phi}{dx} \approx \frac{\text{concentration difference}}{\text{membrane thickness}} = \frac{50 \text{ mM} - 5 \text{ mM}}{3 \text{ nM}} = 15 \text{ mM/nm} \quad \text{Eq. 2.4}$$

Converting the above equation to units with moles and meters, we can state that

$$15 \text{ mM/nm} = \frac{0.015 \text{ moles/L}}{10^{-9}\text{m}} \times \frac{10^3 \text{L}}{\text{m}^3} = 1.5 \times 10^{10} \text{moles/m}^4 \quad \text{Eq. 2.5}$$

Therefore, from Fick's First Law, we can write

$$J = -D\frac{d\phi}{dx} = -\left(10^{10} \text{ m}^2/s\right) \times 15 \times 10^{10} \text{ moles/m} = -1.5 \text{moles}/\text{m}^2/\text{s} \quad \text{Eq. 2.6}$$

A negative sign indicates that diffusion is in the direction opposite to the concentration gradient.

Diffusion through a membrane pore can be illustrated as in Figure 2.5.

We can assume the pore diameter is equal to 1 nm and area = $\pi r^2 = \pi (0.5 \text{ nm})^2 = 7.8 \times 10^{19} \text{ m}^2$ and flux is $J = 1.5$ moles/(sm^2), thus moles per second through the pore is given by 1.5 moles/moles/(sm^2) $\times 7.8 \times 10^{19}$ m^2 = 1.2×10^{18} moles/s.

2.3 FICK'S SECOND LAW

The prediction of change in concentration along with time due to diffusion is given by Fick's Second Law as

$$\frac{\partial\phi}{\partial t} = D\frac{\partial^2\phi}{\partial x^2} \quad \text{Eq. 2.7}$$

where

D = Diffusivity
t = Time
ϕ = Concentration
x = Position in x-direction

The concentration of the compound in question is the dependent variable in the second diffusion law of Fick, which is a linear equation. Individual chemical species spread on their own. The mass transport systems defined by Fick's Second Law are easily numerically simulated, thanks to their characteristics.

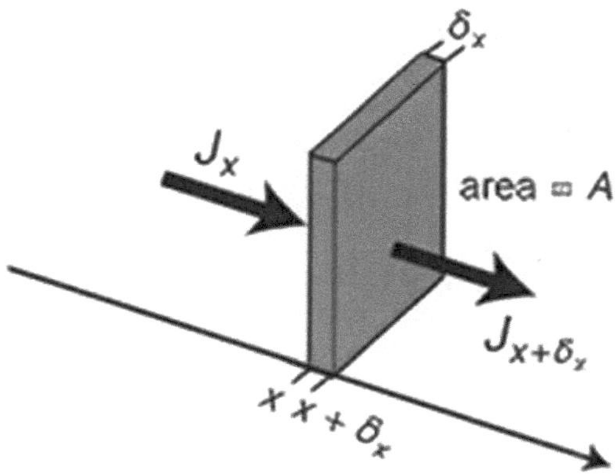

FIGURE 2.6 Diffusion versus time and distance.

How does the concentration at a given point change with time? For this, we use Figure 2.6 and define the net number of molecules moving to the right in Figure 2.6 at two sides of a slice, during interval dt, then we can write

$$\begin{aligned} N_x &= J_x A dt \\ N_{x+\delta_x} &= J_{x+\delta_x} A dt \end{aligned} \qquad \text{Eq. 2.8}$$

Then, the change in the number of molecules in the slice in Figure 2.6 is given as

$$\begin{aligned} dN &= N_x - N_{x+\delta_x} \\ &= AJ_x dt - AJ_{x+\delta_x} dt \\ &= Adt\left(J_x - J_{x+\delta_x}\right) \end{aligned} \qquad \text{Eq. 2.9}$$

The change in concentration in the slice is written as

$$\begin{aligned} d\phi &= \frac{dN}{A\delta_x} = \frac{Adt\left(J_x - J_{x+\delta_x}\right)}{A\delta_x} \\ &= -dt\frac{J_{x+\delta_x} - J_x}{\delta_x} \end{aligned} \qquad \text{Eq. 2.10}$$

In the limit of small dt and small δ_x, we can conclude that

$$\frac{d\phi}{dt} = -\frac{J_{x+\delta_x} - J_x}{\delta_x} = -\frac{dJ}{dx} \qquad \text{Eq. 2.11}$$

and that is how J changes with variable x.

Moreover, the Fick's Second Law of Diffusion, in general, can be expressed as

$$J = -D\frac{d\phi}{dt} \qquad \text{Eq. 2.12}$$

Derivative of J in respect to x, yields as

TABLE 2.1
Fick's First and Second Laws of Diffusion

First Law	Second Law
$J = -D\frac{d\phi}{dx}$	$\frac{d\phi}{dt} = D\frac{d^2\phi}{dx^2}$
Flux J, at position x is proportional to the concentration gradient at that position	Rate of change in concentration at position x is proportional to the derivative of the concentration gradient

$$\frac{dJ}{dx} = -D\frac{d^2\phi}{dx^2} \quad \text{Eq. 2.13}$$

From Equation 2.7, which is Fick's Second Law, we can deduce that

$$\frac{d\phi}{dt} = -\frac{dJ}{dx} = D\frac{d^2\phi}{dx^2} \quad \text{Eq. 2.14}$$

This is also called the diffusion equation. Now the questions arise: How does this diffusion equation work? What is it good for? How do we use it?

In our case, for the purposes of reactor physics and neutronics analysis, we apply neutron diffusion theory for one-group and multi-group calculations related to reactor design and criticality analysis, as discussed in Section 2.4.

A summary of Fick's First and Second laws of diffusion is shown Figure 2.7.

Given Fick's Second law, we can see that is a "second-order differential equation" and the solution to the equation can be found as $\phi = f(x,t)$

$$\frac{d\phi}{dt} = D\frac{d^2\phi}{dx^2}$$
$$\phi = f(x,t) \quad \text{Eq. 2.15}$$
$$\frac{df(x,t)}{dt} = D\frac{d^2 f(x,t)}{dx^2}$$

The trick is to find $\phi = f(x,t)$. The solution depends on the shape of the volume, the initial concentrations, and the Boundary Conditions, as in Figure 2.3.

$$\text{For } x < 0 \ \rightarrow \ \phi(x) = 0, \quad \frac{d\phi}{dx} = 0, \quad \frac{d^2\phi}{dx^2} = 0 \quad \text{Eq. 2.16(a)}$$

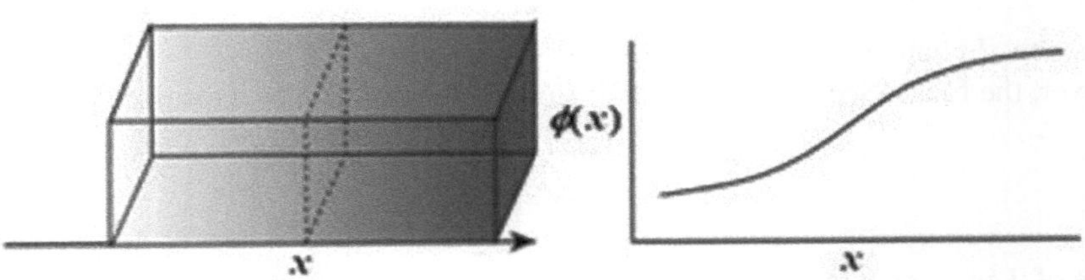

FIGURE 2.7 Fick's Laws presentations.

At $x = 0 \ \dfrac{d\phi}{dx} \to \infty$ Eq. 2.16(b)

For $x \geq 0 \ \to \ \phi(x) = 1, \ \dfrac{d\phi}{dx} = 0, \ \dfrac{d^2\phi}{dx^2} = 0$ Eq. 2.16(c)

Under diffusion from a sharp boundary condition, molecules will begin to diffuse via an RW. How will the molecules initially at position a be distributed after a time t using Figure 2.8?

In this case, results are produced in Figure 2.9 for diffusion from a Sharp Boundary and distribution of molecules originally at position $x = a$.

$$p(x) = \frac{1}{\sqrt{2\pi n\left(\delta_x^2\right)}} e^{-\left((x-a)^2/(2\pi\delta_x^2)\right)} \quad \text{Eq. 2.17}$$

The diffusion coefficient is given by $D = \dfrac{\delta_x^2}{2\tau}$, where τ is average time for each RW step, and after time t, $n = t/\tau$ will be given as

$$n\left(\delta_x^2\right) = \frac{t(\delta_x^2)}{\tau} = 2Dt \quad \text{Eq. 2.18}$$

Thus, the distribution of molecules originally at position $x = a$ is given as

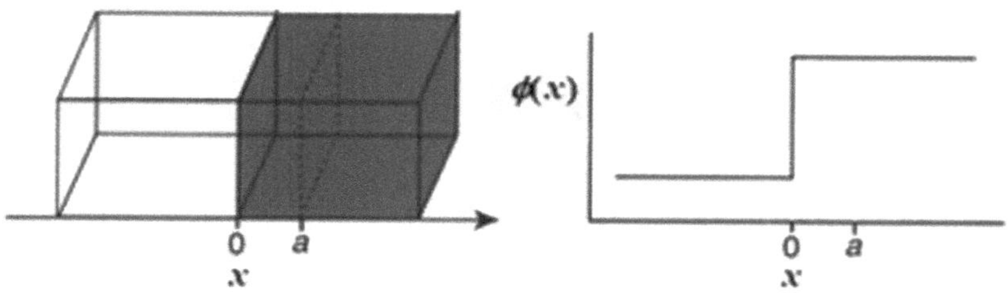

FIGURE 2.8 Sharp boundary conditions.

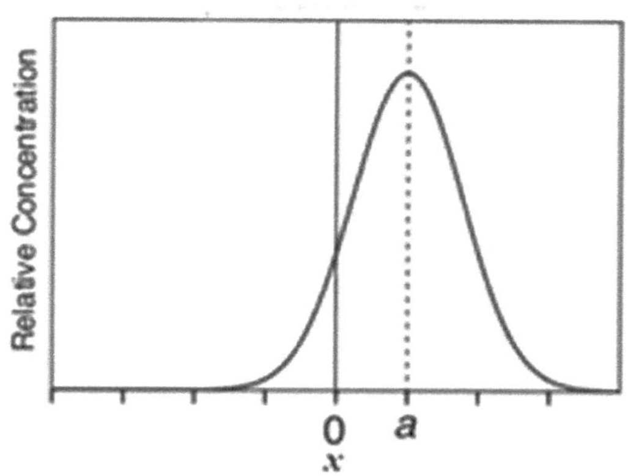

FIGURE 2.9 Distribution of molecules originally at position $x = a$

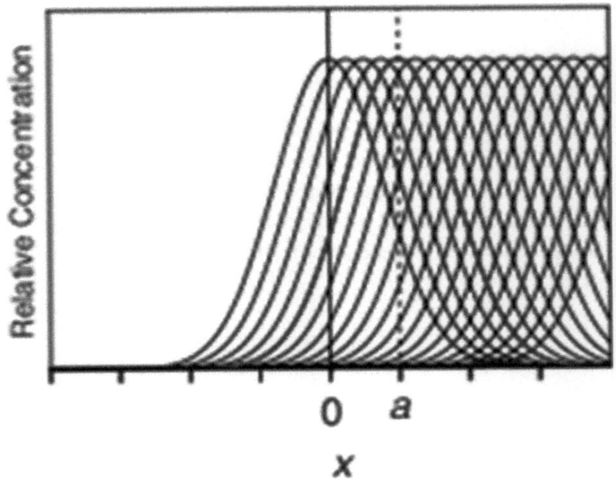

FIGURE 2.10 Distribution of molecules originally at position $\mathbf{a} \geq \mathbf{0}$.

$$p(x) = \frac{1}{\sqrt{4\pi Dt}} e^{-9x-a)^2/(4Dt)} \quad \text{Eq. 2.19}$$

For the distribution of molecules from all starting points for $a \geq 0$ and using Figure 2.10, we state that the resulting concentration profile becomes increasingly uniform over time, indicating that molecular motion leads to the spreading out and eventual equilibration of the system.

At position x, concentration is the sum of molecules that have diffused from $a \geq 0$, given as

$$\phi(x,t) = \frac{1}{\sqrt{4\pi Dt}} \int_0^\infty e^{-(x-a)^2/(4Dt)} da \quad \text{Eq. 2.20}$$

2.4 APPLICATIONS OF FICK'S LAW

In biological applications, Fick's law is given as

Flux = P (C_2 C_1) from Fick's First Law

where

P = Permeability
C_2, C_1 = Difference in concentration

Other applications of Fick's law include:

- Liquids: When two miscible liquids come into contact with one another and macroscopic diffusion occurs, Fick's law applies.
- Fabrication of semiconductor: Integrated circuits are made using Fick's Law of Diffusion equations.
- Pharmaceutical application.
- Food industries.

In the case of a Nuclear Power Plant (NPP) reactor, Fick's Law in reactor theory states the following:

The negative of the neutron flux gradient determines the current density vector $\vec{J}$. The symbol D stands for the proportionality constant, often known as the diffusion coefficient.

In one spatial dimension, Fick's law is, using Figure 2.11, described as

$$\vec{J} = -D\frac{\partial\vec{\phi}(x,t)}{\partial x} \qquad \text{Eq. 2.21}$$

Fick's law, which asserts that solute diffusion (neutron current) from high concentration (high flux) to low concentration is necessary for the development of the diffusion equation. As is evident, we must look at how the flux (ϕ) and the current (J) relate to one another. The law (Fick's law) used to investigate physical diffusion in liquids and gases has the same shape as this connection between flux and current.

In chemistry, Fick's law states that

> Assume that one area of a solution has a higher solute concentration than another. The solute then diffuses with a magnitude proportional to the concentration gradient from the area of greater concentration to the area of lower concentration.

In one spatial dimension, the law is as shown in Equation 2.21 and plotted in Figure 2.12 as well, under Physical Interpretation of the diffusion process.

FIGURE 2.11 Physical interpretation of a diffusion process.

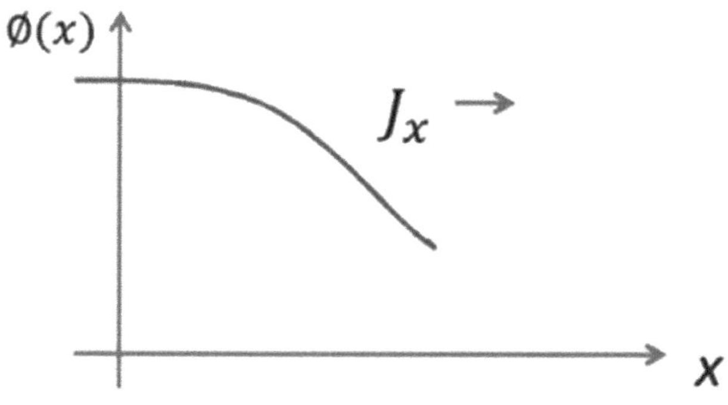

FIGURE 2.12 Diffusion flux approximation in the x-direction.

This law's application to nuclear reactor theory results in the diffusion approximation, which is once more described by Fick's First Law.

$$\vec{J} = -D\frac{\partial \vec{\phi}}{\partial x} \qquad \text{Eq. 2.22}$$

where

$\vec{J}$ = Diffusion flux
D = Diffusion coefficient
$\vec{\phi}$ = Concentration for ideal gas

Its physical interpretation is comparable to that of gas fluxes. There is a net flow of neutrons in the direction with the lowest density. Greater collision densities in locations with higher neutron densities naturally result in this.

Think about neutrons moving from left to right through the plane at $x = 0$ as a result of collisions to the left of the plane. There are more collisions per cubic centimeter on the left because the neutron concentration and flux are higher for negative values of x. As a result, more neutrons are dispersed from left to right than vice versa. As a result, the neutrons naturally go to the right.

2.4.1 Fick's Law Application in Reactor Theory

The Fick's law theory states that

> The negative of the neutron flux gradient determines the current density vector *J*. The symbol *D* stands for the proportionality constant, often known as the diffusion coefficient.

The law is expressed in one spatial dimension as Equation 2.23, which is also comparable to Equation 2.22 for one-dimensional analysis.

$$J = -D\frac{\partial \phi(x)}{\partial x} \qquad \text{Eq. 2.23}$$

where

- Figure 2.6 shows the net flow of neutrons that pass per unit of time across a unit area perpendicular to the x-direction, where J is the neutron current density (i.e., neutrons/cm^2s) along the x-direction.
- D is the diffusion coefficient, it has the unit of centimeter (cm) and is given by

$$D = \frac{1}{3\Sigma_s(1-\bar{\mu})} = \frac{1}{3\Sigma_{tr}} = \frac{\lambda_{tr}}{3}$$

The above relationship has been proved and shown with all its parameters in Section 2.5 of this chapter.

Σ_s = Scattering cross-section of neutron
$\bar{\mu} = 2/(3A)$, where A is the mass number of nuclei in the scattering medium

$\Sigma_{tr} = \frac{1}{\lambda_{tr}} = \Sigma_s\left(1-\bar{\mu}\right)$ = The transport cross-section

λ_{tr} = Transport mean free path

- The quantity of neutrons moving through a given cross-sectional unit area in all directions in a given amount of time is known as the neutron flux, or ϕ.

Furthermore, as we stated before, the use of this law in "Nuclear Reactor Theory" leads to the Diffusion Approximation.

The generalized Fick's First Law in three-dimensional space is given by

$$\vec{J} = -D\nabla\vec{\phi}(x,y,z) \quad \text{Eq. 2.24}$$

where the diffusion flux vector is shown by J. It should be noted that the gradient operator converts the scalar neutron flux into a vector neutron current.

2.4.1.1 Neutron Flux Density in Reactor

The neutron flux $\vec{\phi}$, which represents the total route length taken by all neutrons, is often employed in reactor physics. The overall distance these neutrons may travel in a second is determined by their velocity. As a result, the neutron density (n) times the neutron velocity ($\vec{\mathrm{v}}$) yields the neutron flux density value.

$$\vec{\phi} = n\cdot\vec{\mathrm{v}} \quad \text{Eq. 2.25}$$

where

$\vec{\phi}$ = Neutron flux (neutrons/cm^2s)
n = Neutron density (neutrons/cm^3)
$\vec{\mathrm{v}}$ = Neutron velocity (cm/s)

The number of neutrons that go through a given unit area in all directions in a given amount of time is known as the neutron flux, and it is a scalar quantity. As a result, another name for it is the scalar flux. The total distance covered in a second by all neutrons with energy between E and dE, that are contained inside 1 cm^3, is expressed as $\phi(E)\cdot dE$.

It is clear that the reactor power is related to the Reaction Rate (RR). We may calculate the rate of interactions (such as the rate of fission reactions) by knowing the neutron flux, which is the total route length of all the neutrons in a cubic centimeter in a second, and the macroscopic cross-sections, which is the likelihood of having an interaction per centimeter journey length. They are then multiplied collectively to determine the RR, or the quantity of interactions occurring in that cubic centimeter in a second.

$$\begin{cases} RR = \phi \cdot \Sigma = \phi \cdot (n\sigma) \\ \text{or} \\ RR = \int_0^\infty \Sigma(E) \cdot \phi(E) dE \end{cases} \qquad \text{Eq. 2.26}$$

where

ϕ = Neutron flux (neutrons/cm^{2}s)
σ = Microscopic cross-section (cm^2)
n = Atomic number density (atoms/cm^3)

2.4.1.2 Neutron Current Density

The neutron flow is not described by the neutron flux, φ . Numerous interactions (I) might take place even if there is no neutron flow ($I = \Sigma\phi$). The neutrons may not flow because they travel in random directions. As a result, densities and neutron flux have a closer relationship. When the density of neutrons varies spatially, they will show a net flow. We can thus have a neutron flow!

The neutron current density is the name given to this neutron flow. The neutron current density and the neutron flux must be distinguished. Despite this, both physical values are measured in neutrons. Their physical interpretations differ by cm^{2}s^1. The number of neutrons moving through a given unit area in a single direction in a unit amount of time is known as the neutron intensity, as opposed to the neutron flux (a surface is perpendicular to the direction of the beam). One vector quantity is the neutron current density. Refer to Figure 2.13.

The vector $\vec{J}$ is defined as the following integral

$$\vec{J} \equiv \int_{4\pi} n(\vec{r},\omega)\vec{v}d\Omega \qquad \text{Eq. 2.27}$$

where $\vec{v}$ is velocity in vector quantity instead of speed as scalar quantity as in neutron flux definition and Ω is representation of solid angle.

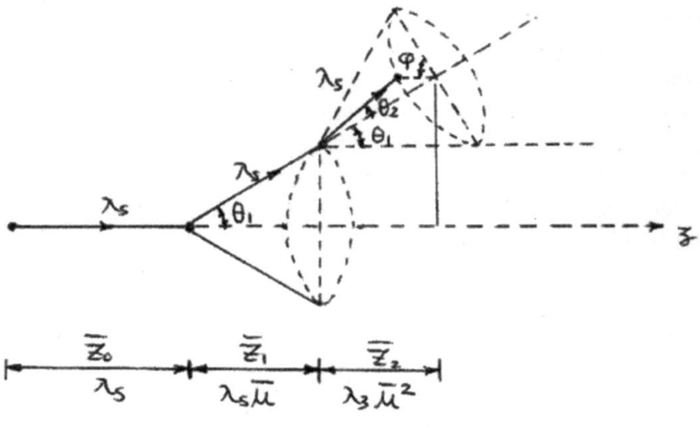

FIGURE 2.13 Geometry for the transport of a particle over three collision.
(Source: Courtesy https://mragheb.com, as cited in [13].)

Neutron Flux 

$$\phi(\mathbf{r},E) \equiv \sum_i vn(\mathbf{r},\boldsymbol{\Omega}_i,E) = \int_{4\pi} vn(\mathbf{r},\boldsymbol{\Omega},E)\,d\boldsymbol{\Omega}$$

$$\mathbf{J}(\mathbf{r},E) \equiv \sum_i \mathbf{v}_i n(\mathbf{r},\boldsymbol{\Omega}_i,E) = \int_{4\pi} \mathbf{v}n(\mathbf{r},\boldsymbol{\Omega},E)\,d\boldsymbol{\Omega}$$

$$\mathbf{v} = v\boldsymbol{\Omega},$$

Neutron Current Density 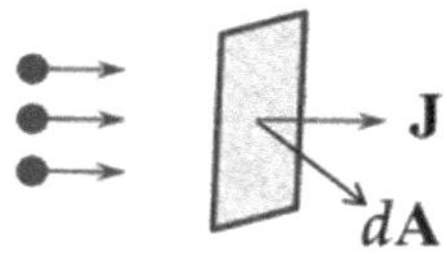

2.4.1.3 Neutrons in a Typical Thermal Reactor

About 100 tons of uranium with an average enrichment of 2% are found in a typical thermal reactor (do not mistake it with the enrichment of the new fuel). Calculate the average core heat flux and the reaction rate if the reactor has a 3,000 MWth power output.

Solution

The total number of reactions taking place in the reactor core per unit time is obtained by multiplying the reaction rate per unit volume ($RR = \phi \cdot \Sigma$) by the total volume of the core (V). However, we also know that each fission process releases around 200 MeV of energy. The rate at which energy is released (power) as a result of the fission process may now be calculated by the following equation.

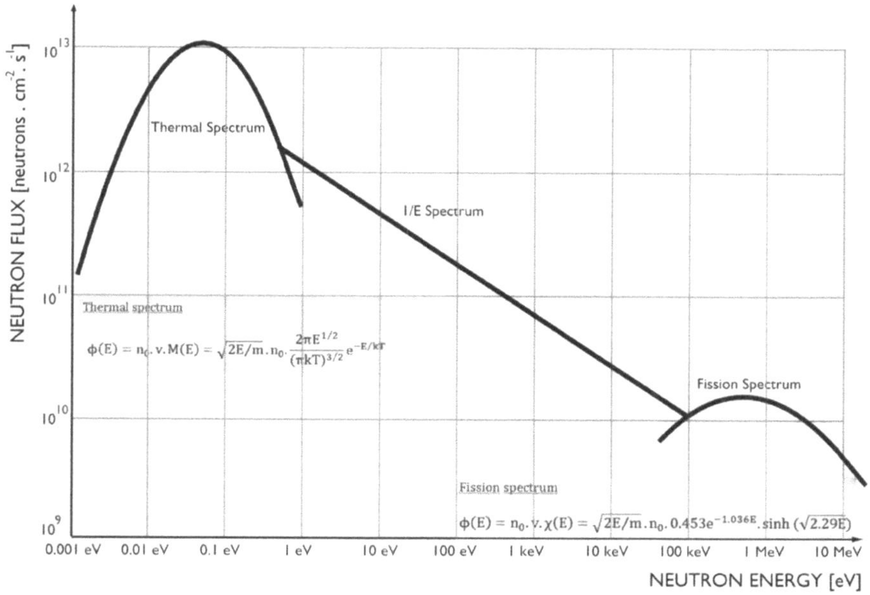

$$P = \phi \cdot \Sigma_f \cdot E_r \cdot V = \phi \cdot N_{U235} \cdot \sigma_f^{235} \cdot E_r \cdot V$$

where

P = Reactor power (MeV.s[1])
ϕ = Neutron flux (neutronscm[2]s[1])
σ = Microscopic cross-section (cm[2])
N = Atomic number density (atomscm[3])
E_r = Average recoverable energy per fission (MeV/fission)
V = Total volume of the core (m3)

The amount of fissile ^{235}U per the volume of the reactor core is a critical parameter that determines the core's reactivity and power generation capability, and it must be carefully controlled to maintain a sustained chain reaction while ensuring operational safety.

M_{235} [g/core] = 100 [metric tons] × 0.02 [g of ^{235}U/g of U] × 10^6 [g/metric ton] = 2 × 10^6 grams of ^{235}U per the volume of the reactor core.

Therefore, the atomic number density of ^{235}U in the volume of the reactor core is

N_{235}. V = m_{235}. N_A/M_{235} = 2 × 10^6 [g 235/core] × 6.022 × 10^{23} [atoms/mol]/235 [g/mol] = 5.13 × 10^{27} atoms/core

The microscopic fission cross-section of ^{235}U (for thermal neutrons) is

σ_f^{235} = 585 barns

The average recoverable energy per ^{235}U fission becomes

E_r = 200.7 MeV/fission

The resulting average neutron flux is

$$\bar{\varphi}_{th} = {}^{P}\!/_{\sigma_f^{235}\,.E_r\,.N_{235}\,.V}$$

$$\bar{\varphi}_{th} = \frac{3000\,[MWth]\,x\,10^6\,[Wth/MWth]}{585\,x\,10^{-24}[cm^2]x\,200.7\,[MeV/fission]x\,5.13\,x\,10^{27}[atoms/core]\,x\,10^6 x\,1.6\,x\,10^{-19}[Wth.s/MeV]}$$

$$\bar{\varphi}_{th} = \frac{3\,x\,10^9}{9.64\,x\,10^{-5}} = 3.11\,x\,10^{13}\,neutrons.cm^{-2}.s^{-1}$$

The resulting reaction rate per entire reactor core [fissions/second] is:

$$RR = \Sigma_f^{235}.\ \bar{\varphi}_{th} = \sigma_f^{235}.N_{235}.\ \bar{\varphi}_{th}$$

$$RR = 585\,x\,10^{-24}[cm^2]x\,5.13\,x\,10^{27}[atoms/core]\,x\,3.11\,x\,10^{13}[neutrons.cm^{-2}.s^{-1}]$$

$$RR = 9.33\,x\,10^{19}\,fissions\,/\,second$$

2.5 NEUTRON DIFFUSION THEORY

Since it is realistic enough to examine many significant design issues and simple enough to allow for scientific insight, the diffusion theory model of neutron transport is essential to reactor theory. The model permits initial design estimations and the neutrons are described by a single energy or speed. Such a model may be analyzed mathematically using the same techniques, as more complex approaches, like transport theory and multigroup diffusion theory.

Although a straight derivation from the transport equation is also feasible, the diffusion equation will be reliant on Fick's law. The constraints of the diffusion equation and the boundary conditions applied to real-world issues are examined, along with the derivation of the Helmholtz equation.

2.5.1 Transport Cross-Section

The transport cross-section is used to examine how the scattering angular distribution affects a neutron's velocity (Lamarsh 2001, cited in other sources).

Using Figure 2.13, let us assume

1. An infinite medium.
2. A purely scattering medium without absorption.
3. When a neutron collides with the medium's nuclei, its energy remains unchanged.
4. The particle follows a Scattering Mean Free Path λ_s and is deflected by an angle θ following each impact.

After injection into the system, the neutron moves a distance, calculated by

$$\bar{Z}_0 = \lambda_s \qquad \text{Eq. 2.28}$$

The distance traveled after the first collision along the z-axis is shown in Figure 2.13.

$$Z_1 = \lambda_s \cos(\theta_1) \qquad \text{Eq. 2.29}$$

The average value of Z_1 is

$$\bar{Z}_1 = \lambda_s \overline{\cos(\theta_1)} = \lambda \bar{\mu} \qquad \text{Eq. 2.30}$$

where $\bar{\mu} = 2/(3A)$ and A = Mass number of the nuclei in the scattering medium.

The distance traveled after the second collision is given by

$$Z_2 = \lambda_s \cos(\theta_1)\cos(\theta_2) \qquad \text{Eq. 2.31}$$

where the azimuthal angle (ϕ) is the scattering isotope.

The average value of Z_2 is given by

$$\bar{Z}_2 = \lambda_s \overline{\cos(\theta_1)\cos(\theta_2)} \approx \lambda_s \left(\overline{\cos(\theta_1)}\right)^2 = \lambda_s \left(\bar{\mu}\right)^2 \qquad \text{Eq. 2.32}$$

In general, for $(n-1)$ collisions, we can state that

$$\bar{Z}_n = \lambda_s (\bar{\mu})^n \quad \text{for} \quad n = 0, 1, 2, \cdots \qquad \text{Eq. 2.33}$$

Given that $\bar{Z} \to 0$ equals $n \to \infty$, the neutron may be equally likely to be dispersed in either direction upon its subsequent encounter. This indicates that following a series of collisions and being transported by the distance of a Transport Mean Free Path λ_{tr}, the neutron loses its initial direction of travel. Equation 2.33 may be used to calculate this.

$$\lambda_{tr} = \bar{Z}_0 + \bar{Z}_1 + \cdots + \bar{Z}_n = \sum \lambda_s \bar{\mu}^n$$
$$= \lambda_s (1 + \bar{\mu} + \bar{\mu}^2 + \cdots + \bar{\mu}^2) \qquad \text{Eq. 2.34}$$
$$= \frac{\lambda_s}{1 - \bar{\mu}} \rightarrow \text{for} \rightarrow \forall \bar{\mu} < 1$$

The transport cross-section Σ_{tr}, using Equation 2.34, can be defined as

$$\Sigma_{tr} = \frac{1}{\lambda_{tr}} = \frac{1}{\frac{\lambda_s}{(1-\bar{\mu})}}$$
$$= \frac{(1-\bar{\mu})}{\lambda_s} = \underbrace{\frac{1}{\lambda_s}}_{\Sigma_s}(1-\bar{\mu}) \qquad \text{Eq. 2.35}$$
$$= \Sigma_s(1-\bar{\mu})$$

If absorption is present, we generalize the definition of the transport cross-section to be (Lamarsh as cited in [14]):

$$\Sigma_{tr} = \Sigma_a + \Sigma_s(1-\bar{\mu})$$
$$= \underbrace{\Sigma_a + \Sigma_s}_{\Sigma_t} - \Sigma_s\bar{\mu} \qquad \text{Eq. 2.36}$$
$$= \Sigma_t - \Sigma_s\bar{\mu}$$

where Σ_t is the total "macroscopic cross-section".

2.5.2 Fick's Law and Diffusion Approximation Derivations

As explained at the beginning of this chapter, there are several basic circumstances under which the neutron flux (ϕ) and current ($\vec{J}$) are connected. The shape of this connection between ϕ and $\vec{J}$ is the same as Fick's law, which is used to examine diffusion processes in liquids and gases.

As previously mentioned, Fick's law in chemistry states: "A solute diffuses from the region of higher concentration to the region of lower concentration if the concentration of a solute in one region of a solution is greater than in another region".

The diffusion approximation is the result of applying this rule to reactor theory. The following presumptions should be made.

1. We consider an infinite medium.
2. The cross-sections indicate a homogenous medium since they are constant regardless of place.
3. Scattering is isotropic in the LAB system. (The LAB system is a frame of reference where the target nucleus is initially at rest, and the incoming particle [e.g., a neutron] is moving toward it.)
4. The neutron flux is a slowly varying function of the position.
5. The neutron density is not a function of time in our one-speed system.
6. A steady state system where the neutron density is not a function of time.
7. No fission source in the system.

Some of these assumptions above can be later relaxed.

In Figure 2.14, we attempt to determine the current density at the coordinate system's center. In a three-dimensional space, the vector is then provided using the following equation.

$$\vec{J} = J_x\hat{i} + J_y\hat{j} + J_z\hat{k}s \qquad \text{Eq. 2.37}$$

Now, we must calculate the components of J_x, J_y , and J_z here. The partial axial current may be used to express these net current components as follows:

$$\begin{aligned} J_x &= J_x^+ - J_x^- p \\ J_y &= J_y^+ - J_y^- \\ J_z &= J_z^+ - J_z^- \end{aligned} \qquad \text{Eq. 2.38}$$

As seen in Figure 2.14, let us focus on the estimation of a single component: $\overline{J}_z$, which crosses the element of area dS_z at the coordinate system's origin in the negative z-direction.

In the x–y plane, every neutron that travels through dS_z is the result of a scattering collision. As a result, neutron scattering above the x–y plane will descend through dS_z.

Using Figure 2.14 to examine the volume element, we can now write

$$dV = r^2 \sin(\theta) dr d\theta d\varphi \qquad \text{Eq. 2.39}$$

Using Equation 2.39, the number of scattering collisions that occur in the volume element dV per unit of time is given as (Lamarsh as cited in [14])

$$\Sigma_s\, \phi(\vec{r}) dV = \Sigma_s\, \phi(\vec{r}) r^2 \sin(\theta) dr d\theta d\varphi \qquad \text{Eq. 2.40}$$

where

$\phi(\vec{r})$ = Particle flux
Σ_s = Macroscopic scattering cross-section

Given that the scattering in the LAB system is isotropic, the percentage that reaches dS_z is subtended by the solid angle $d\Omega$, as shown.

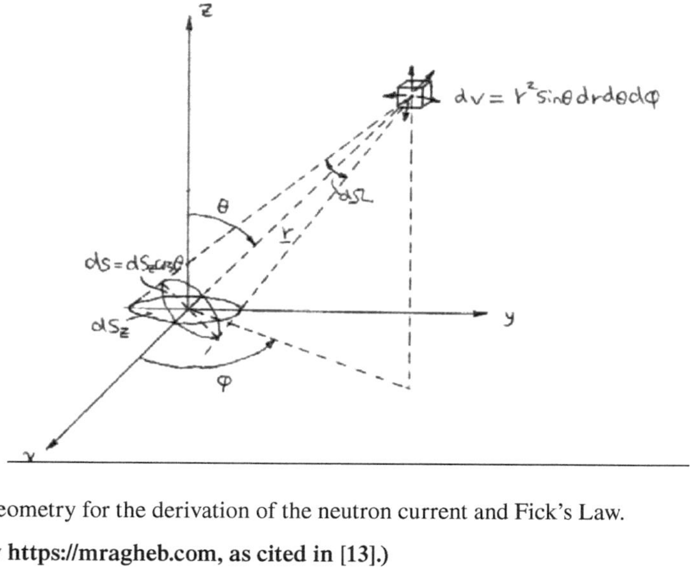

FIGURE 2.14 Geometry for the derivation of the neutron current and Fick's Law.

(Source: Courtesy https://mragheb.com, as cited in [13].)

$$\frac{d\Omega}{\Omega} = \frac{dS/r^2}{4\pi} = \frac{dS_z \cos(\theta)}{4\pi r^2} \qquad \text{Eq. 2.41}$$

Therefore, after being attenuated in the medium by the exponential factor $e^{-\Sigma_t r}$, the number of neutrons dispersed per unit time in dV that reach dS_z is

$$dN = e^{-\Sigma_t r}\phi(\vec{r})\Sigma_s r^2 \sin(\theta) dr d\theta d\varphi \cdot \frac{dS_z \cos(\theta)}{4\pi r^2} \qquad \text{Eq. 2.42}$$

The partial current J_z^- can now be written as (Lamarsh as cited in [14])

$$J_z^- = \frac{\int dN}{dS_z} = \frac{\Sigma_s}{4\pi}\int_0^\infty \int_0^{\pi/2} \int_0^{2\pi} e^{-\Sigma_{tr}} \phi(\vec{r}) \sin(\theta)\cos(\theta) dr d\theta d\varphi \qquad \text{Eq. 2.43}$$

Since $\phi(\vec{r})$ is an unknown function, we expand it in a Taylor's series assuming it varies slowly with position, then we can write

$$\phi(\vec{r}) = \varphi_0 + x\left.\frac{\partial\phi}{\partial x}\right|_0 + y\left.\frac{\partial\phi}{\partial y}\right|_0 + z\left.\frac{\partial\phi}{\partial z}\right|_0 + \cdots \qquad \text{Eq. 2.44}$$

Writing x, y, z in spherical coordinates

$$\varphi(\vec{r}) = \varphi_0 + r\sin(\theta)\cos(\varphi)\left.\frac{\partial\varphi}{\partial x}\right|_0 + r\sin(\theta)\sin(\varphi)\left.\frac{\partial\varphi}{\partial y}\right|_0 + r\cos(\theta)\left.\frac{\partial\varphi}{\partial z}\right|_0 \qquad \text{Eq. 2.45}$$

Substituting Equation 2.45 into Equation 2.43, we obtain (Lamarsh as cited in [14])

$$\begin{aligned} J_z^- &= \frac{\Sigma_s}{4\pi}\int_0^\infty \int_0^{\pi/2} \int_0^{2\pi} e^{-\Sigma_{tr}} \left[\phi_0 + r\sin(\theta)\cos(\varphi)\left.\frac{\partial\varphi}{\partial x}\right|_0\right] \\ &+ \left[r\sin(\theta)\sin(\varphi)\left.\frac{\partial\phi}{\partial y}\right|_0\right] + r\cos(\theta)\left.\frac{\partial\phi}{\partial z}\right|_0 \sin(\theta)\cos(\theta) dr d\theta d\varphi \end{aligned} \qquad \text{Eq. 2.46}$$

The terms containing $\cos(\varphi)$ and $\sin(\varphi)$ integrate to zero over the interval $\varphi \in [0, 2\pi]$, thus Equation 2.46 reduces to the following form (other references [14])

$$J_z^- = \frac{\Sigma_s}{4\pi}\int_0^\infty \int_0^{\pi/2} \int_0^{2\pi} e^{-\Sigma_{tr}} \left[\phi_0 + r\cos(\theta)\left.\frac{\partial\phi}{\partial z}\right|_0\right] \sin(\theta)\cos(\theta) dr d\theta d\varphi \qquad \text{Eq. 2.47}$$

The first term can be evaluated as

$$\begin{aligned} I_1 &= \frac{\Sigma_s}{4\pi}\phi_0 \int_0^\infty \int_0^{\pi/2} \int_0^{2\pi} e^{-\Sigma_{tr}} \sin(\theta)\cos(\theta) dr d\theta d\varphi \\ &= \frac{\Sigma_s}{4\pi}\phi_0 \cdot 2\pi \cdot \frac{1}{\Sigma_t} \cdot \frac{1}{2} \\ &= \frac{1}{4}\frac{\Sigma_s}{\Sigma_t}\phi_0 \end{aligned} \qquad \text{Eq. 2.48}$$

The second term also becomes

$$I_2 = \frac{\Sigma_s}{4\pi} \frac{\partial \phi}{\partial z}\bigg|_0 \int_0^{\infty} \int_0^{\pi/2} \int_0^{2\pi} r e^{-\Sigma_{tr}} \sin(\theta)\cos^2(\theta) dr d\theta d\varphi$$
$$= \frac{\Sigma_s}{4\pi} \frac{\partial \phi}{\partial z}\bigg|_0 \cdot 2\pi \cdot \frac{1}{\Sigma_t} \cdot \frac{1}{3} \qquad \text{Eq. 2.49}$$
$$= \frac{1}{6} \frac{\Sigma_s}{\Sigma_t} \frac{\partial \phi}{\partial z}\bigg|_0$$

By substituting Equations 2.48 and 2.49 into Equation 2.47, we reduce it to the following form:

$$J_z^- = \frac{1}{4} \frac{\Sigma_s}{\Sigma_t} \phi_0 + \frac{1}{6} \frac{\Sigma_s}{\Sigma_t^2} \frac{\partial \phi}{\partial z}\bigg|_0 \qquad \text{Eq. 2.50}$$

Similarly

$$J_z^+ = \frac{1}{4} \frac{\Sigma_s}{\Sigma_t} \phi_0 + \frac{1}{6} \frac{\Sigma_s}{\Sigma_t^2} \frac{\partial \phi}{\partial z}\bigg|_0 \qquad \text{Eq. 2.51}$$

Substituting Equations 2.50 and 2.51 into Equation 2.38, we get the following relationship:

$$J_x = -\frac{1}{3} \frac{\Sigma_s}{\Sigma_t^2} \frac{\partial \phi}{\partial x}\bigg|_0$$
$$J_y = -\frac{1}{3} \frac{\Sigma_s}{\Sigma_t^2} \frac{\partial \phi}{\partial y}\bigg|_0 \qquad \text{Eq. 2.52}$$
$$J_z = -\frac{1}{3} \frac{\Sigma_s}{\Sigma_t^2} \frac{\partial \phi}{\partial z}\bigg|_0$$

Substituting all the terms from Equation 2.52 into Equation 2.37, we obtain the following equation (Lamarsh as cited in [14]).

$$\vec{J} = J_x \hat{i} + J_y \hat{j} + J_z \hat{k}$$
$$= -\frac{1}{3} \frac{\Sigma_s}{\Sigma_t^2} \frac{\partial \phi}{\partial x}\bigg|_0 \hat{i} - \frac{1}{3} \frac{\Sigma_s}{\Sigma_t^2} \frac{\partial \phi}{\partial y}\bigg|_0 \hat{j} - \frac{1}{3} \frac{\Sigma_s}{\Sigma_t^2} \frac{\partial \phi}{\partial z}\bigg|_0 \hat{k}$$
$$= -\frac{1}{3} \frac{\Sigma_s}{\Sigma_t^2} \left[\frac{\partial \phi}{\partial x}\bigg|_0 \hat{i} + \frac{\partial \phi}{\partial y}\bigg|_0 \hat{j} + \frac{\partial \phi}{\partial z}\bigg|_0 \hat{k} \right] \qquad \text{Eq. 2.53}$$
$$= -\frac{1}{3} \frac{\Sigma_s}{\Sigma_t^2} \nabla \phi$$
$$= -\frac{1}{3} \frac{\Sigma_s}{\Sigma_t^2} grad\varphi$$

Here, in Equation 2.53, we define the Diffusion Coefficient D as

$$D = \frac{\Sigma_s}{3\Sigma_t^2} \qquad \text{Eq. 2.54}$$

Thus, Fick's law for neutron diffusion is given by

$$\vec{J} = -D\nabla\vec{\varphi} = -D\nabla\vec{\phi} \text{ Fick's First Law} \qquad \text{Eq. 2.55}$$

Here, in Equation 2.55, in order to keep our notation consistent with the beginning of this chapter, we converted $\varphi = \phi$.

Fick's Law for neutron transport asserts that, given the above assumptions, the current density vector is proportional to the flux's negative gradient and draws a connection between the two. As you can see, the gradient operator transforms the scalar neutron flux into the vector neutron current, represented by $\vec{\nabla} = \frac{\partial}{\partial x}\hat{i} + \frac{\partial}{\partial y}\hat{j} + \frac{\partial}{\partial z}\hat{k}$.

2.6 LIMITATIONS OF DIFFUSION THEORY

The fact that the current density is positive if the flux gradient is negative is expressed by Fick's law. This indicates that by collisions in the medium, the particles will diffuse from the area of higher flux to the area of lower flux. Due to its underlying assumptions, Fick's law places some restrictions on the problems that can be solved. In the following situations, it becomes invalid and requires revisions.

1. **Closeness to Boundaries:**
   An infinite medium was assumed in derivation. Fick's law only applies to points that are more than a few mean free pathways from the medium's edges in a finite medium. This is because only locations of a few number of mean free pathways from the place where the flux is computed significantly contribute to the integral, and the exponential term rapidly dies off with distance.
2. **Proximity to Source or Sink:**
   It was thought that scattering collisions accounted for the majority of the flux contribution. There may be sources. However, if there are more than a few mean free pathways from sources, only a small portion of the source neutrons will contribute to the flux due to the attenuation factor.
3. **Anisotropic LAB System Scattering:**
   Although it is not always the case, isotropic scattering in the LAB system happens at low energy levels. However, if a modified form of the diffusion coefficient based on transport theory is applied, Fick's law remains valid with substantial anisotropy in scattering. For D, such an expression is provided by

$$D = \frac{1}{3\Sigma_s(1-\bar{\mu})\left(1 - \frac{4}{5}\frac{\Sigma_a}{\Sigma_t} + \cdots\right)} \qquad \text{Eq. 2.56}$$

   If $\Sigma_a \ll \Sigma_t$, then $(\Sigma_a/\Sigma_t) \to 0$ and utilization of Equation 2.35, then Equation 2.56 reduces to

$$D = \frac{1}{3\Sigma_s(1-\bar{\mu})} = \frac{1}{3\Sigma_{tr}} = \frac{\lambda_{tr}}{3} \qquad \text{Eq. 2.57}$$

   and we have learned from Equation 2.35 that the following is true.

$$\Sigma_s(1-\bar{\mu}) = \Sigma_{tr} = \frac{1}{\lambda_{tr}} \quad \text{Eq. 2.58}$$

where λ_{tr} is the transport mean free path as we defined earlier in Section 2.5.1 of this chapter.

4. **Highly Absorbing Media:**

   It was anticipated that the flux varied slowly and was enlarged in a Taylor's series. However, the flow varies quickly in medium that absorbs substantially. Fick's law is therefore applicable to systems where

$$\Sigma_a \ll \Sigma_s \quad \text{Eq. 2.59}$$

   When absorption is present, the diffusion coefficient D should be computed using Equation 2.56 but not Equations 2.54 or 2.57.

5. **Proximity to Interfaces:**
   Fick's law was derived under the presumption of a uniform medium. As long as the abrupt shift does not result in a quickly fluctuating flux that renders Taylor's series expansions of the flux in the derivation incorrect, Fick's law remains valid at the boundary between two media with distinct scattering qualities. The second-order derivatives cancel out or integrate to zero in their contribution to J_z. The integral of J is indeed influenced by the third derivative. Therefore, if there is no discernible variation in the flux's second derivative, Fick's law would hold true.
6. **Rapidly Time-Varying Flux:**
   The flow was thought to be time-independent. If there is little change in ϕ throughout the time it takes for a neutron to transit a few mean free routes, this criterion can be relaxed. We may say that the temporal variation of the flux must meet the following requirement if the slow neutrons in a reactor travel along three scattering mean free routes and have a velocity of $v = 10^5$ cm/s.

$$\left[\frac{\phi}{d\phi/dt}\right] \gg \frac{3\lambda_s}{\mathrm{v}} \approx \frac{3\lambda_s}{10^5}(s) \quad \text{Eq. 2.60}$$

   It claims that the typical time for the neutron flux fluctuation is greater than the time required for the neutron to travel three mean free paths.

2.7 DERIVATION OF THE NEUTRON DIFFUSION EQUATION

The following presumptions are made in order to construct the neutron diffusion equation (Lamarsh as cited in [14]).

1. The neutrons are described by a single average kinetic energy in our one-speed or one-group approximation.
2. The particle density $N(\vec{r},t)$ or the number of neutrons per unit volume at a place r at time t, is used to describe the neutron dispersion in the reactor. It has $\phi(\vec{r},t) = \mathrm{v}N(\vec{r},t)$ connections with the flow.

The balancing equation is written as follows, taking into account an arbitrary volume V (Lamarsh as cited in [14]).

$$\begin{aligned}&\left[\textit{Time rate of change of the number of neutrons in } V\right]=\\&\left[\textit{production rate in } V\right]-\left[\textit{Net absorbtion in the surface of } V\right]\\&-\left[\textit{Net leakage from the surface of } V\right]\end{aligned} \quad \text{Eq. 2.61}$$

The first term is expressed mathematically as

$$\frac{d}{dt}\left[\int_V N(\vec{r},t)dV\right]=\frac{d}{dt}\left[\int_V \frac{1}{\mathrm{v}}\phi(\vec{r},t)dV\right]=\frac{1}{\mathrm{v}}\left[\int_V \frac{\partial}{\partial t}\phi(\vec{r},t)dV\right] \quad \text{Eq. 2.62}$$

The Production Rate (PR) can be written as

$$\int_V S(\vec{r},t)dV \quad \text{Eq. 2.63}$$

The Absorption Term (AT) is

$$\int_V \Sigma_a(\vec{r})\phi(\vec{r},t)dV \quad \text{Eq. 2.64}$$

and the Leakage Term (LT) is

$$\int_s \vec{J}(\vec{r},t)\cdot ds=\int_V \nabla\cdot\vec{J}(\vec{r},t)dV \quad \text{Eq. 2.65}$$

where we used Gauss' Theorem or the divergence theorem to transform the surface integral to a volume integral in the generic form shown below (Lamarsh as cited in [14]).

$$\begin{aligned}&\iint_S \vec{F}\cdot\hat{n}\,ds=\iiint_V \nabla\cdot\vec{F}dV\\&\vec{F}=x^3\,\hat{i}+y^3\,\hat{j}+z^3\,\hat{k} \qquad \text{Gauss Theorem}\\&\nabla\cdot\vec{F}=\frac{\partial}{\partial x}(x^3)+\frac{\partial}{\partial y}(y^3)+\frac{\partial}{\partial z}(z^3)\end{aligned} \quad \text{Eq. 2.66}$$

Taking Equations 2.63–2.65 and substituting them for the differential terms in the balance Equation 2.61, we obtain

$$\int_V \frac{1}{\mathrm{v}}\frac{\partial\phi(\vec{r},t)}{\partial t}dV=\int_V S(\vec{r},t)dV-\int_V \Sigma_a(\vec{r})\phi(\vec{r},t)dV-\int_V \nabla\cdot\vec{J}(\vec{r},t)dV \quad \text{Eq. 2.67}$$

or manipulation of Equation 2.67, yields the following result as

$$\int_V\left(\frac{1}{\mathrm{v}}\frac{\partial\phi(\vec{r},t}{\partial t}-S(\vec{r},t)+\Sigma_a\,\phi(\vec{r},t)+\nabla\cdot\vec{J}\right)dV=0 \quad \text{Eq. 2.68}$$

Since the volume V is arbitrary, the integration of Equation 2.69 over such volume results in the following form as

$$\begin{cases}\left(\frac{1}{\mathrm{v}}\frac{\partial\phi(\vec{r},t}{\partial t}-S(\vec{r},t)+\Sigma_a\,\phi(\vec{r},t)+\nabla\cdot\vec{J}\right)=0\\ \frac{1}{\mathrm{v}}\frac{\partial\phi(\vec{r},t}{\partial t}=-\nabla\cdot\vec{J}-\Sigma_a\,\phi(\vec{r},t)+S(\vec{r},t)\end{cases} \quad \text{Eq. 2.69}$$

We now use the relationship between $\vec{J}$ and ϕ (Fick's First Law, Equation 2.55) to write the diffusion equation as a new form here (Lamarsh as cited in [14])

$$\frac{1}{\text{v}}\frac{\partial\phi(\vec{r},t}{\partial t} = \nabla\cdot\left[D(\vec{r})\nabla\phi(\vec{r},t)\right] - \Sigma_a\,\phi(\vec{r},t) + S(\vec{r},t) \qquad \text{Eq. 2.70}$$

A large portion of the advancement in reactor theory employing diffusion theory is based on this equation.

2.8 THE HELMHOLTZ EQUATION

Diffusion Equation 2.71 is a parabolic type partial differential equation. It also explains the physical processes of material diffusion, gas diffusion, and heat conduction (Lamarsh 2001 as cited in other sources).

When the medium is homogeneous or uniform, this equation becomes simpler: D and Σ_a are independent of location $\vec{r}$ as follows:

$$\frac{1}{\text{v}}\frac{\partial\phi(\vec{r},t)}{\partial t} - D\nabla^2\phi(\vec{r},t) + \Sigma_a\,\phi(\vec{r},t) = S(\vec{r},t) \qquad \text{Eq. 2.71}$$

where we applied the knowledge that the Laplacian operator is reached by the gradient's divergence.

$$\nabla\cdot\nabla = \nabla^2 \qquad \text{Eq. 2.72}$$

The Laplacian operator ∇^2, which serves as a memory refresher, is dependent on the coordinate system in use.

$$\text{Cartesian: } \nabla^2 \equiv \frac{\partial^2}{\partial x^2} + \frac{\partial^2}{\partial y^2} + \frac{\partial^2}{\partial z^2}$$

$$\text{Cylindrical: } \nabla^2 \equiv \frac{1}{r}\frac{\partial}{\partial r}\left(r\frac{\partial}{\partial r}\right) + \frac{1}{r^2}\frac{\partial^2}{\partial\theta^2} + \frac{\partial^2}{\partial z^2}$$

$$\text{Spherical: } \nabla^2 \equiv \frac{1}{r^2}\frac{\partial}{\partial r}\left(r^2\frac{\partial}{\partial r}\right) + \frac{1}{r^2\sin(\theta)}\frac{\partial}{\partial\theta}\left(\sin(\theta)\frac{\partial}{\partial\theta}\right) + \frac{1}{r^2\sin^2(\theta)}\frac{\partial^2}{\partial\varphi^2}$$

In case of symmetry around the angular variables θ and φ, these equations simplify the one-dimensional case into

$$\nabla^2 \equiv \frac{1}{r^\alpha}\frac{d}{dr}\left(r^\alpha\frac{d}{dr}\right)$$

where

$\alpha = 0$, for Cartesian coordinates
$\alpha = 1$, for Cylindrical coordinates
$\alpha = 2$, for Spherical coordinates

and the partial derivative has been replaced by the total derivative.

We utilize the scalar Helmholtz equations or the steady state diffusion equation when the flow is not a function of time (Lamarsh as cited in [14]).

$$\begin{cases} D\nabla^2\phi(\vec{r}) - \Sigma_a\,\phi(\vec{r}) + S = 0 \\ \nabla^2\phi(\vec{r}) - \dfrac{\Sigma_a}{D}\phi(\vec{r}) + \dfrac{S}{D} = 0 \end{cases} \quad \text{Eq. 2.73}$$

This is an elliptic-type PDE (Partial Differential Equation).

One way to express the Helmholtz equation is as follows:

$$\nabla^2\phi(\vec{r}) - \frac{1}{L^2}\phi(\vec{r}) = -\frac{S(\vec{r})}{D} \quad \text{Eq. 2.74}$$

where $L^2 = D/\Sigma_a$, and L is the diffusion length.

In contrast, if the neutron flux and density are location-independent, we may get from Equation 2.71, taking div $\vec{J} = 0$ into consideration.

$$\frac{1}{\text{v}}\frac{d\phi(t)}{dt} = S(t) - \Sigma_a\,\phi(t) \quad \text{Eq. 2.75}$$

which is a time-dependent equation in the flux.

2.9 BOUNDARY CONDITIONS FOR THE STEADY STATE DIFFUSION EQUATION

The following boundary conditions must be described mathematically for the Helmholtz equation: either ϕ or the normal derivative $\partial\phi/\partial n$, or a linear combination of the two.

It is not possible to specify ϕ and $\partial\phi/\partial n$ separately. Boundary conditions are often defined using physical reasoning that does not go against this requirement.

1. **Vacuum Boundary Conditions:**

Reactor calculations may consider air as a vacuum since its mean free path is significantly bigger than that of the reactor. Equation 2.51 may be written as follows if we assume that there are no neutrons reflected to the reactor core from the vacuum (Figure 2.15).

$$J^-(x) = \frac{1}{4}\frac{\Sigma_s}{\Sigma_t}\phi_0 + \frac{1}{6}\frac{\Sigma_s}{\Sigma_t^2}\frac{\partial\phi}{\partial x}\bigg|_0 = 0 \quad \text{Eq. 2.76}$$

Recalling from the Fick's law derivation

$$D = \frac{\Sigma_a}{3\Sigma_t^2} \quad \text{Eq. 2.77}$$

Substituting Equation 2.77 into Equation 2.76, we can write

$$J^-(x) = \frac{1}{4}\frac{\Sigma_s}{\Sigma_t}\phi_0 + \frac{D}{2}\frac{\partial\phi}{\partial x}\bigg|_0 = 0 \quad \text{Eq. 2.78}$$

If, at the boundary, the materials are mostly a scatter ($\Sigma_a \ll \Sigma_t$), then $(\Sigma_s/\Sigma_t) \simeq 1$, and we can substitute for $D = \dfrac{\lambda_{tr}}{3}$

$$\frac{1}{\phi_0}\frac{\partial\phi}{\partial x}\bigg|_0 \approx -\frac{1}{2D} \approx -\frac{3}{2}\frac{1}{\lambda_{tr}} \qquad \text{Eq. 2.79}$$

However, from the geometry of Figure 2.15, if the diffusion theory is linearly extrapolated, we can write

$$\frac{\phi_0}{d} = \tan(\theta) = \frac{\partial\phi}{\partial x}\bigg|_0 \qquad \text{Eq. 2.80}$$

where d is the "extrapolated length".

Comparing Equations 2.79 and 2.80, we get the following result

$$\frac{1}{\phi_0}\frac{\partial\phi}{\partial x}\bigg|_0 \approx -\frac{3}{2}\frac{1}{\lambda_{tr}} = -\frac{1}{d} \qquad \text{Eq. 2.81}$$

from which

$$d \approx \frac{2}{3}\lambda_{tr} \qquad \text{Eq. 2.82}$$

In the case of curved borders, more intricate analyses provide more complex formulations for d. For example, it is provided for flat geometry by

$$d = 0.7104\lambda_{tr} \qquad \text{Eq. 2.83}$$

"The flux vanishes at the extrapolated distance beyond the edge of the reactor surface" is the new enunciation of the boundary condition.

$$\phi(R+d) = 0 \qquad \text{Eq. 2.84}$$

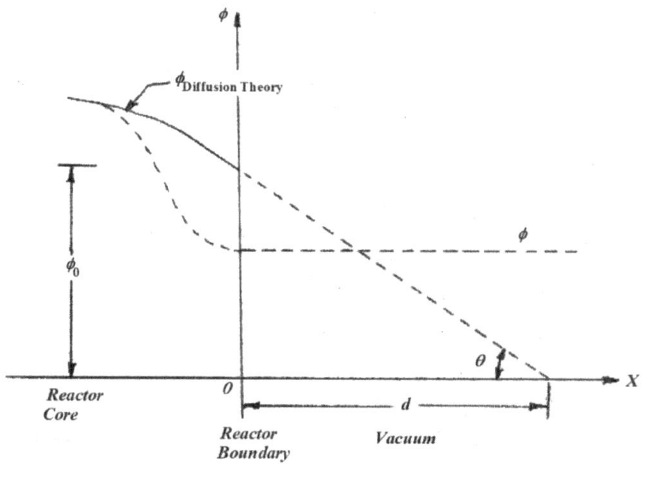

FIGURE 2.15 Geometry for the estimation of the extrapolation length in diffusion theory.

(Source: Courtesy https://mragheb.com, as cited in [13].)

where R is the reactor radius.

2. **Interface Boundary Conditions:**

Across the border between two distinct media, A and B, the flow is constant.

$$\phi_A\big|_{\text{interface}} = \phi_B\big|_{\text{interface}} \quad \text{Eq. 2.85}$$

Therefore, the current density normal components at the border need to be equal.

$$(J_A)_n\big|_{\text{interface}} = (J_B)_n\big|_{\text{interface}} \quad \text{Eq. 2.86}$$

3. **Physical Requirement Boundary Conditions:**

The diffusion equation solution needs to be single-valued, real, and non-negative.

$$\phi \geq 0 \quad \text{Eq. 2.87}$$

Additionally, the diffusion equation's solution needs to be finite in the areas where it applies, with the exception of singular points in the source distribution; however, this is not always the case.

$$\phi < \infty \quad \text{Eq. 2.88}$$

Unnecessary functions can be removed from the solutions using these requirements. This opens the door for flux and power distribution calculations in real-world reactor design scenarios (other references [14]).

BIBLIOGRAPHY

Nuclear and Reactor Physics

1. J. R. Lamarsh, *Introduction to Nuclear Reactor Theory*, 2nd edition, Addison-Wesley, Reading, MA, 1983.
2. J. R. Lamarsh, A. J. Baratta, *Introduction to Nuclear Engineering*, 3rd edition, Prentice-Hall, 2001.
3. W. M. Stacey, *Nuclear Reactor Physics*, John Wiley & Sons, 2001.
4. S. Glasstone, *Nuclear Reactor Engineering: Reactor Systems Engineering*, 4th edition, Springer, 1994.
5. W. S. C. Williams, *Nuclear and Particle Physics*, 1st edition, Clarendon Press, 1991.
6. G. R. Keepin, *Physics of Nuclear Kinetics*, 1st edition, Addison-Wesley Pub. Co, 1965
7. R. R. Burn, *Introduction to Nuclear Reactor Operation*, 1988.
8. U.S. Department of Energy, Nuclear Physics and Reactor Theory. *DOE Fundamentals Handbook*, Volume 1 and 2, 1993.

Advanced Reactor Physics

1. K. O. Ott, W. A. Bezella, *Introductory Nuclear Reactor Statics*, Revised edition, American Nuclear Society, 1989.
2. K. O. Ott, R. J. Neuhold, *Introductory Nuclear Reactor Dynamics*, American Nuclear Society, 1985.
3. D. L. Hetrick, *Dynamics of Nuclear Reactors*, American Nuclear Society, 1993.
4. E. E. Lewis, W. F. Miller, *Computational Methods of Neutron Transport*, American Nuclear Society, 1993.

Other References

1. M. Ragheb, *Lecture Notes on Fission Reactors Design Theory*, FSL-33, University of Illinois, 1982.
2. J. R. Lamarsh, *Introduction to Nuclear Engineering*, Addison-Wesley Publishing Company, 1983. http://www.mragheb.com/NPRE%20402%20ME%20405%20Nuclear%20Power%20Engineering/index.htm.

3 Nuclear Reactor Physics and Theory

3.1 INTRODUCTION

The branch of physics known as "nuclear reactor physics" is concerned with the engineering and practical study of chain reactions to produce energy in nuclear reactors at a controlled rate of fission [1]. A nuclear power plant's (NPP) core is the nuclear reactor, which, in the case of fissionable nuclear reactors, as shown in Figure 3.1, contains and regulates nuclear chain reactions that generate heat through a physical process known as fission.

This heat is used to make steam through a steam generator system that spins a turbine to generate electricity. A reactor's primary function is to contain and regulate nuclear fission processes, which produce energy as atoms break.

Nuclear power is one of the biggest sources of dependable, carbon-free electricity available, with over 440 operational reactors globally, including 94 in the United States*.

Uranium is used as nuclear fuel in reactors. Small ceramic pellets made from uranium are piled in enclosed metal tubes known as fuel rods. To create a fuel assembly, more than 200 of these rods are usually grouped together. Depending on the power level, a reactor core usually consists of a few hundred components.

The water that surrounds the fuel rods within the reactor vessel serves as a moderator and coolant. In order to maintain the chain reaction, the moderator aids in slowing down the neutrons created during fission. Thereafter, the control rods can be removed to speed up the reaction or put back into the reactor core to slow it down. The water is converted to steam by the heat produced by fission, and this steam drives a turbine to provide carbon-free power.

All commercial nuclear reactors in the United States today are Generation-III (GEN-III) reactors, which are classified as Light Water Reactors (LWRs) and use ordinary water as a coolant and neutron moderator. The two categories of LWRs that are now in operation in the USA are as follows:

1. **Pressurized Water Reactors**

Pressurized water reactors, or PWRs, make up over 65% of all commercial reactors in the United States. In order to keep the water from boiling, as shown graphically in Figure 3.2, these reactors push water into the reactor core under tremendous pressure.

In this type of reactor, the water in the core is heated by nuclear fission and then pumped into tubes inside a heat exchanger. These tubes heat a separate water source to create steam. The steam is then turned into an electric generator to produce electricity.

The cycle is repeated with the core water cycling back to the reactor for reheating.

2. **Boiling Water Reactors**

Boiling Water Reactors (BWRs) make up around one-third of all reactors in operation in the United States. BWRs generate steam and heat water inside the reactor vessel. Fission heats the water as it is

* IAEA

DOI: 10.1201/9781003590194-3

FIGURE 3.1 Large pool containing reactor core.

(Source: Department of Energy [14])

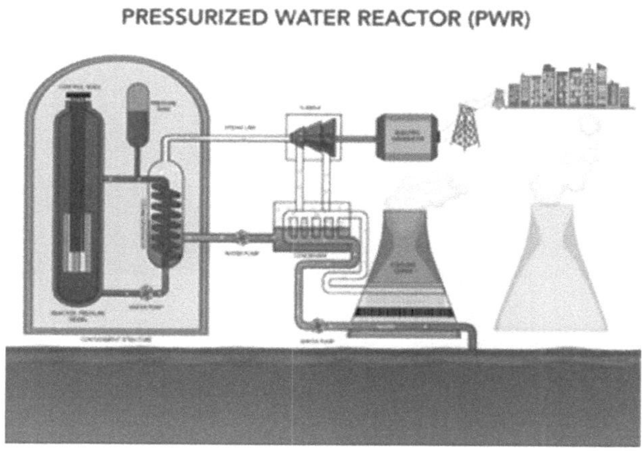

FIGURE 3.2 Pressurized water reactor depiction.

(Source: Department of Energy [14])

forced up through the reactor core. As Figure 3.3 illustrates graphically, pipes subsequently supply the steam straight to a turbine, which generates power.

The unused steam is then condensed back to water and reused in the heating process.

In general, research is done by the Light Water Reactor Sustainability (LWRS) program to create technologies and other solutions that will increase the economy and dependability, maintain the safety, and prolong the lifespan of our country's fleet of NPPs. In terms of long-term operations, it has two goals:

1. To aid the industry in overcoming the existing labor-intensive business model and related practices by offering science and technology-based solutions.
2. To control the aging of Systems, Structures, and Components (SSCs) in order to maintain the economical and safe operation of NPPs.

The following three objectives are the main focus of the LWRS program.

1. Establishing the basic scientific foundation for comprehending, forecasting, and quantifying aging-related changes in materials and SSCs in settings linked to the ongoing, long-term operations of current NPPs (i.e., life cycle).
2. Using this foundational information to create and illustrate techniques and technologies that promote long-term, safe, and cost-effective operation of nuclear power reactors that are now in operation.
3. Investigating novel technologies to improve the safety, economics, and performance of NPPs.

Over the last 20 years, nuclear power has consistently and profitably contributed to around 19% of electricity output in the United States. With over 56% of the nation's non-greenhouse gas-emitting electric power generation, it continues to be the single greatest source.

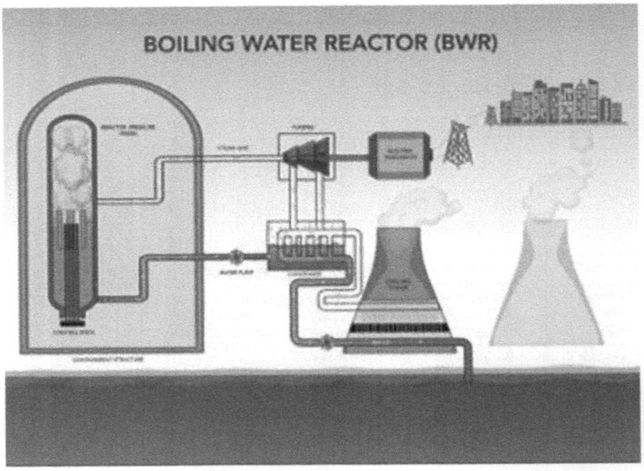

FIGURE 3.3 Boiling water reactor depiction.

(Source: Department of Energy [14])

FIGURE 3.4 Typical light water reactor of Generation-III (GEN-III) in operation.

(Source: Department of Energy [14])

To fulfill the nation's energy needs – supply, dependability, and diversity – it is imperative that the current fleet of plants be operated for 60 years, their working lives be extended beyond 60 years, and, where feasible, additional productivity gains be made. An overhead view of an existing LWR undergoing a review of the LWRS program is shown in Figure 3.4.

In order to start the process of renewing their licenses and extending their operational license duration beyond 60 years from the date of their first licensing, domestic nuclear utilities started applying to the Nuclear Regulatory Commission (NRC) in 2018 or declaring plans to do so. Actually, Florida Power and Light's request to extend the licenses for its Turkey Point Nuclear reactors 3 and 4 was granted by NRC staff on December 5, 2019, enabling the company to continue operating the reactors until 2052 and 2053, respectively. Reactor operating licenses have been extended from 60 to 80 years for the first time by the NRC. In the history of commercial nuclear power operations in the United States, this is a significant, planned milestone that highlights the long-term reliability of these plant designs and the dedication of the companies operating them to their long-term performance.

There are other advanced reactor technologies in the pipe that we will talk about later and the ongoings with Advance Reactor Demonstration Program (ARDP) under auspices United States Department of Energy (DOE) and its Office of Nuclear Energy (ONE).

The neutron population of a nuclear reactor is generally determined by the rate of neutron generation (fission processes) and the rate of neutron losses (leakage from the system and non-fission absorption mechanisms). The fission chain reaction is self-sustaining when the neutron population of a reactor stays constant from generation to generation (producing as many new neutrons as are lost), and the reactor is said to be in "critical" condition. The reactor is said to be "supercritical" when its neutron output surpasses losses, as shown by a growing power level, and "subcritical" when losses predominate, as indicated by a declining power level.

Reactor criticality exists when there is a balance between the number of neutrons produced in fission and the number lost, either by absorption in the reactor or leakage taking place from its surface. One of the main driving factors in design of a reactor core is the calculation of the size and composition of the system required to maintain this balance requirement. The criticality condition of reactor core is driven by solving neutron diffusion equations as we described in the previous chapter and gets carried out by the group diffusion method that will be introduced in this chapter going forward. Thus, we start our discussion with one-group reactor equation and move on to two-groups and eventually to multi-groups with modified form of computation of some thermal reactors.

In conclusion, neutrons in a nuclear reactor can break certain very heavy nuclei (such as ^{235}U) into two pieces, releasing a significant amount of energy and possibly some new neutrons that can lead to other fissions. A steady state chain reaction, in which energy is continually released, is started if, on average, one of the neutrons produced in a fission triggers another fission. About 200 MeV is released during a fission (1 eV = 1.6×10^{-19} J). This is a huge quantity of energy when compared to a chemical process, which releases a few electron Volts (eV; for instance, compare the burning of 1 m^3 of natural gas, which has an energy content of 30 MJ). For details, please refer to Figure 3.5.

Even if a ^{235}U ball, which is slightly bigger than a tennis ball, may maintain a chain reaction, such a structure is not appropriate for producing a significant quantity of energy. In order to do this, a substantially greater quantity of uranium is put in the reactor core, and a coolant that can efficiently remove the heat is applied. Furthermore, pure ^{235}U is not found in the natural world.

Table 3.1 lists the weight percentages of the isotopes ^{234}U, ^{235}U, and ^{238}U that make up natural uranium, which is present in different ore levels in the earth's crust.

This will make it clear why, for example, ^{238}U is significantly less suitable for neutron fission, making the proportion of ^{235}U, or the degree of enrichment, crucial.

A reactor's nuclear fuel is enriched to a level of around 3% to 5% for economic reasons. As shown in Figure 3.6, a reactor core is created by assembling several fuel pieces in a large reactor vessel.

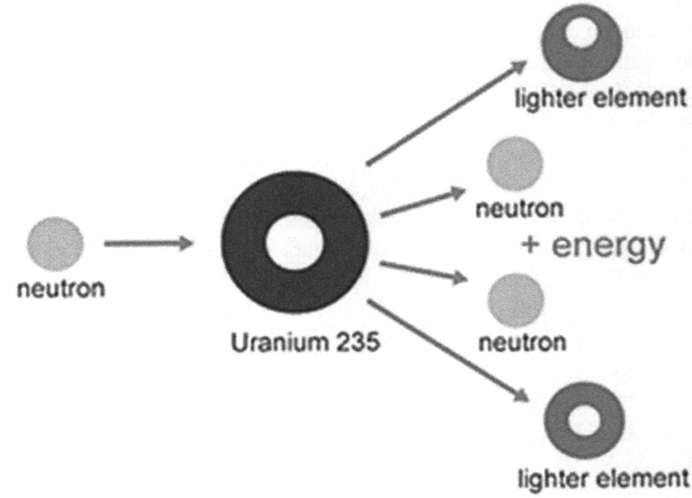

FIGURE 3.5 The Uranium ($_{92}U^{235}$) atom split by energetic neutron particle.

TABLE 3.1
Composition of Natural Uranium

Isotope	Weight Percentage	Mass Number	Radioactive Half-Life
$_{92}U^{234}$	0.006	234.0409	2.4×10^5 a
$_{92}U^{235}$	0.712	235.0439	7.0×10^8 a
$_{92}U^{238}$	99.282	238.0508	4.5×10^9 a

The coolant, which is primarily water, travels between a number of long, thin rods that contain uranium (often in the form of UO_2) in a rectangular fuel lattice. The nuclear fuel is enclosed by a metal covering to stop chemical interactions between it and the coolant.

Figure 3.6 is the presentation of a typical and possible shape of a fuel assembly and as we can see in an LWR-type reactor, the heat generated in the nuclear fuel induced by a fissionable process as depicted in Figure 3.5, and this heat is transferred to the cooling water, which is pumped upward along the rods as depicted in Figures 3.2 and 3.3. In this process, the water can start to boil in the case of BWR, then the steam is subsequently driven to a steam turbine via the Intermediate Heat Exchanger (IHX), where the blades of the turbine begin to rotate.

The turbine shaft, where the blades are attached, in return, drives an electrogenerator, which produces the electric energy to be supplied to electrical power network grids.

The water is circulated through the core again once the turbine's expanded steam has condensed. The steam is condensed using cooling water, which is obtained from rivers or seas, or a cooling tower. A schematic illustration of one of these so-called BWRs is shown in Figure 3.6.

In PWR (refer to Figure 3.7), the water pressure is maintained at a significantly greater level to prevent boiling in the core. A steam generator receives the heated water and transfers heat to the secondary side, which is under lower pressure, causes the water to begin boiling. The steam then makes its way back to the turbine.

In a BWR or PWR, the water slows down the energy of the neutrons in addition to acting as a coolant. As we shall see in the following section, the neutrons emitted during fission have a high energy, whereas the neutrons with a low energy have a greater likelihood of triggering a new fission.

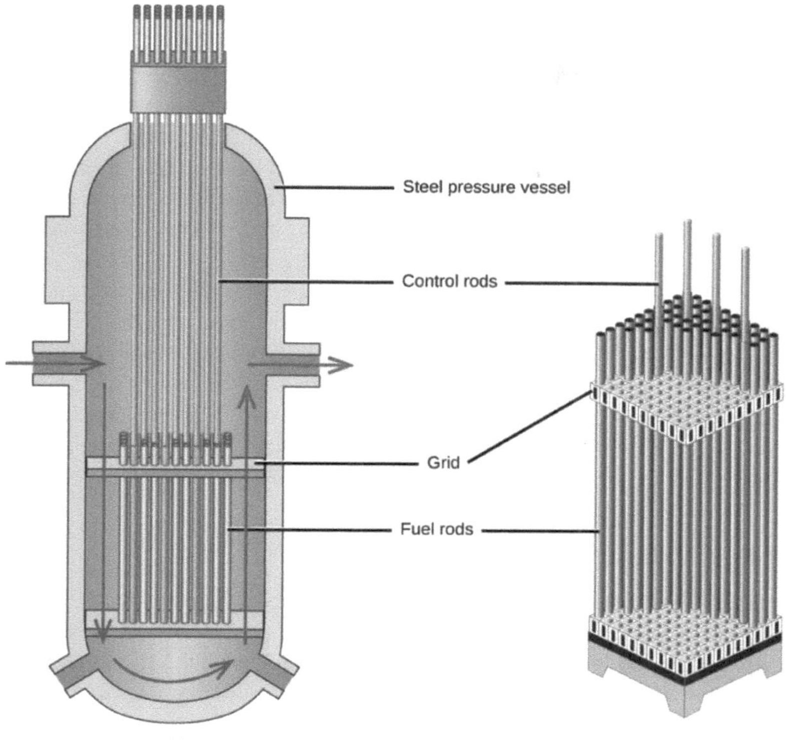

FIGURE 3.6 The nuclear reactor core shown in (a) contains the fuel and control rod assembly shown in (b).
(Source: Modification work by E. Generalic, http://glossary.periodni.com/glossary.php?en=control+rod)

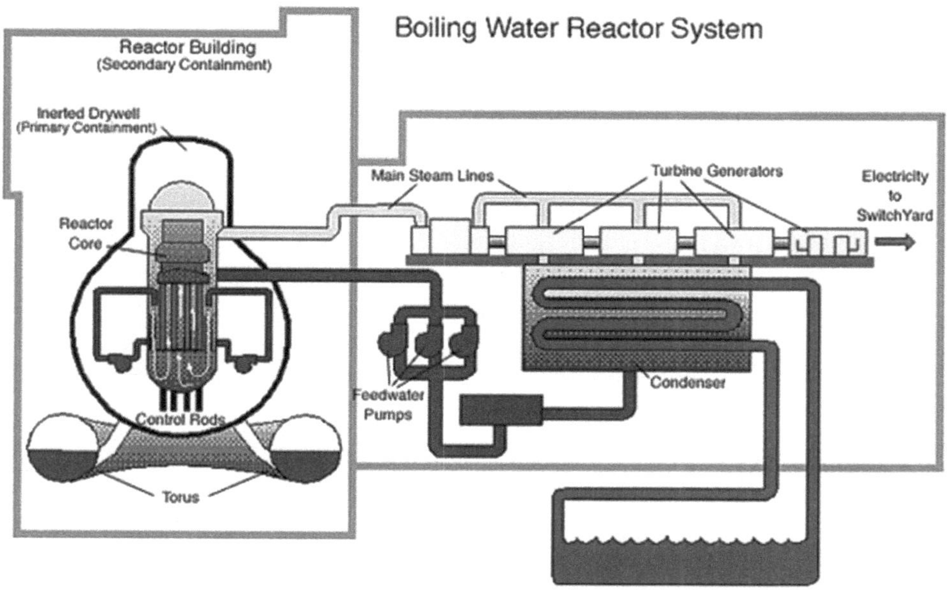

FIGURE 3.7 Principal sketch of a boiling water reactor (BWR).
(Source: Department of Energy [14])

Water also acts as a moderator because neutrons lose energy when they collide with light nuclei like hydrogen in water.

Ultimately, the neutrons receive an energy distribution that matches the atomic nuclei's heat movement. As a result, thermal reactors are another name for reactors that need a moderator. Neutrons in reactors without a moderator primarily maintain a high energy. We refer to these reactors as fast reactors. To understand why energy is released in nuclear fission and why only specific nuclei may be used, we shall first examine the fission process in the following sections. Additionally, the likelihood of interactions will be measured, and additional potential neutron–nucleus interactions will be explored.

The movement of neutrons in a reactor core is, therefore, described by an equation that may be obtained in the next chapter. The requirements that a reactor core must meet in order to facilitate a self-sustaining chain reaction of fissions may be inferred from the solution of this equation (or a simplified version).

Again, as we stated before, their technology reactor design is getting pushed and thriving to Generation IV (GEN-IV) and Advance Reactor Concepts (ARCs) that are more modularized, smaller in size, and are known as Small Module Reactors (SMRs).

3.2 ONE-GROUP REACTOR THEORY

The steady state neutron diffusion was considered; the theory was presented in Chapter 1 as Fick's First Law and was specialized to the situation of multiplying media. Specifically, the source term in the Helmholtz Equation 2.74 is expressed as a function of the fission medium's multiplication factor. By using the Helmholtz Equation 2.74 in this one-group reactor theory, one important assumption is that the reactor of this model is a "bare reactor", where we consider a critical fast reactor containing a homogeneous mixture of fuel and coolant. It is also assumed that the reactor consists of only one region and has neither a blank nor a reflector.

Thus, this reactor is described in a one-group calculation by the one-group time-dependent diffusion Equation 2.72 in form of the new equation.

$$D\nabla^2\phi - \Sigma_a\phi + S = \frac{1}{\mathrm{v}}\frac{\partial\phi}{\partial t} \qquad \text{Eq. 3.1}$$

where D [cm] is the neutron diffusion coefficient and is not a function of position, Σ_a [1/cm] is called macroscopic absorption cross-section, S is the source and the rate at which neutrons are emitted from sources per cm^3 in volume V [cm^3] of the given reactor internal space under consideration, and the rate at which neutrons are produced throughout volume is given by Equation 2.62, and finally v is the neutron speed. For the purpose of this analysis, here in this section, only time-independent problems are considered, thus, Equation 3.1 becomes

$$D\nabla^2\phi - \Sigma_a\phi + S = 0 \qquad \text{Eq. 3.2}$$

This implies a steady state diffusion equation and it is often convenient. Dividing Equation 3.2 by parameter D, we get

$$\nabla^2\phi - \frac{1}{L^2}\phi = -\frac{S}{D} \qquad \text{Eq. 3.3}$$

where the parameter L^2 is defined as

$$L^2 = \frac{D}{\Sigma_a} \qquad \text{Eq. 3.4}$$

Quantity L shows up frequently in nuclear engineering problem, called the "diffusion length" and thus, the quantity L^2 is called the "diffusion area". Since parameter D and Σ_a have dimension of [cm] and [1/cm], respectively, it forces the dimension of L^2 to become [cm^2] and has unit or dimension of area and L holds dimension or unit of centimeter. A physical interpretation of L and L^2 is given later on this chapter in Section 3.5.

3.2.1 Steady State Neutron Diffusion Equation

The one-group diffusion equation can be written as neutron balance equation over an arbitrary volume V of a multiplying or non-multiplying medium as follows [8–10]

$$\frac{\partial n}{\partial t}=\left[\frac{\text{Neutron Produced}}{\text{Unit Volume}\times\text{Unit Time}}\right]-\left[\frac{\text{Neutron Lost by Absorption}}{\text{Unit Volume}\times\text{Unit Time}}\right]-\left[\frac{\text{Neutron Lost by Leakage}}{\text{Unit Volume}\times\text{Unit Time}}\right] \quad \text{Eq. 3.5}$$

where n is the density of neutrons at any point and time in volume V.

If the system is just critical, then

Loss = Gain,
and

$$\frac{\partial n}{\partial t}=0 \quad \text{Eq. 3.6}$$

for a steady state reactor operational condition.

The diffusion Equation 3.5 can be symbolically written as below [10–12]

$$\frac{\partial n}{\partial t}=0=S-\Sigma_a(\phi+D\nabla^2\phi) \quad \text{Eq. 3.7}$$

where ϕ is the neutron flux $\left[\frac{\text{n}}{\text{cm}^3\times\text{s}}\right]$, Σ_a is the macroscopic absorption coefficient [1/cm], and D is the diffusion coefficient [cm].

3.2.2 Diffusion Coefficient

Diffusion theory provides the following expression for the diffusion coefficient in terms of the macroscopic scattering cross-section [8–9].

$$D=\frac{1}{3\Sigma_s} \quad \text{Eq. 3.8}$$

where Σ_s is the macroscopic scattering cross-section.

Nonetheless, the macroscopic transport Σ_{tr}, scattering Σ_s, and absorption Σ_a cross-sections may be used to represent them using the more sophisticated transport theory as follows:

$$\begin{aligned}D&=\frac{1}{3[\Sigma_a+\Sigma_s(1-\bar{\mu})]}\\&=\frac{1}{3[\Sigma_a+\Sigma_{tr}]}\end{aligned} \quad \text{Eq. 3.9}$$

where again, Σ_{tr} is the macroscopic transport cross-section, which is equal to

$$\Sigma_{tr} = \Sigma_s(1-\bar{\mu}_0)$$

$$\lambda_{tr} = \frac{1}{\Sigma_{tr}} = \frac{1}{\Sigma_s(1-\bar{\mu}_0)} \qquad \text{Eq. 3.10}$$

where

λ_{tr} = Transport mean free path
Σ_a = Macroscopic absorption cross-section
Σ_s = Macroscopic scattering cross-section
$\bar{\mu}_0$ = Average cosine of the scattering angle in the laboratory system

In a weakly absorbing medium where $\Sigma_a \ll \Sigma_s$, and after the utilization of Equation 2.35, the neutron coefficient D becomes as before (i.e., see Equations 2.56 and 2.57).

$$D = \frac{1}{3\Sigma_{tr}} = \frac{\lambda_{tr}}{3} \qquad \text{Eq. 3.11}$$

The transport mean free route (λ_{tr}) is the average distance a neutron will travel in its original direction following an unlimited number of scattering collisions, and it may alternatively be expressed as follows:

$$\bar{\mu} = \overline{\cos \vartheta} \qquad \text{Eq. 3.12}$$

As shown in Figure 3.7, taking into account the average value of the cosine of the angle in the lab system at which neutrons are dispersed in the medium. For the majority of neutron energies, it may be computed as follows (where A is the target nucleus mass number).

$$\bar{\mu} = \frac{2}{3A} \qquad \text{Eq. 3.13}$$

We may say that the diffusion coefficient is a crucial parameter in thermal reactors and that its magnitude may be altered during reactor operation, as part of the operational changes that impact the diffusion length. Previously, we examined the effects of operational modifications on the diffusion coefficient as it is reliant on the macroscopic scattering cross-section, Σ_s.

The diffusion coefficient D is sensitive to changes in the moderator temperature, particularly when the moderator temperature changes. In other words, the diffusion coefficient gradually rises as the moderator temperature does [15].

The reduction in the macroscopic scattering cross-section, $\Sigma_s = \sigma_s \cdot N_{H2O}$, brought on by the thermal expansion of water (a drop in the atomic number density) is specifically responsible for the rise in the diffusion coefficient ($\Sigma_s = \sigma_s . N_{H2O}$).

Given the above condition, the calculation of diffusion coefficient for the scattering cross-section of carbon at 1 eV is 4.8 barn (4.8×10^{-24}). One can calculate the diffusion coefficient and the transport mean free path as well. The solution can be easily provided as illustrated in Equation 3.10, and we can calculate the diffusion according to this advanced formula and write it as

$$D = \frac{1}{3\Sigma_s(1-\bar{\mu})} \qquad \text{Eq. 3.14(a)}$$

where

$$\bar{\mu} = \frac{2}{3A} = \frac{2}{36} = 0.0555 \qquad \text{Eq. 3.14(b)}$$

First, we have to determine the atomic number density of carbon and then the scattering macroscopic cross-section as follows:

Density

$M_C = 12$
$N_C = \rho \bullet (N_a/M_C) = (2.2\ \text{g/cm}^3) \times (6.022 \times 10^{23}\ \text{nuclei/mol})/(12\ \text{g/mol}) = 1.1 \times 10^{23}\ \text{nuclei/cm}^3$

Microscopic cross-section

$\sigma_s^{12C} = 4.8$ b

Macroscopic cross-section

$\Sigma_s^{12C} = 4.8 \times 10^{-24} \times 1.1 \times 10^{23} = 0.528\ \text{cm}^{-1}$

Diffusion coefficient then becomes

$D = 1/(3 \times 0.528 \times 0.9445) = 0.668$ cm

Transport mean free path

$\lambda_{tr} = 3 \times D = 2.005$ cm

Additionally, the weighted sum is used to get an average value of $(1 - \bar{\mu}_0)$ in a reactor that contains a variety of materials.

$$(1 - \bar{\mu}_0)_{\text{reactor}} = \sum_{i=1}^{n} \frac{(1 - \bar{\mu}_0)_i \Sigma_{si}}{\Sigma_{s\,\text{reactor}}} \quad i = 1, \cdots, n \qquad \text{Eq. 3.15}$$

where n represents all of the elements that are in the reactor. Additionally, the scattering angle average cosine is determined by

$$\bar{\mu}_0 = \overline{\cos\theta} = \frac{2}{3A} \qquad \text{Eq. 3.16}$$

where θ is the neutron scattering angle and A is the mass number of the scattering isotope [amu].

Moreover, since we derived result as shown in Equation 3.11, it is worthwhile to evaluate the "boundary conditions", when the neutron flux can be found by solving the diffusion equation by imposing certain boundary conditions that must be satisfied by the solution.

Some of these are determined from obvious requirements for a reasonable physical flux. For example, since a negative or imaginary flux has no meaning, it follows that ϕ must be a real, non-negative function. The flux must also be finite, except perhaps at artificial singular points of a source distribution.

In many reactor problems related to neutron diffusion equation in steady state [2], neutrons diffuse in a medium that has an outer surface, which is a surface between the medium and the atmosphere. This was pointed out in Chapter 1 that Fick's Law is not valid in the immediate vicinity of such a surface, and it follows that diffusion equation is not valid there either. Higher-order techniques demonstrate that the flux derived from the diffusion equation is almost exactly equal to the flux in the medium's interior if it is assumed that the flux will disappear at a small distance d, known as the extrapolation distance, as shown in Figure 3.8 beyond the surface.

Keeping in mind that a reactor's outer surface simulates a condition in which the atmosphere is actually a vacuum and the mean free path of neutrons in air is significantly greater than in non-gaseous materials. In reactor analysis, it is usually possible to treat it as a vacuum. However, as we previously mentioned, the related diffusion theory is not valid near such a surface, and in order to further analyze such a problem, we must take extra care and handle the solution of the following, assuming that, given the physical problem in hand, the above definition or requirements must not and will not be violated under any circumstances. Now, if we assume a boundary condition *at the surface* of the medium of the form, we get [2]:

$$\frac{1}{\phi}\frac{d\phi}{dn} = -\frac{1}{d} \text{ Boundary condition} \qquad \text{Eq. 3.17}$$

Given that Equation 3.14 may be expressed as $\left[d/dn(\ln\phi)\right] = -1/d$, commonly known as the logarithmic boundary condition where $d\phi/dn$ is the normal derivative of the neutron flux and d is a parameter known as the reciprocal logarithm derivative. These kinds of boundary requirements often apply to reactor issues with basically vacuous media and somewhat dense media [2].

Again, the assumption is that the flux vanishes at a small distance of d beyond the surface is clearly non-physical, and it is a convenient mathematical approximation that provides a high degree of accuracy for estimating flux in the medium. In the case of this special handling of diffusion equation, the solution can be as close as possible to the rigorous solution that is given by neutron transport theory in the interior of the reactor media and satisfy the boundary condition of Equation 3.14 and this equation should be written as [2]:

$$\frac{d\phi}{dn} + \frac{\phi}{d} = 0 \qquad \text{Eq. 3.18}$$

Utilizing Figure 3.8, the extrapolation distance or extrapolation length for most cases of interest is given by the simple formula as

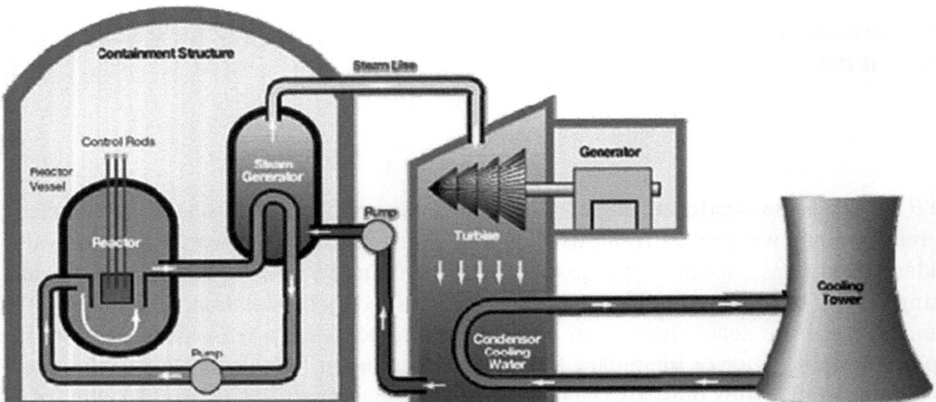

FIGURE 3.8 Principal sketch of a pressurized water reactor (PWR).

(Source: Department of Energy [14])

$$d = 0.71\lambda_{tr} \quad \text{Eq. 3.17}$$

where again λ_{tr} is the transport mean free path of the medium that is given by Equation 3.11, thus Equation 1.16 in terms of d becomes

$$d = 2.13D \quad \text{Eq. 3.18}$$

However, measured values of D for some non-gaseous materials are given in Table 3.2 for thermal neutron diffusion parameters of moderators at 20°C [3].

Thus, from Equation 3.17, it can be seen that d is usually small compared with most reactor dimensions. It is often possible while solving the diffusion equation to assume that the flux vanishes at the actual surface of the system [4].

In cases where d is not eligible, we must establish a mathematical boundary for the problem. This boundary is referred to as the extrapolated boundary. It is located at a distance d from the real physical boundary. This boundary condition is referred to as the vacuum boundary condition [4].

It is typically imposed in problems where there is a medium having a low density or in a region at the edge of the problem where there are few neutrons that re-enter the regions of interest. A typical case is in the air-filled annulus outside the reactor vessel. In such cases, the dimensions of the problem are increased by d. For our case, if the distance from the center of the diffusing media in Figure 3.8 is a, then the distance to the extrapolated boundary is$a + d$, and we define the quantity $\tilde{a}$ as [4]:

$$\tilde{a} = a + d \quad \text{Eq. 3.19}$$

The flux is thought to disappear on the border at $\tilde{a}$. It is important to note the assumption that the flux disappears is only a mathematical one. Beyond the boundary, the flux is non-zero and fluctuates according to its shape and material characteristics.

At an interface between two distinct diffusing media, like the interface between the reactor core and the reflector, boundary conditions must also be specified. It is easy to understand that the flux and component of the current normal to the surface must be continuous across the boundary since neutrons pass through an interface without encountering any obstacles. Equations 1.85 and 1.86 demonstrate that the following relations must be met at the interface between two regions, A and B.

$$\begin{aligned} \phi_A &= \phi_B \\ (J_A)_n &= (J_B)_n \end{aligned} \quad \text{Eq. 3.20}$$

where ϕ_A and ϕ_B are, respectively, the fluxed in regions A and B evaluated at the interface, and $(J_A)_n$ and $(J_B)_n$ are the normal components of the neutron current evaluated at the interface. Equation 3.19 is sometimes called interface boundary condition.

TABLE 3.2
Thermal Neutron Diffusion Parameters of Common Moderators 20°C*

Moderator	Density g/m³	$\bar{D}$, cm	Σ_n, cm⁻¹	L_T^2, cm²	L_T, cm
H_2O	1.00	0.16	0.0197	8.1	2.85
D_2O†	1.10	0.87	9.3×10^{-5}	9.4×10^{3}	97
Be	1.85	0.50	1.04×10^{-3}	480	21
Graphite	1.60	0.84	2.4×10^{-4}	3,500	59

*Based on *Reactor Physical Constants* [3].

†D_2O containing 0.25 weight/percent H_2O. These values are very sensitive to amount of H_2O impurity.

3.3 THE DIFFUSION EQUATION SOLUTION

Some simple diffusion problems in simple reactor design are now considered to demonstrate how the diffusion equation may be solved subject to the prior defined boundary conditions.

3.3.1 The Infinite Medium Multiplication Factor and the Source Term

Before we tackle this subject here, we need to look at the fundamental aspect of nuclear fission reactor, and consequently, the nuclear power that is driven by the fission chain reaction in this case.

Fission chain reactions are what propel the nuclear energy emission from a nuclear fission reactor. Neutrons are released by fissioning nuclei that induce fissile or fissionable nuclei in this process, which is depicted in Figure 3.9. The neutrons generated from these fissions then cause fissions in still more fissile or fissionable nuclei, and so on.

Keep in mind that if each fission generates enough neutrons to trigger at least one additional fission, a chain reaction can result in self-sustained fission. This depends on several variables, such as the average number of neutrons generated and the ease of creating a specific nuclide fission.

Not all fission-produced neutrons cause fission. While some neutrons interact with a nucleus without causing it to fission, others escape the fissionable material. Having a lot of fissionable material allows us to increase the number of fissions that neutrons create. The critical mass of a given nuclide is the smallest quantity required for self-sustained fission. Certain nuclides, such as ^{239}Pu

- produce more neutrons per fission than others, such as ^{235}U.
- Additionally, nuclides are easier to fission than others. ^{235}U and ^{239}Pu are easier to fission than the much more abundant ^{238}U.
- Both factors affect critical mass, which is the smallest for ^{239}Pu.

Such above fissionable chain reaction can be described qualitatively in terms of the factor that is known as multiplication factor, denoted by the symbol k, which is presented by the definition in Equation 3.21 below [9].

In an infinite medium, we assume that the flux has a constant value, no gradient exists, and hence there is no neutral leakage, and we can define an infinite medium multiplication factor k as

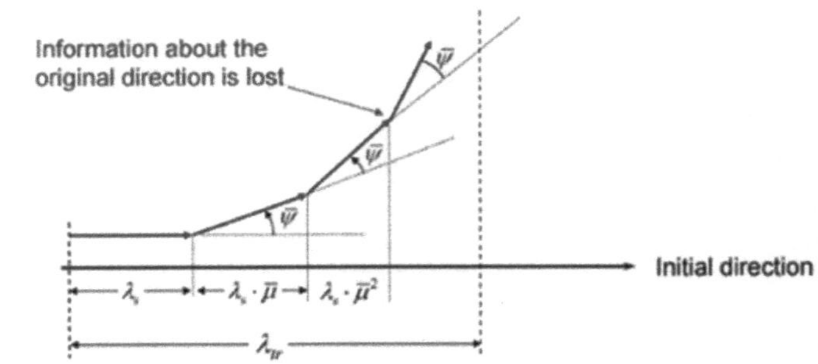

No absorption:

$$\lambda_{tr} = \lambda_s + \lambda_s\bar{\mu} + \lambda_s\bar{\mu}^2 + \lambda_s\bar{\mu}^3 + \ldots = \frac{\lambda_s}{1-\bar{\mu}}$$

With absorption:

$$\lambda_{tr} \equiv \frac{1}{\Sigma_t - \bar{\mu}\cdot\Sigma_s} = \frac{1}{\Sigma_t\left(1-\bar{\mu}\cdot\Sigma_s/\Sigma_t\right)} = \frac{\lambda_t}{1-\bar{\mu}\cdot\Sigma_s/\Sigma_t}$$

FIGURE 3.9 The transport mean free path concept [11, 12].

$$k \equiv k_{\infty} = \frac{\text{Neutrons produced in current fission generation}}{\text{Neutrons absorbed in previous fission generation}} \quad \text{Eq. 3.21}$$

Now, it is appropriate that we clarify the role of the multiplication factor as the formal calculation of reactor criticality [4].

According to the definition of the multiplication factor k given above, if $k = 1$, the reactor is in critical mode at this point, and the quantity of neutrons in any two successive fission generations is equal, we may claim the chain reaction to be time-independent. Similarly, if $k < 1$, the reactor is in subcritical condition, and if $k > 1$, the core is working in supercritical mode.

In summary, the reactor core can operate under three following conditions:

$k < 1$ Subcritical
$k = 1$ Critical
$k > 1$ Supercritical

To achieve our main goal of reactor design, which is to have a reactor that can be in critical mode, all of the selected materials and associated subsystems inside the core (such as fuel rods) should be taken into account as part of the critical analysis. Occasionally, we must restate our analysis and documentation that the reactor ultimately reaches criticality when the criticality requirement is satisfied and $k = 1$.

It should be noted that the word "lifecycle" refers to the multiplication factor k in terms of subsequent fission neutron production. Additionally, keep in mind that during a chain reaction, some neutrons may not even cause fission reactions; instead, they may be absorbed in a non-productive capture or leak out of the core system. In these situations, we can define multiplication factor k in a way that is somewhat more accurate in order to achieve neutron balance, as follows.

$$k \equiv \text{Multiplication Factor} \equiv \frac{\text{Rate of neutron production in reactor}}{\text{Rate of neutron loss or absorption plus capture in reactor}} \quad \text{Eq. 3.22}$$

The aforementioned description makes it clear that the pace of production and loss is dependent on time and fuel consumption, and it may change as a result [5].

At this stage, the neutron lifetime L may alternatively be defined as follows [2]:

$$L \equiv \frac{N(t)}{L(t)} \quad \text{Eq. 3.23}$$

where $L(t)$ is the reactor's loss rate at time t and $N(t)$ is the entire neutron population, or the number of neutrons counted. Additionally, we may examine a basic kinetics of chain reactions utilizing the following analysis to have a better understanding of $N(t)$ in a nuclear reactor at time t. Therefore,

$$\frac{dN(t)}{dt} = \text{Production Rate} - \text{Loss Rate} = P(t) - L(t) \quad \text{Eq. 3.24}$$

Equation 3.24 may also be enlarged to the following version of Equation 3.25 using the definition of the multiplication factor k for balancing neutrons as follows [10]

$$\frac{dN(t)}{dt} = \left[\frac{P(t)}{L(t)} - 1\right] L(t) = (k-1)L(t) \quad \text{Eq. 3.25}$$

To proceed further, using Equation 3.23, we can write

$$\frac{dN(t)}{dt} = \frac{(k-1)}{l} N(t) \quad \text{Eq. 3.26}$$

Assuming that both k and L are time-independent, which they typically are not, and setting the initial condition for the aforementioned ordinary differential Equation 3.26 that there are N_0 neutrons in the reactor at time $t = 0$, we may formulate a solution for Equation 3.26 as follows:

$$N(t) = N_0 \exp\left[\left(\frac{k-1}{l}\right)t\right] \qquad \text{Eq. 3.27}$$

The straightforward method for modeling a nuclear reactor's kinetics described above is also shown in Figure 3.10 below and agrees with reactor criticality in terms of k.

Since a nuclear reactor's power level is proportional to its neutron population, we may consider the reactor power level's temporal behavior to be exponential, with a time constant, or reactor period T, provided by the following equation [16]:

$$T = \frac{l}{k-1} \qquad \text{Eq. 3.28}$$

We may observe that if $k \to 1$, then $T \to \infty$, which corresponds to a time-independent neutron population or react power level, by analyzing Equation 3.28. This is consistent with our earlier assertions in the section above. However, the topic of nuclear reactor kinetic analysis is outside the purview of this book and must be studied if we wish to extend our basic kinetic model.

We must look at what is happening with the neutrons in each fission reaction or generation before moving on with our study and calculations. These fission neutrons and the likelihood that they may leak out of the reactor, be lost to the chain reaction, and then be absorbed – possibly as a result of a non-productive capture event in the fuel or other materials – can be seen in Figure 3.11.

To make it more formal, we define the probabilities for each of these possible events presented by Figure 3.10 as follows:

$P_{NL} \equiv$ Possibility that a neutron will not escape the system prior to absorption
$P_{AF} \equiv$ Conditional potential that the neutron will be absorbed in the fuel if it is absorbed
$P_f \equiv$ Conditional likelihood that a fission reaction will occur if a neutron is absorbed in the fuel

We can easily calculate the last two probability conditions and the conditional probability for absorption in the fuel P_{AF} can be expressed as the ratio of the macroscopic absorption cross-section for fuel Σ_a^F, and for the fuel plus the rest of the material in the core Σ_a [5].

Please note that the general shape of the macroscopic cross-section was specified as Σ in Chapter 1. But in this case, we use the superscript to denote the material we are interested in (fuel), and its

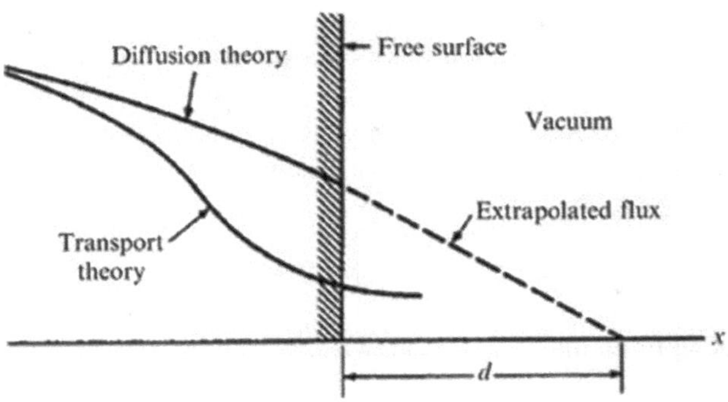

FIGURE 3.10 The extrapolation distance at a surface [2].

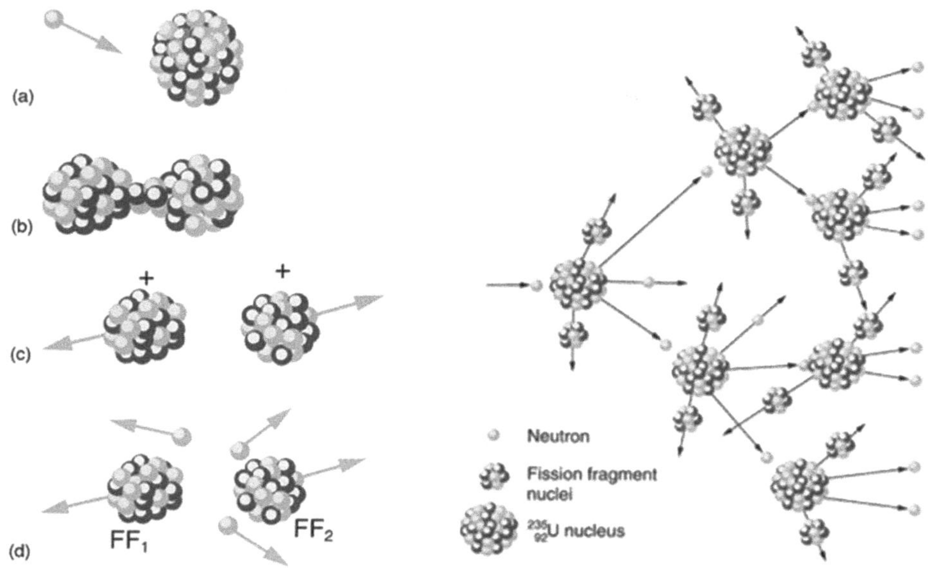

FIGURE 3.11 A nuclear fission chain reaction.

(Source: www.wikipedia.org)

absence will show that the macroscopic cross-section is the sum of the materials in the system. We use the symbol Σ_t, so we can write the probability of absorption as follows:

$$P_{AF} = \frac{\Sigma_a^F}{\Sigma_a} \qquad \text{Eq. 3.29}$$

Only when the reactor's composition is uniform does Equation 3.29 hold true. Even though the majority of reactor design engineers actually deal with non-uniform compositions in modern reactor analysis, which vary from point to point because of fuel elements, coolant channels, support structures, etc.; in more general cases, they use macroscopic cross-sections, which are represented by the symbol Σ_a in relation to the spatial average over the entire reactor. Equation 3.55 from the Reference [2] is relevant for correctly averaging over energy, just as they are over space, as we must evaluate the neutron energy at which these cross-sections are under consideration [17].

It is well-known that the likelihood of absorption is known as the reactor's thermal utilization factor, and it is represented by the letter f as shown below. This will be covered in more detail in Section 3.23 of the reference [2].

$$P_{AF} \equiv f \qquad \text{Eq. 3.30}$$

Historically, the term in Equation 3.29 originates from early thermal reactor analyses, where thermal neutrons caused nearly all of the fuel's fissions. The cross-sections in f would be measured at thermal neutron energy in these conditions and would show how well the fuel competes with other materials in the reactor. The efficiency with which the reactor used the thermal neutrons in the fuel is known as the absorption of thermal neutrons, and Equation 3.30 typically applies to reactors of any kind [2].

We are now returning to our notion of conditional probability for causing a fission reaction in the fuel from above. As a result, it may be expressed in terms of cross-sections. In this scenario, we simply consider the ratio of the absorption cross-section resulting from both fission and radiative capture in the fuel material to that of the fission cross-section.

$$P_f = \frac{\Sigma_f^F}{\Sigma_a^F} = \frac{\sigma_f^F}{\sigma_a^F} \qquad \text{Eq. 3.31}$$

We can now use all of these probabilities to get the multiplication factor *k*, which is described above, in order to track the neutrons.

Equation 3.30 and Figure 3.11 allow us to calculate the number of neutrons N_2 in the subsequent step generation, starting with the N_1 neutrons that are present in the reactor during a certain fission generation process.

$$N_2 = v\, P_f P_{AF} P_{NL} N_1 \qquad \text{Eq. 3.32}$$

So, if we substitute both Equations 3.30 and 3.31 into Equation 3.32 and use the definition of number of fission neutrons produced per absorption in the fuel as $\eta = v\left(\sigma_t^F / \sigma_a^F\right)$, then, we get

$$N_2 = \eta f P_{NL} N_1 \qquad \text{Eq. 3.33}$$

The majority of the neutrons received in the fuel induce fission, while some do not. η is sometimes referred to as the reproduction factor. The ratio of fast neutrons generated by thermal fission to thermal neutrons absorbed in the fuel is known as the reproduction factor, or η. The reproduction factor may be shown as [2] below:

$$\eta = \frac{\text{Number of fast neutrons produced by thermal fission}}{\text{Number of thermal neutrons absorbed in the fuel}}$$

As seen below, the reproduction factor may also be expressed as a ratio rating.

$$\eta = \frac{\text{Rate of production of fast neutrons by thermal fission}}{\text{Rate of absorption of thermal neutrons by the fuel}}$$

The product of the fission reaction rate ($\Sigma_f^U \phi^U$) and the average number of neutrons generated per fission (*n*) may be used to calculate the rate at which thermal fission produces fast neutrons. Uranium-235 undergoes thermal fission with an average emission of 2.42 neutrons. The fuel absorbs thermal neutrons at a rate of $\Sigma_f^U \phi^U$. The following equation is produced when these terms are substituted into the definition above.

$$\eta = \frac{\Sigma_f^U \phi^U v}{\Sigma_f^U \phi^U} \qquad \text{Eq. 3.34}$$

Table 3.3 lists values of v and η for fission of several different materials by thermal neutrons and fast neutrons.

When there are many fissionable components in the fuel, each one must be taken into consideration. A reactor core that contains both uranium-235 (^{235}U) and uranium-238 (^{238}U) would have the reproduction factor (η) computed as follows [2]:

$$\eta = \frac{N^{U-235}\, \sigma_f^{U-235}\, \upsilon^{U-235}}{N^{U-235} \sigma_\alpha^{U-235} N^{U-238} \sigma_\alpha^{U-238}} \qquad \text{Eq. 3.35}$$

TABLE 3.3
Average Number of Neutrons Liberated in Fission [2].

Fissile Nucleus	Thermal Neutrons ν	Thermal Neutrons η	Fast Neutrons ν	Fast Neutrons η
Uranium-233	2.49	2.29	2.58	2.40
Uranium-235	2.42	2.07	2.51	2.35
Uranium-239	2.93	2.15	3.04	2.90

The computation can be further reduced to the one below in order to find the reproduction factor for a single nuclide instead of a combination.

$$\eta = \nu\left(\sigma_t^F / \sigma_a^F\right) \quad \text{Eq. 3.36}$$

Note that the superscript refers to Uranium.

Now, once again, we write the following equation using the definition of multiplication factor k, which is the ratio of the amount of neutrons in two subsequent fission generations.

$$k = \frac{N_2}{N_1} = \frac{\eta f P_{NL} N_1}{N_1} = \eta f P_{NL} \quad \text{Eq. 3.37}$$

It is quite challenging to compare the non-leakage probability P_{NL} and quantify neutron leakage from the reactor core in this equation; more precise analysis is needed, and more studies should be conducted. The corresponding multiplication factor then becomes known as the infinite medium multiplication factor and is expressed by the following equation by substituting this value of P_{NL} in Equation 3.34. However, since no neutrons could leak out, we can quickly conclude that we should set the non-leakage probability $P_{NL} = 1$.

$$k_\infty = \eta f \quad \text{Eq. 3.38}$$

Although no reactor is infinitely large, $k \equiv k_\infty$ is a helpful parameter in reactor analysis because it distinguishes the geometry of the reactor core from the multiplication characteristics of the material utilized in the reactor. Naturally, we need $k_\infty > 1$ to have any likelihood of a critical chain reaction in the reactor, which is in-line with what we previously said [2], as $P_{NL} < 1$, more generally, for a finite reactor from which any neutron leakage can occur [2].

One more parameter that we must define and be familiar with in nuclear reactor analysis is the fast fission factor ε. As previously said, we need to know another term in order to continue our discussion of neutronic analysis. Fast fission is one of the initial processes that neutrons of one generation may go through. Neutrons in the fast energy range that slow down to thermal energies are the source of rapid fission. The reactor core's fast neutron population increases net as a result of rapid fission. This component of the four-factors formula accounts for the fact that, while thermal neutrons will cause the majority of fissions in fissile materials, fast neutrons will cause some fissions in both fissile and fissionable materials. Equation 3.37 illustrates this relationship. Only a limited number of fast neutrons generate fission in uranium-235 (^{235}U) or uranium-238 (^{238}U) due to the narrow cross-section for rapid fission. Thus, the fast fission factor increases the fast neutron population in a single generation. The ratio of the net number of fast neutrons generated by all fissions to the number of fast neutrons produced by thermal fissions is known as the fast fission factor (ε). This ratio's mathematical expression is shown below [13].

$$\varepsilon = \frac{\text{Total number of fission neutron (fromboth fast and thermal fission)}}{\text{Number of fission neutrons from thermal fissions}}$$

Furthermore, the above definition can be presented by the following equation.

$$\varepsilon \equiv \frac{\int_{fuel} dV \int_0^{\infty} \nu \Sigma_f(\vec{r},E) dE}{\int_{fuel} dV \int_0^{E_2} \nu \Sigma_f(\vec{r},E)\phi_1(\vec{r},E) dE} \quad \text{Eq. 3.39}$$

where $\phi(\vec{r},E)$ is the flow and will be shown for an infinite lattice of cells. These cells will be identical to any other cell and are for a particular square fuel rod and its associated moderator. Therefore, if we know $\phi(\vec{r},E)$ for one cell, we automatically know it for all of them [6].

The same is true for infinite homogeneous media: If we know $\phi(E)$ at one point, we know it at every point. The rapid fission factor ε is defined by the relationship provided by the fraction in Equation 3.37. The aforementioned relationship is the ratio of the lattice's overall rate of fission neutron production to the rate of fission-related production in the thermal energy range. Equation 3.34 makes it evident from the definition that ε solely depends on $v^j\sigma_f^j(E)n^j(\vec{r},E)$ of the different fissionable isotopes present in the fuel rod [6].

Additionally, a neutron must travel near enough to a fuel nucleus when it is a fast neutron in order to be absorbed by it. The fuel and moderator concentrations and arrangement will have an impact on the value. In a homogeneous reactor with moderator atoms surrounding the fuel atoms, the value of ε is practically 1.00. All of the fuel atoms, on the other hand, are densely packed in components like pellets, rods, or pins in a heterogeneous reactor. It is quite likely that neutrons released when one fuel atom fissions will travel close to another fuel atom before experiencing a major slowdown. In the majority of heterogeneous reactors, the core element configuration yields a ε value of around 1.03. Neutron poison concentrations, enrichment, temperature, and pressure have no discernible effects on the value of ε. We recommend that the reader consult Reference [2] for a more thorough discussion of poisons, which are non-fuel compounds that readily absorb neutrons.

Resonance escape probability is another statistic that should be understood, which is the likelihood that neutrons will continue to diffuse through the reactor after their numbers have increased due to rapid fission. As they travel, the neutrons hit with the reactor's moderator and fuel and non-fuel material nuclei, losing some of their energy and slowing down. It is possible that some neutrons will be trapped when they slow down via Uranium-238's resonance zone, which spans from around 6 eV to 200 eV. The chance of resonance escape is the likelihood that a neutron will not be absorbed by a resonance peak. The ratio of neutrons that attain thermal energy to the number of fast neutrons that begin to slow down is known as the resonance escape probability (*P*). The ratio is displayed below.

$$p = \frac{\text{Number of neutron that reach thermal energy}}{\text{Number of fast neutrons that start to slow down}}$$

The fuel-moderator configuration and the degree of uranium-235 (^{235}U) enrichment (if any) have a significant impact on the resonance escape probability value. A neutron must approach a Uranium-238 (^{238}U) nucleus sufficiently to be absorbed while slowing down in order to experience resonance absorption. The neutron slows down in the vicinity of the fuel nuclei in a homogeneous reactor, and this need is readily satisfied. This indicates that a neutron's escape probability is decreased since it has a higher chance of being absorbed by Uranium-238 (^{238}U) as it slows down. However, the neutron slows down in the moderator of a heterogeneous reactor, which is devoid of uranium-238 (^{238}U) atoms. As a result, its escape probability is larger and its likelihood of experiencing resonance absorption is low.

Pressure and poison concentration have no discernible effects on the resonance escape probability value. The Doppler effect, which is an apparent broadening of the typically narrow resonance peaks owing to thermal motion of nuclei, causes the resonance absorption in Uranium-238 (^{238}U) to increase as the fuel temperature is raised in water moderation, low Uranium-235 (^{235}U) enrichment reactors. The likelihood of a resonance escape is decreased by an increase in resonance absorption, and the fuel temperature coefficient for resonance escape is negative (described in more detail below). For the moderator temperature, the temperature coefficient of resonance escape probability is likewise negative. The density of water reduces as its temperature rises. More neutrons with resonance energy can enter and be absorbed by the fuel due to the reduction in water density. The resonance escape probability, which is typically between 0.95 and 0.99, is always a little less than 1.

The ratio of the number of fast neutrons that survive slowing down (thermalization) to the number of fast neutrons that initially started the production is the product of the fast fission factor and the resonance escape probability ($\varepsilon \cdot p$).

Lastly, as a practical step, we adjust our definition of the non-leakage probability to account for the possibility of two separate neutron leakage phases, which will necessitate two rather different kinds of analysis in our further work. As it slows down, the neutron may first seep out.

Such rapid neutron leakage may be rather sluggish down to thermal energy; in fact, the neutron mean free path is relatively wide for high energies. The neutron may continue to disperse after slowing down, finally leaking out before it has had a chance to be absorbed. We will deconstruct our previous non-leakage probability P_{NL} as follows [2, 6] to account for these two processes.

$$P_{NL} = P_{FNL} P_{TNL} \qquad \text{Eq. 3.40}$$

where P_{FNL} and P_{TNL} are defined as

$P_{FNL} \equiv$ Probability that fast neutron will not leak out (fast non-leakage)
$P_{TNL} \equiv$ Probability that thermal neutron will not leak out (thermal non-leakage)

Equations 3.35 and 3.37, together with the two definitions mentioned above, are inserted into Equation 3.37 to get the following new form for the infinite medium multiplication factor.

$$k_\infty = \eta f p \varepsilon \qquad \text{Eq. 3.41}$$

This formula, sometimes referred to as the four-factor formula, is described in Section 2.15 of the Reference [2]. In addition, the multiplication factor k of Equation 3.35 can be written as follows:

$$k = \eta f p \varepsilon P_{FNL} P_{TNL} \qquad \text{Eq. 3.42}$$

Unexpectedly, this version of Equation 3.40 is referred to as the six-factor formula.

It should be noted that altering the reactor's shape or the materials around it with a bigger scattering cross-section can change the non-leakage probability P_{NL} by causing some of the neutrons that leak out to scatter back into the reactor.

Since the parameters in the four-factor formula are averages of the different neutron energies in the reactor, they will also change when leakage is altered. This results in a distribution of energy that varies with the degree of leakage. This leads to somewhat different notation for multiplication factors that describe a finite system. The effective multiplication factor, sometimes known as k_{eff}, is represented as follows:

$$k_{eff} = k_\infty P_{FNL} P_{TNL} \qquad \text{Eq. 3.43}$$

We advise readers to see Henry [6] for further information, specifics, and mathematical analysis.

Returning to the subject of this subsection, which is the infinite medium multiplication factor and the source term, we can say that in an infinite medium, the flux takes on a constant value, there is no gradient, and therefore no neutron leakage. We can define an infinite medium multiplication factor as Equation 3.20, which can be rewritten as follows:

$$k \equiv k_\infty = \frac{\text{Neutrons produced in current fission generation}}{\text{Neutrons absorbed in previous fission generation}}$$

Equation 3.40, the four-factor formula, can be used to represent the infinite medium multiplication factor as follows:

$$k = \eta f p \varepsilon P_{FNL} P_{TNL}$$

where

η = Regeneration factor
ε = Fast fission factor
p = Resonance escape probability
f = Fuel utilization factor

In a homogeneous medium, the fuel utilization factor is defined as follows for a constant flow value.

$$\begin{aligned} f &= \frac{\text{Rate of neutrons absorptions in the fuel}}{\text{Total rate of absorptions in fuel, moderator, structure, etc.}} \\ &= \frac{\Sigma_{aF}\,\phi}{\Sigma_{aF}\,\phi + \Sigma_{aM}\,\phi + \Sigma_{aS}\,\phi + \cdots} \\ &= \frac{\Sigma_{aF}}{\Sigma_{aF} + \Sigma_{aM} + \Sigma_{aS} + \cdots} \end{aligned} \qquad \text{Eq. 3.44}$$

where

M = Moderator
F = Fuel
S = Structure

As previously mentioned, the resonance escape probability P is the percentage of fast neutrons that slow down to thermal energies after the cross-section energy resonance area without being absorbed.

The ratio of the total fissions produced by thermal and fast neutrons to the total fissions caused by thermal neutrons is known as the fast fission factor.

The regeneration factor η is

$$\eta = \frac{\text{Neutrons produced from fission}}{\text{Neutrons absorbed in fuel}}$$

For ^{235}U,

$$\begin{aligned}\eta &= v\frac{\Sigma_f \phi}{\Sigma_{aF} \phi} \\ &= v\frac{\Sigma_f}{\Sigma_f + \Sigma_\gamma} \\ &= v\frac{N\sigma_f}{N\sigma_f + N\sigma_\gamma} \\ &= v\frac{\sigma_f}{\sigma_f + \sigma_\gamma}\end{aligned} \qquad \text{Eq. 3.45}$$

where

v = Average number of neutrons emitted per fission

Σ_f, Σ_γ = Macroscopic fission and radiative capture cross-section, respectively.

If we designate the ratio of the radiative capture to fission microscopic cross-sections as the capture to fission ratio

$$\alpha = \frac{\sigma_\gamma}{\sigma_f} \qquad \text{Eq. 3.46}$$

Then,

$$\eta = \frac{v}{1+\alpha} \qquad \text{Eq. 3.47}$$

Since Equation 3.7 applies to the neutron of a single energy that are thermal, we can define for the thermal neutrons source term s as

$$s = k_\infty \ell_f \Sigma_a \phi \qquad \text{Eq. 3.48}$$

where ℓ_f is the fast neutron non-leakage probability. For the transient state, Equation 3.7 can be written as

$$\frac{\partial n}{\partial t} = \frac{1}{\text{v}}\frac{\partial(n\text{v})}{\partial t} = \frac{1}{\text{v}}\frac{\partial \phi}{\partial t} = D\nabla^2\phi - \Sigma_a \phi + k_\infty \ell_f \Sigma_a \phi \qquad \text{Eq. 3.49}$$

Dividing by Σ_a, we have

$$\frac{1}{\text{v}\Sigma_a}\frac{\partial \phi}{\partial t} = L^2\nabla^2\phi - \phi + k_\infty \ell_f \phi \qquad \text{Eq. 3.50}$$

Where $L^2 = \frac{D}{\Sigma_a}$ is the thermal diffusion area [cm^2], and L is the diffusion length [cm]. See next section for definition of diffusion length L.

Rearranging Equation 3.48, after dividing by $L^2\phi$, we get

$$\begin{cases} \dfrac{1}{vL^2\Sigma_a}\dfrac{1}{\phi}\dfrac{\partial\phi}{\partial t} - \dfrac{\nabla^2\phi}{\phi} = \dfrac{k_\infty \ell_f - 1}{L^2} \\ \dfrac{1}{vL^2\Sigma_a}\dfrac{1}{\phi}\dfrac{\partial\phi}{\partial t} + (B_g)^2 = (B_m)^2 \end{cases} \qquad \text{Eq. 3.51}$$

where

$(B_m)^2 = \dfrac{k_\infty \ell_f - 1}{L^2}$ is denoted as the "material buckling" since it is a function of the materials composition.

Because it depends on the composition of the material, $(B_g)^2 = -\dfrac{\nabla^2\phi}{\phi}$ is referred to as the "geometrical buckling". Additionally, the term "geometrical buckling" originates in structural mechanics, where it refers to the extent of deformational buckling that occurs in a column when external forces are applied to its ends [7].

3.4 THE CRITICALITY EQUATION

The following requirement, where the geometric buckling is equivalent to the material buckling, must be satisfied for criticality and a self-sustaining chain reaction in the steady state, according to $(\partial\phi/\partial t) = 0$ and Equation 3.50.

$$\begin{cases} -\dfrac{\nabla^2\phi}{\phi} = \dfrac{k_\infty \ell_f - 1}{L^2} \\ (B_g)^2 = (B_m)^2 \end{cases} \qquad \text{Eq. 3.52 (a)}$$

In general

$$B^2 = (B_g)^2 = (B_m)^2 \qquad \text{Eq. 3.52 (b)}$$

where B^2 is called the buckling, in short.

Therefore, for a nuclear reactor to achieve critical and be able to produce a steady state, self-sustained neutron chain reaction, the geometrical and material buckling must be equal.

This criticality criterion indicates that the geometry of the system, as shown by its geometrical buckling, must be matched with the material characteristics of the medium, as stated in the equation for the material buckling.

Rearranging the expression $(B_m)^2 = \dfrac{k_\infty \ell_f - 1}{L^2}$ and taking advantage of Equation 3.52(b) for the material buckling, we get

$$1 = \frac{k_\infty \ell_f}{(1 + L^2B^2)} \qquad \text{Eq. 3.53}$$

The last equation represents a just critical system. In general, though, we can define an effective multiplication factor as

$$k_{eff} = k_\infty \ell_f \left[\frac{1}{1 + L^2B^2}\right] = k_\infty \ell_f \ell_{th} \qquad \text{Eq. 3.54}$$

where ℓ_{th} is the thermal neutrons non-leakage probability as

$$\ell_{th} = \frac{1}{1+L^2B^2} \quad \text{Eq. 3.55}$$

where again ℓ_f is the fast neutrons non-leakage probability as before and the following criteria can be met as it was stated in Section 3.3.1 above.

$k_{eff} < 1$, for a subcritical reactor
$k_{eff} > 1$, for a supercritical reactor
$k_{eff} = 1$, for a just critical reactor

All these above conditions are considered equations of criticality.

3.5 THE DIFFUSION LENGTH

As we go through one-group theory in Section 3.2 above, we discussed this theory by driving Equations 3.1 through 3.4, whereby we established that the quantity L [cm] as $L^2 = D/\Sigma_a$, where D [cm] is neutron diffusion and Σ_a [1/cm] is presenting the macroscopic absorption cross-section. The quantity L is called the diffusion length, and thus L^2 is called the diffusion area with unit dimension of [cm^2].

Furthermore, it would be interesting at this stage to investigate the physical interpretation of the diffusion length, which appears in the diffusion Equation 3.3 and its many solutions. So, if we consider a monoenergetic point source S emitting neutrons per second in an infinite homogeneous moderator, and these neutrons diffuse about in the medium, then the individual neutrons move in complicated, zigzag paths due to successive collisions as illustrated in Figure 3.12.

Under these conditions, ultimately every neutron is absorbed in the medium and none can escape since we assumed the medium is an infinite form, thus if the number of neutrons, dn that are absorbed per second at a distance from the source between r and $r+dr$ can be written as

$$dn = \Sigma_a \phi(r)dV \quad \text{Eq. 3.56}$$

where again Σ_a is defined as macroscopic absorption cross-section, $\phi(r) = Se^{-r/L}/4\pi Dr$ (from point source S emitting neutron per second isotopically in an infinite medium) as neutron flux in element volume dV and in case of spherical geometry [4].

From this statement and usage of Equations 3.55 and 3.56, we can deduce the following equation:

$$dn = \frac{S\Sigma_a}{D} re^{-r/L}dr = \frac{S}{L^2} e^{-r/L}dr \quad \text{Eq. 3.57}$$

Since S neutron per second are emitted by the source and dn are absorbed per second between r and $r+dr$, it follows the probability $p(r)dr$ that a source neutron is absorbed in dr is given by

$$p(r)dr = \frac{1}{L^2} re^{-\left(\frac{r}{L}\right)}dr \quad \text{Eq. 3.58}$$

With all this information in hand, it is possible new to compute the average distance from the source at which a neutron is absorbed by averaging r over the probability distribution $p(r)dr$.

For somewhat obscure reasons, it is more usual in nuclear engineering to compute the average of the square of this distance rather than the average of the distance itself. Thus, we can express that

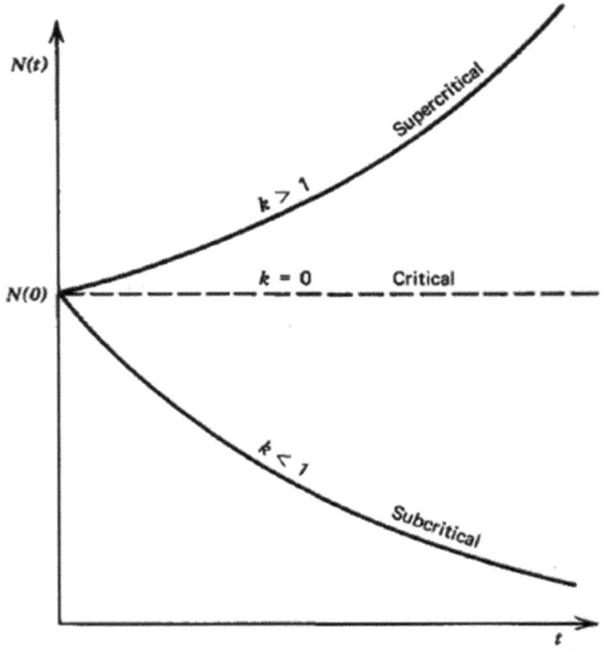

FIGURE 3.12 Time behavior of the neutrons in a reactor [2].

$$\overline{r}^2 = \int_0^\infty r^2 p(r) dr$$
$$= \frac{1}{L^2} \int_0^\infty r^3 e^{-\left(\frac{r}{L}\right)} dr = 6L^2 \qquad \text{Eq. 3.59}$$

Solving for L^2, we obtain

$$L^2 = \frac{1}{6}(\overline{r})^2 \qquad \text{Eq. 3.60}$$

Equation 3.60 expresses that L^2 is equal to one-sixth the average of the square of the vector (crow-flight, i.e., the expression as the crow flies is an idiom for the most direct path between two points, rather similar to "in a beeline". This meaning is attested from the early 19th century and appeared in Charles Dickens's 1838 novel "Oliver Twist": "We cut over the fields at the back with him between us – straight as the crow flies – through hedge and ditch.") distance that a neutron travels from the point where it is emitted to the point where it is finally absorbed.

It follows from this result that greater the value of L, the further neutrons travel and move on the average, before they are absorbed; thus the more diffusive and less absorptive the medium is. The measured values of L and L^2 for thermal neutrons are discussed and can be found in Section 5.9 of Lamarsh book [4].

3.6 THE SOLUTION OF THE UNREFLECTED REACTOR EIGENVALUE EQUATION

Using the geometrical buckling definition from Section 3.3.1 above, which is based on Equation 3.49's findings, we may write

$$B_g^2 = -\frac{\nabla^2 \phi}{\phi} \qquad \text{Eq. 3.61}$$

Thus, we get

$$\nabla^2 \phi = -B_g^2 \phi \qquad \text{Eq. 3.62}$$

In spherical geometry, expressing the Laplacian operator, we obtain

$$\frac{1}{r^2}\frac{d}{dr}\left(r^2 \frac{d\phi}{dr}\right) = \frac{d^2\phi}{dr^2} + \frac{2}{r}\frac{d\phi}{dr} = -B_g^2 \phi \qquad \text{Eq. 3.63}$$

The solution of this equation is given as

$$\phi(r) = \frac{1}{r}\left[A\cos(B_g r) + C\sin(B_g r)\right] \qquad \text{Eq. 3.64}$$

For $r \to 0$ and $A = 0$, we see that $(\cos(0)/0) = (1/0) = \infty$ and such a condition would result in the unphysical scenario of an infinite value for the flux, and if the flux is limited throughout the reactor, we are forced to select the following as a physical solution.

$$\phi(r) = \frac{C\sin(B_g r)}{r} \qquad \text{Eq. 3.65}$$

When L'Hôpital's rule is applied as the initial boundary condition, the value of the flux at the reactor's center, represented by $r \to 0$, shows that

$$\phi_0 = \lim_{r \to 0}\left(\frac{CB_g \cos(B_g r)}{1}\right) = CB_g \qquad \text{Eq. 3.66}$$

L'Hôpital's rule states that for functions f and g which are differentiable on an open interval I except possibly at a point c contained in I, if $\lim_{x \to c} f(x) = \lim_{x \to c} g(x) = 0 \text{or} \pm \infty$ and for all x in I with $x \neq c$, and $\lim_{x \to c} \frac{f'(x)}{g'(x)}$ exists, then

$$\lim_{x \to c} \frac{f(x)}{g(x)} = \lim_{x \to c} \frac{f'(x)}{g'(x)}$$

The differentiation of the numerator and denominator often simplifies the quotient or converts it to a limit that can be evaluated directly.

Consequently, we obtain

$$C = \frac{\phi_0}{B_g} \qquad \text{Eq. 3.67}$$

from which we can get Equation 3.64 as a reduced form.

$$\phi(r) = \phi_0\left[\frac{\sin(B_g r)}{(B_g r)}\right] \qquad \text{Eq. 3.68}$$

where

ϕ_0 = Flux at the reactor center

A second boundary condition is flux at the extrapolated radius, thus, we can write

$$R_{extrapolated} = R + d \qquad \text{Eq. 3.69}$$

where from Equation 2.82 or 2.83, the extrapolated length d is written as

$$d = \frac{2}{3}\lambda_{tr}$$

As we can observe from transport theory, a better estimate of d is usually written as

$$d = 0.71\lambda_{tr}$$

Substituting for d in Equation 3.67, for this boundary condition, we get

$$B_g R_{extrapolated} = n\pi \text{ for } n = 1, 2, \cdots \qquad \text{Eq. 3.70}$$

For the main harmonic solution or $n = 1$ eigenvalue, we can write

$$B_g = \frac{\pi}{R_{extrapolated}} \qquad \text{Eq. 3.71}$$

Thus, Equation 3.68 in terms of the following equation becomes

$$\phi(r) = \phi_0 \left[\frac{\sin\left(\dfrac{\pi r}{R_{extrapolated}}\right)}{\left(\dfrac{\pi r}{R_{extrapolated}}\right)} \right] \qquad \text{Eq. 3.72}$$

3.7 THE EFFECT OF REACTOR POWER, SEMI-INFINITE SLAB REACTOR

The effect of the reactor power level can be included in the magnitude constant of integration to the reactor criticality equation. Mathematically, for the eigenvalue equation,

$$\nabla^2\phi = -B_g^2\phi \qquad \text{Eq. 3.73}$$

Equation 3.73 may still be solved using Figure 3.13, which shows the flow multiplied by any constant. In practical terms, this means that, as long as there is adequate cooling to remove the heat produced, we may run the nuclear reactor at whatever power level we want. Otherwise, the reactor core would melt and possibly evaporate if the emitted energy was not captured.

Moreover, to determine the flux within the reactor, Equation 3.73 must be solved subject to the boundary condition that ϕ vanishes at the extrapolated faces of the slab – that is at $x = +\frac{\tilde{a}}{2}$ and $x = -\frac{\tilde{a}}{2}$, where $\tilde{a} = a + 2d$. Then, the boundary condition becomes

$$\phi\left(+\frac{\tilde{a}}{2}\right) = \phi\left(-\frac{\tilde{a}}{2}\right) = 0 \qquad \text{Eq. 3.74}$$

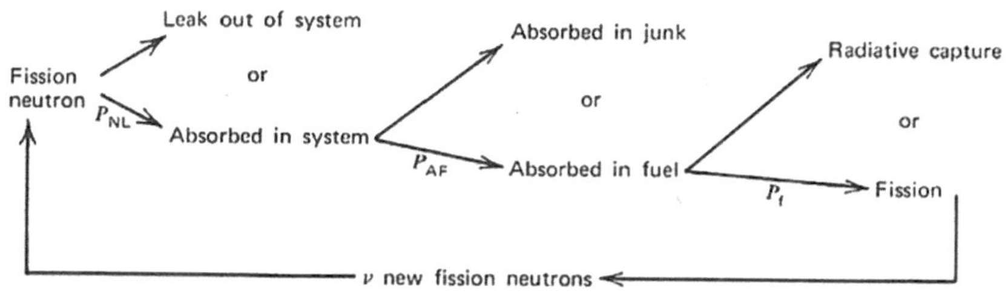

FIGURE 3.13 Schematic of new fission neutron generation and possible events sequence [5].

Furthermore, it may also be noted that because of the symmetrical problem, there can be no net flow of neutron at the center of slab. Since the neutron current density is proportional to the derivative of ϕ, this means

$$\frac{d\phi}{dx} = 0 \text{ at } x = 0 \qquad \text{Eq. 3.75}$$

The condition given in Equation 3.75 is equivalent to requiring that ϕ be an even function – that is

$$\phi(-x) = \phi(+x) \qquad \text{Eq. 3.76}$$

and has a continuous derivative within the reactor. Furthermore, in any problem where the flux is a well-behaved function, Equation 3.76 is often easier to apply than Equation 3.73 [4].

In any case, the general solution to Equation 3.73 that is presented below as Equation 3.77 is Equation 3.78.

For a hypothetical semi-infinite slab reactor of thickness a in the x direction with the origin taken at the center of the slab, and infinite in extent in the y and $z$directions, with a power produced per unit area P, the reactor equation can be written in one-dimensional Cartesian coordinates as

$$\nabla^2\phi = -B_g^2\phi \qquad \text{Eq. 3.77}$$

Its solution is harmonic as

$$\phi(x) = A\cos(Bx) + C\sin(Bx) \qquad \text{Eq. 3.78}$$

Since the flux cannot be zero and reaches a maximum at the origin reject the $\sin(Bx)$ solution implying that $C = 0$

$$\phi(x) = A\cos(Bx) \qquad \text{Eq. 3.79}$$

Applying the boundary condition of a vanish flux at the boundaries $+a/2$ and $-a/2$, and neglecting the extrapolated length for a large reactor.

$$B\left(\frac{a}{2}\right) = n\left(\frac{\pi}{2}\right)$$
$$Ba = n\pi \qquad \text{for } n = 1,3,5,\cdots \qquad \text{Eq. 3.80}$$
$$B = n\left(\frac{\pi}{a}\right)$$

Taking into account the fundamental mode or principal harmonic $n = 1$, we may express the equation as follows:

$$\phi(x) = A\cos\left(\frac{\pi}{a}\right)x \qquad \text{Eq. 3.81}$$

The reactor power per unit area can be expressed as

$$P = \int_V \Sigma_f \phi E_f dV$$

$$= \Sigma_f E_f A \int_{x=-\frac{a}{2}}^{x=+\frac{a}{2}} \cos\left(\frac{\pi}{2}\right)x \times 1 \times dx$$

$$= \Sigma_f E_f A\left(\frac{a}{\pi}\right)\left[\sin(\frac{\pi}{2})x\right]_{x=-\frac{a}{2}}^{x=+\frac{a}{2}} \qquad \text{Eq. 3.82}$$

$$= \Sigma_f E_f A\left(\frac{a}{\pi}\right)\left[\sin(\frac{\pi}{2}) - \sin(-\frac{\pi}{2})\right]$$

$$= \Sigma_f E_f A\left(\frac{a}{\pi}\right) \times 2$$

where 190 [MeV/fission] is the extractable energy release per fission event, denoted by E_f.

The constant of integration A can be expressed as

$$A = \frac{\pi P}{2aE_f \Sigma_f} \qquad \text{Eq. 3.83}$$

and the power per unit area P will determine the flux distribution as follows:

$$\phi(x) = \frac{\pi P}{2aE_f \Sigma_f}\cos\left(\frac{\pi}{a}\right)x \qquad \text{Eq. 3.84}$$

implying that the reactor may run at any required power level P as long as sufficient cooling is generated. This is a limiting characteristic of the materials utilized at high temperatures that it shares with any other power-producing engine, such as an automated engine or a boiler from a fossil fuel power plant.

3.8 OTHER REACTOR SHAPES

So far what we have learned and analyzed based on infinite slab bare reactor in above and presented as well was a simplest scenario that in real world is not going to happen. It is pretty much impossible to construct a reactor of such shape. Therefore, it is necessary to generalize the results of preceding section for reactors of more realistic shapes. This can easily be done for the following reactor as illustrated in Figure 3.13 such as sphere, infinite cylinder, rectangular parallelepiped, and finite cylinder.

These reactors, together with the coordinated systems used to describe them, as we stated, are shown in Figure 3.14. It should be noted that each of these reactors is bare, reflected reactors are discussed here briefly and more detail can be found in reference by Lamarsh and Baratta [4].

For first analysis of different shape reactor, we take under consideration a critical spherical reactor of radius R, thus a related flux equation of neutronic diffusion equation in one-dimensional

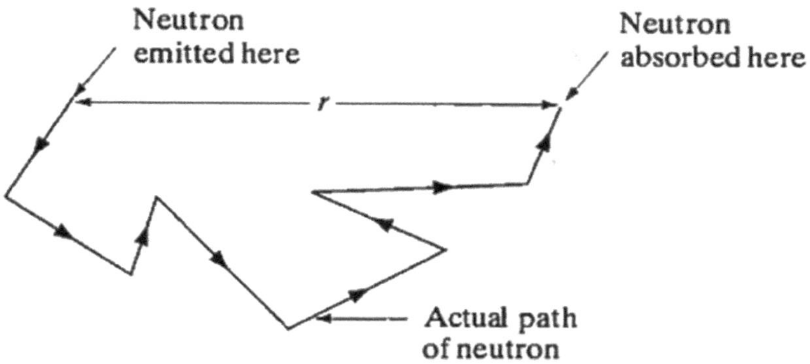

FIGURE 3.14 Trajectory of a neutron in a moderating medium [2].

circumstances as a function of r utilizing the Laplacian operator in a spherical geometry is given by the following equation:

$$\frac{1}{r^2}\frac{d}{dr}r^2\frac{d\phi}{dr}+B^2\phi=0 \qquad \text{Eq. 3.85}$$

In this case, the flux boundary condition must satisfy $\phi(R)=0$ on average and remain finite throughout the interior of the reactor. We solve the ordinary differential equation (ODE) of Equation 3.83 by substitution method and assuming $w(r)$ as a new introductory variable for $w(r)=r\phi(r)$, thus Equation 3.83 structures the following form and it can be written as

$$\frac{d^2w(r)}{dr^2}-\frac{1}{L^2}w(r)=0 \qquad \text{Eq. 3.86}$$

The general solution of Equation 3.85 through help from Equation 3.86 can be easily found as the following:

$$\phi(r)=A\frac{\sin(Br)}{r}+C\frac{\cos(Br)}{r} \qquad \text{Eq. 3.87}$$

where A and C are constants.

We are forced to select the following as a physical solution if the flux is limited throughout the reactor, as shown by the second term in Equation 3.87: $A=0$ would result in an unphysical scenario where the flux has an infinite value due to $[\cos(0)/0]=(1/0)=\infty$.

$$\phi(r)=\frac{C\sin(Br)}{r} \qquad \text{Eq. 3.88}$$

As $r\to 0$, then from L' Hopital's rule, the value of the flux at the center of the reactor is

$$\phi_0=\lim_{r\to 0}\frac{CB\cos(Br)}{r}=CB \qquad \text{Eq. 3.89}$$

and

$$C=\frac{\phi}{B} \qquad \text{Eq. 3.90}$$

from which

$$\phi(r) = \phi_0 \frac{\sin(Br)}{(Br)} \qquad \text{Eq. 3.91}$$

where ϕ_0 is the flux at the reactor center.

A second boundary condition is that of the flux vanishing at the extrapolated radius will be provided the following:

$$R_{ex} = R + d \qquad \text{Eq. 3.92}$$

where d is the extrapolated length given as

$$d = \frac{2}{3}\lambda_{tr} \qquad \text{Eq. 3.93}$$

From the transport theory, a better estimate of d is usually used as

$$d = 0.71\lambda_{tr} \qquad \text{Eq. 3.94}$$

Applying this condition, we get

$$BR_{ex} = n\pi \quad n = 1, 2, \cdots \qquad \text{Eq. 3.95}$$

For the main harmonic solution or $n = 1$ eigenvalue, we come to the conclusion that

$$B_g = \frac{\pi}{R_{ex}} \qquad \text{Eq. 3.96}$$

and the flux distribution becomes

$$\phi(r) = \phi_0 \left[\frac{\sin\left(\frac{\pi r}{R_{ex}}\right)}{\frac{\pi r}{R_{ex}}} \right] \qquad \text{Eq. 3.97}$$

3.9 THE MODIFIED ONE-GROUP CRITICALITY EQUATION

Determining the diffusion area of fast neutrons, commonly known as neutron age τ [cm^2], then using the calculation for the thermal non-leakage probability as an analogy, we obtain

$$\ell_{th} = \frac{1}{1 + L^2B^2} \qquad \text{Eq. 3.98}$$

We can write the expression for the fast non-leakage probability as

$$\ell_f = \frac{1}{1 + \tau B^2} \qquad \text{Eq. 3.99}$$

For the critical system, Equation 3.54 for the effective multiplication factor may, thus, be expressed as follows:

$$k_{eff} = k_\infty \left[\frac{1}{(1+L^2B^2)} \cdot \frac{1}{(1+\tau B^2)} \right] = k_\infty \ell_{th} \ell_f$$
$$k_{eff} = k_\infty \left[\frac{1}{1+(L^2+\tau)B^2 + \tau L^2 B^4} \right] \quad \text{Eq. 3.100}$$

Now, $B_g^2 = (\pi/R_{ex})^2$, so that we can neglect term of B^4 with respect to B^2 in the case of a large reactor with a large radius R, and write the modified one-group theory criticality equation as

$$k_{eff} = \frac{k_\infty}{1+M^2B^2} \quad \text{Eq. 3.101}$$

where $M^2 = L^2 + \tau$ is designated as the migration area, then $B^2 = B_g^2 = B_m^2$ is the buckling.

3.10 DETERMINATION OF REACTOR'S CRITICAL DIMENSION

If we write the buckling expression and rearrange it, we obtain

$$B_m^2 = \frac{k_\infty - 1}{M^2} = \left(\frac{\pi}{R_{ex}} \right)^2 \quad \text{Eq. 3.102}$$

from which,

$$R_{ex} = \pi \sqrt{\frac{M^2}{k_\infty - 1}} \quad \text{Eq. 3.103}$$

and

$$R_c = R_{ex} - d = \frac{\pi M}{\sqrt{k_\infty - 1}} - d \quad \text{Eq. 3.104}$$

is obtained as the reactor critical radius.

3.11 THE CRITICAL RADIUS OF AN UNREFLECTED AND UNMODIFIED CRITICAL FISSILE SPHERE

Since there is no moderator to moderate the neutrons' energy to thermal energy, the neutron spectrum would be rapid if we were looking at an unmoderated fissile reactor.

In the case of moderated systems, the cross-section data must be weighted using a fast neutron spectrum rather than a thermal neutron spectrum. The flux weighting approach used in the literature causes these quick one group constants to vary from one source to another. Table 3.4 displays one of these sets.

3.12 CRITICALITY OF BARE, UNREFLECTED, AND UNMODERATED FAST REACTOR

As an example, we shall calculate the critical radius, volume, and mass of a sphere of ^{235}U, given the following fast spectrum data Wirtz [8].

TABLE 3.4
One Group Fast Group

Nuclide	Neutrons per Fission Event ν	Fission Cross-Section σ_f [barn]	Capture Cross-Section σ_c [barn]	Transport Cross-Section σ_{tr} [barn]
Plutonium^{239}	2.98	1.85	0.260	6.8
Uranium^{235}	2.6	1.40	0.250	6.8
Uranium^{238}	2.6	0.095	0.16	6.9
Fe	–	–	0.006	2.7
Na	–	–	0.0008	3.3
Al	–	–	0.002	3.1

(Source: ANL-5800)

$$\rho\left({}^{235}U\right) = 18.75 \quad \left[gm/cm^3\right]$$
$$\nu\sigma_f = 5.297 \text{ neutron} \cdot \text{b}$$
$$\sigma_a = 2.844 \text{ b}$$
$$\sigma_{tr} = 8.246 \text{ b}$$

Since there is no moderator to indicate the neutron slow down, these data are selected for a fast neutron spectra. We will disregard the extrapolation distance and employ one-group theory. The following is the next step in the solution.

Equating the material and geometrical buckling for a sphere in one-group theory, we get

$$B_g^2 = \left(\frac{\pi}{R_c}\right)^2 = \frac{k_\infty - 1}{L_f^2} = B_m^2 \qquad \text{Eq. 3.105}$$

For the modified two-group theory, L_f^2 is to be replaced by the migration area M^2.

From the last equation, the critical radius is

$$R_c = \pi\left[\frac{\sqrt{L_f^2}}{k_\infty - 1}\right] \qquad \text{Eq. 3.106}$$

where

$k_\infty = \eta\varepsilon pf$, Infinite medium multiplication factor

$L_f^2 = \dfrac{D}{\Sigma_a}$, Fast neutrons diffusion area

$D = \dfrac{1}{3\Sigma_{tr}}$, Fast neutrons diffusion coefficient

For a pure ^{235}U sphere, the fuel utilization factor is

$$f = \frac{\Sigma_{aF}}{\Sigma_a} \approx \frac{\Sigma_a}{\Sigma_a} = 1 \qquad \text{Eq. 3.107}$$

Also, $p \approx \varepsilon \approx 1$ if we neglect fast fissions and resonance absorptions.

Thus,

$$k_\infty \approx \eta \approx \frac{\nu\sigma_f}{\sigma_a} = \frac{5.297}{2.844} = 1.86$$

$$\Sigma_{tr} = N\sigma_{tr} = \sigma_{tr}\frac{\rho \cdot Av}{M} = \left[\frac{18.75 \times 0.602 \times 10^{+24}}{235}\right] \times 8.246 \times 10^{-24}$$

$$= 0.396\,[cm^{-1}]$$

$$\Sigma_a = N\sigma_a = \sigma_a\left(\frac{\rho \cdot Av}{M}\right) = \left[\frac{18.75 \times 0.602 \times 10^{+24}}{235}\right] \times 2.844 \times 10^{-24}$$

$$= 0.137\,[cm^{-1}]$$

$$L_f^2 = \frac{D}{\Sigma_a} = \frac{1}{3\Sigma_{tr}\Sigma_a} = \frac{1}{3 \times 0.396 \times 0.137} = 6.144\,[cm^2]$$

Therefore, the critical radius is

$$R_c = \pi\sqrt{\frac{6.1441}{1.86 - 1}} = \pi\sqrt{7.144} = 8.397\,[cm]$$

The corresponding critical volume is

$$V_c = \frac{4\pi R_c^3}{3} = 2480.054\,[cm^3]$$

and the critical mass becomes

$$M_c = V_c\rho = 2480.054 \times 18.75 = 46501.013\ gm = 46.5\ kg$$

This compares to the radius of 8.71 cm and the critical mass of 48.8 kg for the Godiva critical experimenter, which is a 93.9% of ^{235}U-enriched system or alloy composition. The core radius of different fast critical assemblies is shown in Table 3.5.

In the zero power reactor ZPR-III 48 experiment, a fast reactor with a core made of uranium-235 and plutonium-239 carbides was simulated. Iron served as the structural element, uranium as a reflector, and sodium as a coolant.

This reactor, formerly located in the building adjacent to Experimental Breeder Reactor No. 1 (EBR-I), was used for reactor physics experiments. Fuel and other materials in the form of flat plates were placed in trays and loaded into the two faces of the reactor. See Figure 3.15.

The two halves were brought together remotely to form a critical mass in which uranium atoms were split to release energy. It produced negligible heat – hence the name "Zero Power Reactor". ZPR-3 is now retired, having been replaced by larger and more advanced reactors of its type.

EBR-I Atomic Museum, is located on U.S. Highway 20/26 between Idaho Falls and Arco (A National Historic Landmark at Idaho National Laboratory).

3.13 THE JEZEBEL CRITICAL PLUTONIUM SPHERE

Critical experiments are conducted to experimentally check the calculated values of the critical masses. Figure 3.16 shows a critical experiment for a ^{239}Pu sphere, designated as Jezebel, where the sphere is divided into three parts remotely brought together along sliding wires.

TABLE 3.5
Effective Multiplication Factor of Unreflected and Reflected Fast Reactor Assemblies [19].

Assembly Core	Reflector	Core Radius [cm]	k_{eff}
U^{233}	–	5.965	1.0115
Godova, U^{235}	–	8.710	0.9912
Jezebel, Pu^{239}	–	6.285	1.0039
37.5% U^{235}	–	14.57	0.9855
16.7% U^{235}	7.6 cm U	20.32	0.9893
U^{235}	1.8 cm U	7.725	0.9907
U^{235}	9.9 cm U	6.391	0.9939
Topsy, U^{235}	U	6.045	0.9907
U^{235}	5.1 cm Fe	7.3900	0.9756
U^{235}	4.6 cm Th	7.800	0.9905
ZPR-III 48	30 cm U	47.4200	1.0160

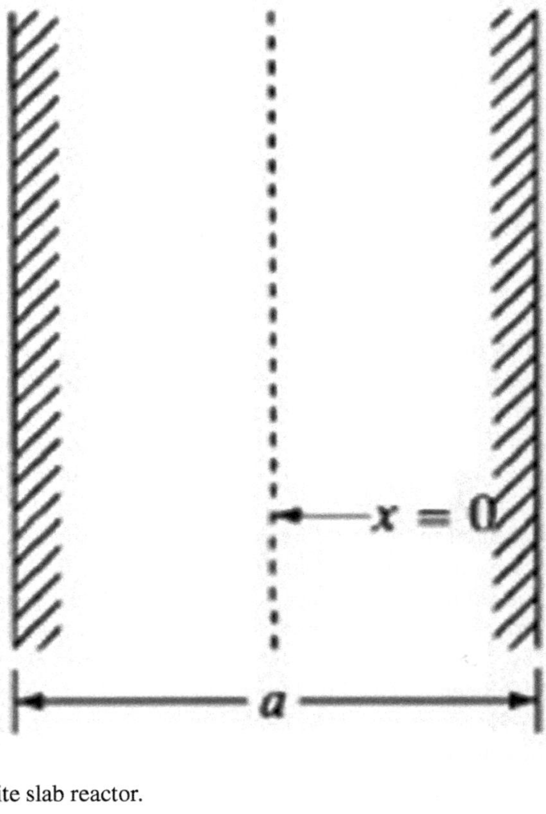

FIGURE 3.15 The infinite slab reactor.

A spherical, naked, unreflected, homogeneous construction of plutonium (Pu) stabilized by 3 to 4.5 atomic percent rare earth element gallium into the delta phase at ambient temperature was the Jezebel experiment, which was conducted between 1945 and 1955.

In 1956, its measured critical mass was determined as 16.45 ± 0.04 kg. This value was updated in 1969 60 16.57 ± 0.10 kg. A later estimation by the X-Division of the Los Alamos National Laboratory (LANL) in 2016 established a value of 16.624 ± 0.065 kg.

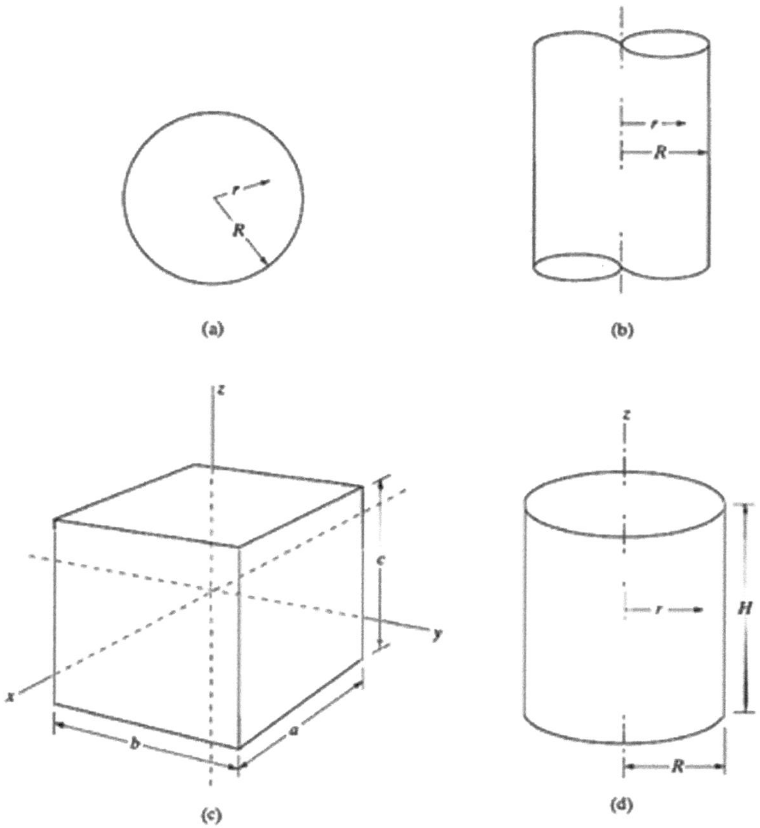

FIGURE 3.16 (a) The spherical reactor, (b) the infinite cylindrical reactor, (c) the parallelepiped reactor, and (d) the finite cylindrical reactor [5].

The delta phase is when solid plutonium is most pliable. Its manufacture is best suited for this phase. Pure plutonium's delta phase is only stable at temperatures between 600 and 700 Kelvin, which is significantly higher than the ambient temperature of 293 Kelvin. Gallium may stabilize the delta phase at ambient temperature, providing greater corrosion resistance, better machinability, and a melting point of around 650°C.

At a relatively low temperature of around 640°C, unalloyed plutonium melts to generate a liquid that is denser than the solid form it melts from. Plutonium's delta face centered cubic (FCC) phase possesses extremely directed or anisotropic elastic characteristics. This indicates that the metal's elasticity varies significantly by a factor of 6 to 7 along various crystallographic axes.

By adding 2 atomic percent of an alloy rare earth metal, such as gallium, the highly symmetric FCC structure may be maintained at normal temperature while the delta FCC form of plutonium remains stable at high temperatures.

3.14 EFFECT OF EXTRAPOLATION DISTANCE

If the extrapolation distance is not neglected in the treated, as it was explained in previous section for a bare reactor with transport cross-section of $\Sigma_{tr} = 0.396 \text{ cm}^{-1}$, then from Equation 3.15, we can write

$$d = 0.71\lambda_{tr} = \frac{0.71}{\Sigma_{tr}} = \frac{0.71}{0.396} = 1.793 \text{ cm}$$

Since, in this case, the extrapolated radius for $R_c = 8.397$ cmis

$$R_{\text{extrapolated radius}} = R_c + d = 10.19 \text{ cm}$$

Thus, the critical assembly radius is

$$R_c = 8.397 - 1.793 = 6.604 \text{ cm}$$

The critical volume will be

$$V_c = \frac{4\pi(6.604)^3}{3} = 1206.451 \text{ cm}^3$$

and the critical mass can be calculated as

$$M_c = 1206.451 \times 18.75 = 22620.962 \text{ g } = 22.6 \text{ kg}$$

We will show all these values in an example of bare, unreflected, and unmoderated fast reactor later on in this chapter.

This highlights the necessity of taking the extrapolation distance into account in tiny reactors since it differs by nearly a factor of 2 from the figure that was calculated in the example, which will be presented later. However, it may be disregarded in massive reactors. Furthermore, the precise size of critical assemblies must be theoretically determined using more complex techniques than one-group diffusion theory, such as transport theory.

3.15 CRITICALITY OF MODERATED HOMOGENEOUS REACTORS WITH DIFFERENT GEOMETRIES

Let us consider a large reactor that is composed of a homogeneous mixture of pure ^{235}U and graphite as carbon. There is one atom of ^{235}U for every 10,000 atoms of carbon, or

$$\frac{N_C}{N_U} = 10^4$$

The migration area is $M^2 = 3040$ cm^2. We ask ourselves the following questions:

1. What is the material buckling?
2. If the reactor is a bare cylinder with a height equal to twice the radius, what is the critical radius?
3. If the reactor is a bare sphere, what is its critical radius?
4. If the reactor is a cube, what is its critical radius?

We will estimate the ^{235}U mass needed for criticality in each of the three final situations by comparing the necessary critical volumes. The following is how we move forward with the solution.

From the modified one-group theory, the material buckling is given by

$$B_m^2 = \frac{k_\infty - 1}{M^2} \quad k_\infty = \eta\varepsilon pf$$

where M^2 is the migration area.

Neglecting fast fission, then fast fission factor $\varepsilon \approx 1$, and neglecting resonance absorption, then resonance escape probability $p \approx 1$ and regeneration factor η can be written as

$$\eta = \nu \frac{\sigma_f}{\sigma_a} = \frac{\nu}{1+\alpha} = 2.08$$

The thermal utilization factor f is given as

$$f = \frac{\Sigma_{aF}}{\Sigma_{aF} + \Sigma_{aC}} = \frac{1}{1 + \frac{N_C}{N_F} \frac{\sigma_{aC}}{\sigma_{aF}}}$$

Substituting the following values in the above equation

$$\sigma_{aC} = 3.4 \times 10^{-6}\ b$$
$$\sigma_{aF} = 2.844\ b$$
$$\frac{N_C}{N_U} = \frac{N_C}{N_F} = 10^4$$

we obtain

$$f = 0.988$$

From which, for k_∞, we have

$$k_\infty \approx 2.08 \times 1 \times 1 \times 0.988 = 2.06$$

The material buckling becomes

$$B_m^2 = \frac{2.06 - 1}{3040} = 3.49 \times 10^{-4}\ \text{cm}^{-2}$$

For an unreflected, bare cylindrical core neglecting the extrapolation length, the geometrical buckling becomes

$$B_g^2 = \left(\frac{\pi}{H}\right)^2 + \left(\frac{2.405}{R}\right)^2$$

Since the height of the reactor $H = 2R$, we get

$$B_g^2 = \left(\frac{\pi}{H}\right)^2 + \left(\frac{2.405}{R}\right)^2 = \frac{8.25}{R^2}$$

Equating the geometrical buckling to the material buckling, the critical radius becomes

$$R_c = \sqrt{\left(\frac{8.25}{3.49 \times 10^{-4}}\right)} = 153.75\ \text{cm}$$

For an unreflected, bare sphere

$$B_g^2 = \left(\frac{\pi}{R}\right)^2$$

and

$$R_c = \sqrt{\left(\frac{\pi}{3.49\times10^{-4}}\right)} = 168.17 \text{ cm}$$

Thus, for cube,

$$B_g^2 = 3\times\left(\frac{\pi}{a}\right)^2$$

and

$$a_c = \sqrt{\frac{3\pi^2}{3.49\times10^{-4}}} = 291.8 \text{ cm}$$

The critical volumes are

$$V_c(\text{Cylinder}) = \pi R^2 H = \pi R^2 H\cdot 2R_c = 2.28\times10^7 \text{ cm}^3$$

$$V_c(\text{Sphere}) = \frac{4\pi R_c^3}{3} = 1.99\times10^7 \text{ cm}^3$$

$$V_c(\text{Cube}) = a_c^3 = 2.47\times10^7 \text{ cm}^3$$

We can notice that even though

$$R_c(\text{Cylinder}) < R_c(\text{Sphere}) < a_c(\text{Cube})$$

Then, we have

$$V_c(\text{Sphere}) < V_c(\text{Cylinder}) < V_c(\text{Cube})$$

The mass of uranium needed for criticality in each case can be calculated as follows:

$$\frac{N_C}{N_U} = 10^4 = \frac{\dfrac{\rho_C(V-V_U)A_V}{12}}{\dfrac{\rho_U V_U A_V}{235}}$$

From which,

$$10^4\cdot\frac{\rho_U}{\rho_C}\cdot\frac{12}{235}\cdot V_U = V - V_U$$

and

$$g_U = \rho_U V_U = \frac{\rho_U V}{1+10^4\cdot\dfrac{\rho_U}{\rho_C}\cdot\dfrac{12}{235}}$$

where g_U is the weight of ^{235}U in volume V in grams, while V_U is the volume occupied by ^{235}U, $\rho_C = 1.60$ g/cm^3, and $\rho_U = 19.1$ g/cm^3.

Substituting for the constants for g_U, we get

$$g_U = \left(3.13\times10^{-3}\right)\cdot V$$

From which,

$$(g_U)_{\text{Cylinder}} = 7.14\times10^4 \text{ gm}$$
$$(g_U)_{\text{Sphere}} = 6.23\times10^4 \text{ gm}$$
$$(g_U)_{\text{Cube}} = 7.73\times10^4 \text{ gm}$$

We can notice that

$$(g_U)_{\text{Sphere}} < (g_U)_{\text{Cylinder}} < (g_U)_{\text{Cube}}$$

This suggests that, in terms of the minimal fissile mass needed to achieve criticality, the sphere is the ideal shape.

3.16 CRITICALITY OF REACTOR CORE WITH AN INFINITE REFLECTOR

Let us look at a spherical reactor with radius R that is encircled by an endless reflector. To determine the relevant criticality requirements and the flux distribution in the core and reflector, we will apply the modified one-group theory. If the thickness of a reflector is equal to the number of neutrons' mean free routes within it, the reflector is said to be "infinite".

The core and reflector's modified one-group thermal diffusion equations are as follows:

$$\begin{aligned} D_C\nabla^2\phi_C - \Sigma_{ac}\phi_C + k_\infty\Sigma_{ac}\phi_C &= 0 \\ D_r\nabla^2\phi_r - \Sigma_{ar}\phi_r &= 0 \end{aligned} \qquad \text{Eq. 3.108}$$

The solution in the core is

$$\phi_C = \frac{C\sin(Br)}{r} = \phi_0\frac{\sin(Br)}{Br} \qquad \text{Eq. 3.109}$$

where

$$B^2 = \frac{k_\infty - 1}{M^2} \qquad \text{Eq. 3.110}$$

Then, the final solution in the reflector is

$$\phi_r = \frac{Ae^{-r/L_r}}{r} + \frac{Fe^{+r/L_r}}{r} \qquad \text{Eq. 3.111}$$

where

$$L_r = \frac{D_r}{\Sigma_{ar}} \qquad \text{Eq. 3.112}$$

For an infinite reflector and a finite flux, $F = 0$.

Using Fick's Law to apply the continuity of the flux and current at the contact between the core and the reflector $r = R$, we obtain

$$\begin{aligned} &\phi_0\frac{\sin(BR)}{BR} = \frac{Ae^{-r/L_r}}{R} \\ &-D_c\frac{\phi_0}{B}\left(\frac{B\cos(BR)}{R} - \frac{\sin(BR)}{R^2}\right) = D_rAe^{-R/L_r}\left(\frac{1}{L_rR} + \frac{1}{R^2}\right) \end{aligned} \qquad \text{Eq. 3.113}$$

Dividing both equations, we get

$$-D_C\left(B\cot(BR)-\frac{1}{R}\right)=D_r\left(\frac{1}{L_r}+\frac{1}{R}\right)$$
$$BR\cot(BR)=1-\frac{D_r}{D_c}\left(\frac{R}{L_r}+1\right) \quad \text{Eq. 3.114}$$

This transcendental equation may be solved visually or iteratively numerically to obtain the critical mass, volume, and radius as well as the flux distribution in the core and reflector.

3.17 CRITICALITY OF REACTOR CORE WITH A FINITE SIZE REFLECTOR

The neutrons seeping from the surface will be reflected by a reflector of a scattering material encircling the fissile core. As a result, the fissile core's critical mass decreases. It is more practical to formulate the solution in the reflector as follows for a finite reflector of thickness *T*, the magnitude of $C \neq 0$:

$$\phi_r=\frac{A\cosh\left(\frac{r}{L_r}\right)}{r}+\frac{C\sinh\left(\frac{r}{L_r}\right)}{r} \quad \text{Eq. 3.115}$$

The following results are obtained from applying the criterion for the flux's disappearing at $r = R + T + d$:

$$C=-A\frac{\cosh\left(\frac{R+T+d}{L_r}\right)}{\sinh\left(\frac{R+T+d}{L_r}\right)}=-A\coth\left(\frac{R+T+d}{L_r}\right) \quad \text{Eq. 3.116}$$

Thus, we have

$$\phi_r=A\frac{\cosh\left(\frac{r}{L_r}\right)}{r}-A\frac{\cosh\left(R+T+d\right)}{L_r}\cdot\frac{\sinh\left(\frac{r}{L_r}\right)}{r}$$
$$=\frac{A}{r\sinh\left(\frac{R+T+d}{L_r}\right)}\left[\sin\left(\frac{R+T+d}{L_r}\right)\cdot\cosh\left(\frac{r}{L_r}\right)-\cosh\left(\frac{R+T+d}{L_r}\right)\cdot\sinh\left(\frac{r}{L_r}\right)\right] \quad \text{Eq. 3.117}$$
$$=\frac{A}{r\sinh\left(\frac{R+T+d}{L_r}\right)}\sinh\left(\frac{R+T+d-r}{L_r}\right)$$

When we apply the flux and current continuity requirements at $r = R$, we obtain

$$\begin{cases} \phi_0 \dfrac{\sin(BR)}{BR} = \dfrac{A}{R\sinh\left(\dfrac{R+T+d}{L_r}\right)} \sinh\left(\dfrac{T+d}{L_r}\right) \\ -D\left(\dfrac{\phi_0}{B}\right)\left(\dfrac{B\cos(BR)}{R} - \dfrac{\sin(BR)}{R^2}\right) = \dfrac{AD_r}{\sinh\left(\dfrac{R+T+d}{L_r}\right)}\left(\dfrac{\cos\left(\dfrac{(T+d)}{L_r}\right)}{L_rR} + \dfrac{\sinh\left(\dfrac{(T+d}{L_r}\right)}{R^2}\right) \end{cases} \quad \text{Eq. 3.118}$$

Dividing both equations, we get

$$-D_c\left(B\coth(BR) - \frac{1}{R}\right) = D_r\left(\frac{\coth\left(\dfrac{T+d}{L_r}\right)}{L_r} + \frac{1}{R}\right) \quad \text{Eq. 3.119}$$

Therefore, we obtain the following transcendental equation to be solved visually or numerically for the critical dimensions and flux distribution as the z criticality condition for a reflector of thickness T.

$$\coth(BR) = \frac{1}{BR}\left(1 - \frac{D_r}{D_c}\right) - \left(\frac{D_r}{D_c}\right) \cdot \left(\frac{1}{BL_r}\right) \cdot \coth\left(\frac{T+d}{L_r}\right) \quad \text{Eq. 3.120}$$

3.18 STANDING-WAVE FAIL-SAFE REACTOR CORE

"If a system is not designed to be fail-safe, tested under all combinations of extreme conditions, and operated perfectly, it will fail", goes the adage of system engineers. Therefore, it is imperative that nuclear reactor designs adhere to this principle during both the design and operation phases.

Murphy's Law, which states that "if anything can go wrong, it will" or "Anything that can go wrong, will go wrong", compels us to conclude that if our engineering systems were not designed to prevent failure in the first place, they would ultimately fail. Therefore, we try to think of a reactor design that is unquestionably fail-safe.

On examining Figure 3.17, it shows a cylindrical or spherical reactor core with a core radius R encircled by an infinite reflector. If, precisely, unity is used as the core infinite medium multiplication factor, then

$$k_\infty = 1 \quad \text{Eq. 3.121}$$

Then, even with the reflector present, the system would be practically subcritical due to leakage from the core to the reflector, which would result in an effective multiplication factor value below unity [13]:

$$k_{eff} < 1 \quad \text{Eq. 3.122}$$

For example, when a reactor is being transported or launched into space by rocket, this would be a desired, naturally fail-safe scenario. If a neutron source of fissile material, such ^{235}U, is added to the system to displace a void or an absorbent layer between the core and the reflector for further safety, the system can be rendered critical or armed after the launch is successfully finished.

FIGURE 3.17 Zero power reactor-III (ZPR-3) photo.

(Source: United States Department of Energy [14])

Additionally, this can serve as the foundation for a fail-safe fission reactor design that uses an accelerator-driven neutron source, a neutron source from a D–D or D–T fusion reactor, or a fission area neutron source.

The neutron source introduces a net current at the interface equal to the following, assuming a thin thickness T of fissile material with a macroscopic absorption cross-section Σ_a and macroscopic fission cross-section Σ_f.

$$\begin{aligned} J_{\substack{neutron\\source}} &= \nu\, \Sigma_f \phi_c(R)\cdot T - \Sigma_{aF}\phi_c(R)\cdot T \\ &= \nu\, \frac{\Sigma_f}{\Sigma_{aF}} \Sigma_{aF}\phi_c(R)\cdot T - \Sigma_{aF}\phi_c(R)\cdot T \\ &= \left(\eta_{\substack{neutron\\source}} - 1\right)\Sigma_{aF}\phi_c(R)\cdot T \end{aligned} \qquad \text{Eq. 3.123}$$

where $\phi_c(R)$ is the flux value when the neutron source is inserted at the core–reflector contact.

Diffusion equations for the core and reflector areas may be expressed as follows:

$$\begin{aligned} &\text{Core}: && D_c\nabla^2\phi_c - \Sigma_{ac}\phi_c + \eta_c\Sigma_{aFc}\phi_c = 0 \\ &\text{Reflector}: && D_r\nabla^2\phi_r - \Sigma_{ac}\phi_r = 0 \end{aligned} \qquad \text{Eq. 3.124}$$

We can propose the following for a fast unmoderated reactor material in the core with a fast fission factor ε and a unity resonance escape probability P:

$$k_\infty = \eta_c \varepsilon p f \simeq \eta_c \cdot 1 \cdot 1 \cdot f = \eta_c f \qquad \text{Eq. 3.125}$$

and

$$\eta_c \Sigma_{aF} = \eta_c \frac{\Sigma_{aF}}{\Sigma_{ac}} \Sigma_{ac} = \eta_c f \Sigma_{ac} = k_\infty \Sigma_{ac} \quad \text{Eq. 3.126}$$

Substituting in the core and reflector diffusion theory equations, we obtain

$$\begin{aligned} \nabla^2 \phi_c + \frac{(k_\infty - 1)}{L_c^2} \phi_c &= 0 \quad L_c^2 = \frac{D_c}{\Sigma_{ac}} \\ \nabla^2 \phi_r - \frac{1}{L_r^2} \phi_r &= 0 \quad L_r^2 = \frac{D_r}{\Sigma_{ar}} \end{aligned} \quad \text{Eq. 3.127}$$

If the material is chosen with an infinite medium multiplication factor of unity, the material buckling in the core looks like this.

$$B_c^2 = \frac{k_\infty - 1}{L_c^2} = \frac{1-1}{L_c^2} = 0 \quad \text{Eq. 3.128}$$

The core exhibits a flat flux distribution due to its zero-material buckling, indicating a consistent power distribution. This is a desirable characteristic that leads to consistent fuel burnup and heat production.

Section 3.19 of this chapter goes into further detail on the neutron absorption cross-section.

3.19 SPHERICAL CORE

We may sum up mathematical connections in spherical geometry as follows:

$$\begin{aligned} &\nabla^2 \phi_c(r) = 0 \\ &\frac{1}{r^2} \frac{d}{dr} \left(r^2 \frac{d\phi_c(r)}{dr} \right) = 0 \\ &\frac{d}{dr} \left(r^2 \frac{d\phi_c(r)}{dr} \right) = 0 \\ &\int d \left(r^2 \frac{d\phi_c(r)}{dr} \right) = \int 0 dr \\ &\left(r^2 \frac{d\phi_c(r)}{dr} \right) = C \\ &\frac{d\phi_c(r)}{dr} = \frac{C}{r^2} \quad \forall r \neq 0 \\ &\int d\phi_c(r) = \int \frac{C}{r^2} dr \\ &\phi_c(r) = -\frac{C}{r} + F \end{aligned} \quad \text{Eq. 3.129}$$

For a finite flux in the core, $C = 0$, and

$$\phi_c(r) = F = \text{constant} \quad \text{Eq. 3.130}$$

3.20 SEMI-INFINITE SLAB GEOMETRY

The simplified form for a semi-infinite slab in slab Cartesian geometry is as follows:

$$\nabla^2\phi_c(X)=0$$

$$\frac{d^2\phi_c(X)}{dx^2}=0$$

$$\frac{d}{dx}\left(\frac{d\phi_c(X)}{dx}\right)=0$$

$$\int d\left(\frac{d\phi_c(X)}{dx}\right)=\int 0dx \qquad \text{Eq. 3.131}$$

$$\frac{d\phi_c(X)}{dx}=C$$

$$\int d\phi_c(X)=\int Cdx$$

$$\phi_c(X)=Cx+F$$

For a finite flux in the core, $C=0$, and again:

$$\phi_c(X)=F=\text{constant} \qquad \text{Eq. 3.132}$$

3.21 SEMI-INFINITE CYLINDRICAL GEOMETRY

The simplified form for a semi-infinite cylinder in cylindrical geometry is as follows:

$$\nabla^2\phi_c(r)=0$$

$$\frac{1}{r}\frac{d}{dr}\left(r\frac{d\phi_c(r)}{dr}\right)=0$$

$$\int d\left(r\frac{d\phi_c(r)}{dr}\right)=\int r\cdot 0dr$$

$$r\frac{d\phi_c(r)}{dr}=C \qquad \text{Eq. 3.133}$$

$$\frac{d\phi_c(r)}{dr}=\frac{C}{r} \quad \forall r\neq 0$$

$$\int d\phi_c(r)=\int\frac{C}{r}dr$$

$$\phi_c(r)=C\ln r+F$$

For a finite flux in the core, $C=0$, and again:

$$\phi_c(r)=F=\text{constant} \qquad \text{Eq. 3.134}$$

The flux solution in the in the infinite reflector is

$$\phi_r(r) = A\frac{e^{-\frac{r}{L_r}}}{r} + G\frac{e^{+\frac{r}{L_r}}}{r} \quad \text{Eq. 3.135}$$

For finite flux, $G = 0$ and

$$\phi_r(r) = A\frac{e^{-\frac{r}{L_r}}}{r} \quad \text{Eq. 3.136}$$

Since the neutron source shell is regarded as thin, we can now apply the thin interface boundary conditions. At the contact, the flux and current continuity results in

$$\begin{aligned} \phi_c(R) &= \phi_r(R) \\ J_{cn}(R) &= J_{rn}(R) + J_{\substack{neutron\\source}} \end{aligned} \quad \text{Eq. 3.137}$$

The flux continuity at the boundary implies that

$$F = A\frac{e^{-\frac{r}{L_r}}}{r} \quad \text{Eq. 3.138}$$

Therefore, the current boundary condition becomes

$$D_c\nabla\phi_c(R) = D_r\nabla\phi_r(R) + J_{\substack{neutron\\source}}$$

$$0 = A\frac{D_r}{R}\left[-\frac{e^{-\frac{R}{L_r}}}{R^2} - \frac{e^{+\frac{r}{L_r}}}{L_r R}\right] + \left(\eta_{\substack{neutron\\source}} - 1\right)\Sigma_{aF} A\frac{e^{-\frac{R}{L_r}}}{R} T \quad \text{Eq. 3.139}$$

When the constant A is eliminated, the assembly's critical condition is as follows:

$$D_r\left[\frac{1}{R} + \frac{1}{L_r}\right] = T\left(\eta_{\substack{neutron\\source}} - 1\right)\Sigma_{aF} \quad \text{Eq. 3.140}$$

3.22 DIFFUSION EQUATION – NON-MULTIPLYING SYSTEMS

As previously mentioned, Figure 3.17 illustrates how the diffusion theory is frequently employed in the fundamental design of modern PWRs or BWRs. This section offers instructive insights into how the neutron flux can be dispersed in any diffusion medium, however, it is not about such computations. We shall solve diffusion equations in this section.

$$D\Delta\phi - \Sigma_a\phi + S = \frac{1}{v}\frac{\partial\phi}{\partial t} \quad \text{Eq. 3.141}$$

In general form, where Equation 3.141 is concerned, neither D nor Σ_a is dependent on the position and time–homogeneous system. However, Equation 3.141 may be expressed as follows in a variety of geometries that meet the boundary requirements established in earlier sections of this chapter.

$$D\Delta\phi(\vec{r},t)-\Sigma_a\phi(\vec{r},t)+S(\vec{r},t)=\frac{1}{v}\frac{\partial\phi(\vec{r},t)}{\partial t} \quad \text{Eq. 3.142}$$

We will begin with basic systems and work our way up to more complicated ones. The idea that all neutrons belong to a single energy group is the most crucial one. At thermal energy (0.025 eV), they are released and dispersed.

Moreover, neutron diffusion in a non-multiplying system – that is, a system in which fissile isotopes are absent – is covered in the first subsection of this article. In this case, the fission cross-section is zero. An external neutron source emits the neutrons. We will suppose that D and Σ_a are constants and that the system is uniform outside the source.

3.22.1 Solution for the Infinite Planar Source

Let us assume the neutron source (with strength) as an infinite plane source (in $y-z$ plane geometry). In this geometry, as illustrated in Figure 3.18, the flux varies so slowly in y and z allowing us to eliminate the y and z derivatives from ∇^2. The flux is then a function of x only, and therefore, the Laplacian and diffusion equations (outside the source) can be written as (everywhere, except $x = 0$)

$$\nabla^2=\frac{d^2}{dx^2}$$

$$D\frac{d^2\phi(x)}{dx^2}-\Sigma_a\phi(x)=0 \quad \text{Eq. 3.143}$$

$$\frac{d^2\phi(x)}{dx^2}-\frac{\phi(x)}{L^2}=0$$

FIGURE 3.18 The jezebel critical plutonium sphere experiment.

(Source: United States Department of Energy [14])

For $x > 0$, this diffusion equation as written in form of Equation 3.141 has two possible solutions of type $\exp(x/L)$ and $\exp(-x/L)$, which give a general solution as

$$\phi(x) = A\exp(x/L) + C\exp(-x/L) \qquad \text{Eq. 3.144}$$

B is not usually used as a constant because B is reserved for a buckling parameter. To determine the coefficients A and C, we must apply boundary conditions. See Figure 3.18.

It may be inferred that A must equal zero from the finite flux requirement ($0 \le \phi(x) < \infty$), which only allowed for realistic flux values. A physically valid solution for $x > 0$ cannot include the term $\exp(x/L)$, as it goes to ∞ as $x \to \infty$. Refer to Figure 3.19.

The physically acceptable solution for $x > 0$ must then be

$$\phi(x) = Ce^{-x/L} \qquad \text{Eq. 3.145}$$

where C is a constant that can be determined from the source condition at $x \to 0$.

The number of neutrons passing outwards per unit area in the positive x-direction must trend to $S_0/2$ as $x \to 0$ if S_0 is the source intensity per unit area of the plane.

$$\begin{aligned}
&\lim_{x\to 0} J(x) = \frac{S_0}{2} \Rightarrow \\
&\lim_{x\to 0}\left[-D\frac{d\phi}{dx}\right] = \frac{S_0}{2} \Rightarrow \\
&\lim_{x\to 0}\left[-De^{-x/L}\right] = \frac{S_0}{2} \Rightarrow \\
&\Rightarrow \frac{CD}{L} = \frac{S_0}{2} \Rightarrow C = \frac{S_0 L}{2D}
\end{aligned} \qquad \text{Eq. 3.146}$$

So that the solution may be written as

$$\phi(x) = \frac{S_0 L}{2D} e^{-|x|/L} \qquad \text{Eq. 3.147}$$

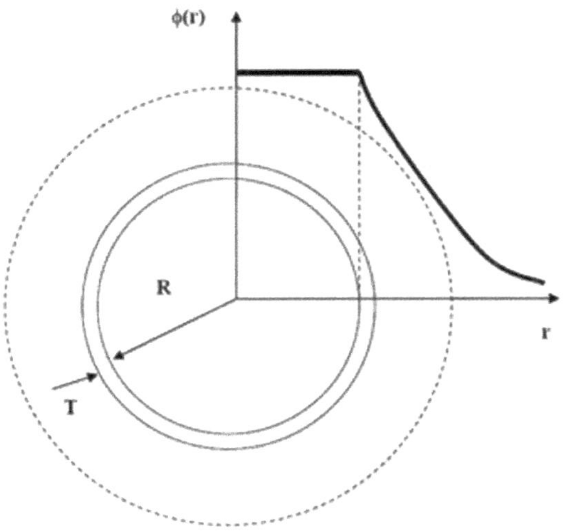

FIGURE 3.19 Geometry of a spherical or cylindrical core with an infinite reflector and neutron source reflector interface.

Assume that Figure 3.20's neutron source (with strength S_0) is an isotropic point source located in a spherical shape.

The origin of coordinates is where this point source is located. The Laplacian must be substituted with its spherical version in order to solve the diffusion equation.

3D Laplacian

$$\nabla^2 \equiv \frac{1}{r^2}\frac{\partial}{\partial r}\left(r^2\frac{\partial}{\partial r}\right)+\frac{1}{r^2\sin(\theta)}\frac{\partial}{\partial\theta}\left(\sin(\theta)\frac{\partial}{\partial\theta}\right)+\frac{1}{r^2\sin^2(\theta)}\frac{\partial^2}{\partial\varphi^2}$$

1D Laplacian

$$\nabla^2 = \frac{1}{r^2}\frac{\partial}{\partial r}\left(r^2\frac{\partial}{\partial r}\right)=\frac{\partial^2}{\partial r^2}+\frac{2}{r}\frac{\partial}{\partial r}$$

Since there is no reliance on an angle (polar or azimuthal), we may substitute the 1D spherical version of the 3D Laplacian. Because of the spherical symmetry, the source is thought to be isotropic. The diffusion equation (outside the source) may thus be expressed as follows (anywhere except $r = 0$) since the flow is a function of radius – r only.

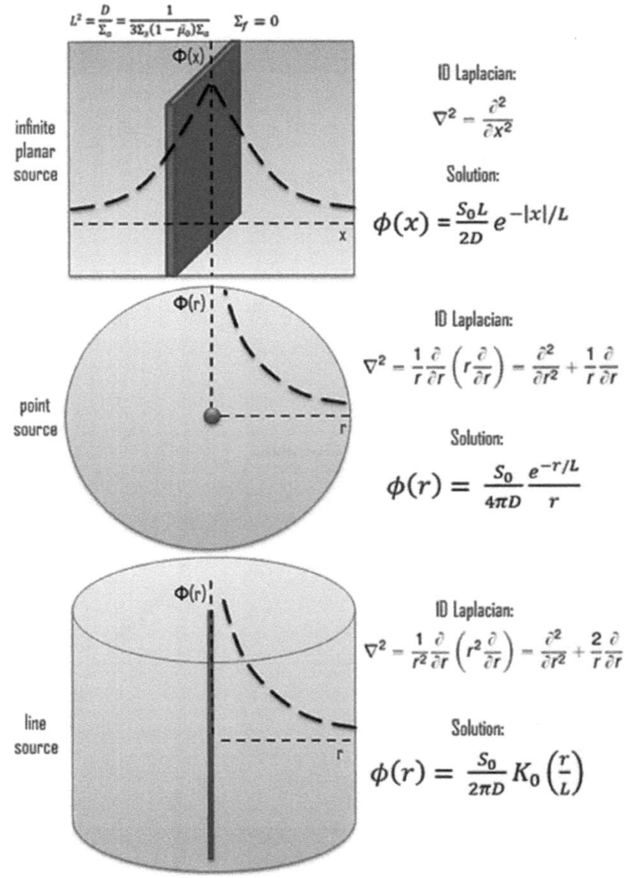

FIGURE 3.20 Neutron flux distribution in various geometries of non-multiplying systems [5].

$$D\nabla^2\phi(r) - \Sigma_a \phi(r) = 0 \quad \text{Eq. 3.148 (a)}$$

i.e.,

$$D\left[\frac{d^2\phi(r)}{dr^2} + \frac{2}{r}\frac{d\phi(r)}{dr}\right] - \Sigma_a \phi(r) = 0 \quad \text{Eq. 3.148 (b)}$$

or

$$\frac{d^2\phi(r)}{dr^2} + \frac{2}{r}\frac{d\phi}{dr} - \frac{\phi(r)}{L^2} = 0 \quad \text{Eq. 3.148 (c)}$$

If we make the substitution $\phi(r) = \frac{1}{r}\psi(r)$, the Equation 3.148(c) will be simplified to

$$\frac{d^2\psi(r)}{dr^2} - \frac{\psi(r)}{L^2} = 0 \quad \text{Eq. 3.149}$$

For $r > 0$, this differential equation has two possible solutions $\exp(r/L)$ and $\exp(-r/L)$, which give a general solution.

$$\begin{cases} \psi(r) = Ae^{r/L} + Ce^{-r/L} \\ \text{or} \\ \phi(r) = A\left(\dfrac{e^{r/L}}{r}\right) + C\left(\dfrac{e^{-r/L}}{r}\right) \end{cases} \quad \text{Eq. 3.150}$$

Remember that B is reserved for a buckling parameter, therefore it is not often used as a constant. We must use boundary conditions in order to ascertain the coefficients A and C.

We follow the same process as for the infinite planar source to determine constants A and C. It may be inferred that A must equal zero from the finite flux requirement $(0 \le \phi(r) < \infty)$, which is only allowed for realistic flux values. $\exp(r/L)/r$ cannot be a part of a physically valid solution for X4 since it refers to ∞ as $r \to \infty$. Then, $r > 0$ physically feasible solution has to be

$$\phi(r) = C\left(e^{-r/L}\middle/r\right) \quad \text{Eq. 3.151}$$

where the source condition at $x \to 0$ may be used to derive the constant C.

The number of neutrons that pass a sphere outward in the positive r-direction must trend to S_0 as $r \to 0$ if S_0 is the source strength.

$$\begin{aligned} &\lim_{x\to 0} 4\pi r^2 J(x) = S_0 \Rightarrow \\ &\lim_{x\to 0}\left[-4\pi r^2 D\frac{d\phi}{dr}\right] = S_0 \Rightarrow \\ &\lim_{x\to 0}\left[-4\pi r^2 CD\frac{-\frac{r}{L}e^{-r/L} - e^{-r/L}}{r^2}\right] = S_0 \\ &\Rightarrow 4\pi CD = S_0 \Rightarrow C = \frac{S_0}{4\pi D} \end{aligned} \quad \text{Eq. 3.152}$$

So that the solution may be written as

$$\phi(r) = \frac{S_0}{4\pi D}\left(\frac{e^{-r/L}}{r}\right) \quad \text{Eq. 3.153}$$

As shown in Figure 3.21, let us suppose that the neutron source (with intensity S_0) is an isotropic line source located in an infinite cylindrical shape. The location of this line source is $r = 0$. We must substitute the cylindrical form of the Laplacian in order to solve the diffusion equation.

3D Laplacian

$$\nabla^2 \equiv \frac{1}{r}\frac{\partial}{\partial r}\left(r\frac{\partial}{\partial r}\right) + \frac{1}{r^2}\frac{\partial^2}{\partial\theta^2} + \frac{\partial^2}{\partial z^2}$$

1D Laplacian

$$\nabla^2 = \frac{1}{r}\frac{\partial}{\partial r}\left(r\frac{\partial}{\partial r}\right) = \frac{\partial^2}{\partial r^2} + \frac{1}{r}\frac{\partial}{\partial r}$$

We may solve the issue just in the radial direction by substituting the 1D version of the 3D Laplacian since it is independent of the angle Θ and the z-coordinate. It is believed that the source is isotropic. The diffusion equation (outside the source) may be expressed as follows (anywhere except $r = 0$) since the flow is a function of radius – r only.

$$D\nabla^2\phi(r) - \Sigma_a\phi(r) = 0 \quad \text{Eq. 3.154 (a)}$$

that is,

$$D\left[\frac{d^2\phi(r)}{dr^2} + \frac{1}{r}\frac{d\phi(r)}{dr}\right] - \Sigma_a\phi(r) = 0 \quad \text{Eq. 3.154 (b)}$$

or

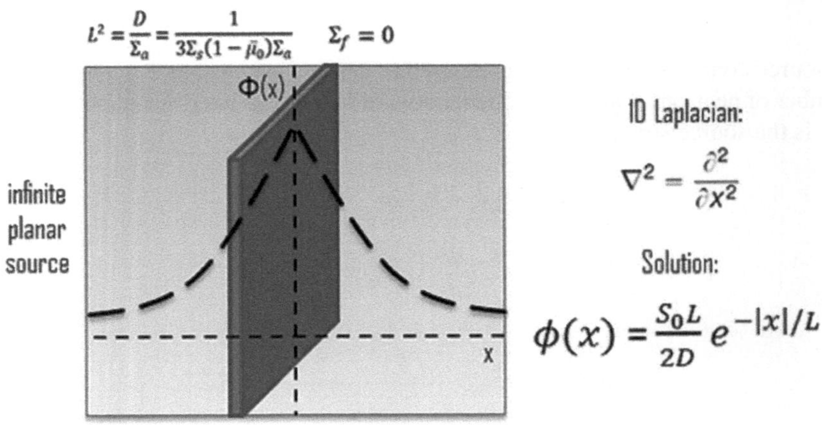

FIGURE 3.21 Infinite plane source configuration [5].

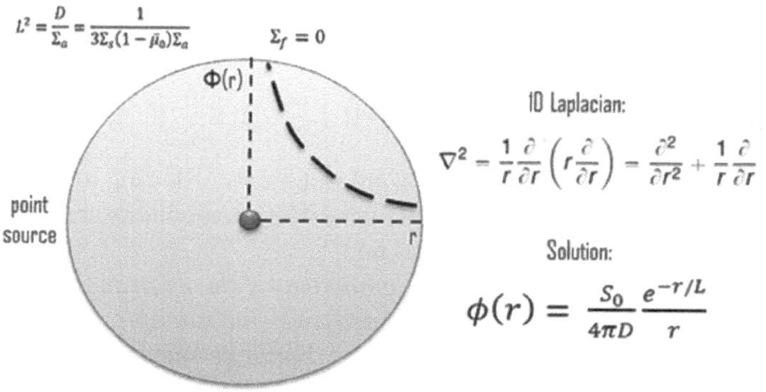

FIGURE 3.22 Isotropic point source configuration [5].

$$\frac{d^2\phi(r)}{dr^2} + \frac{1}{r}\frac{d\phi(r)}{dr} - \frac{\phi(r)}{L^2} = 0 \qquad \text{Eq. 3.154 (c)}$$

Mathematicians are familiar with this differential equation, which is known as Bessel's equation (Equation 3.154(c)). Here, the solutions to Bessel's equation are referred to as the first and second class modified Bessel functions, or sometimes the hyperbolic Bessel functions, $I_a(X)$ and $K_a(r/L)$, respectively. Refer to Figure 3.22.

The two potential solutions to this differential equation for $r > 0$ are $I_0(r/L)$ and $K_0(r/L)$, which are modified Bessel functions of order zero that provide a generic solution.

$$\phi(r) = AI_0\left(\frac{r}{L}\right) + CK_0\left(\frac{r}{L}\right) \qquad \text{Eq. 3.155}$$

We follow the same process as for the infinite planar source to determine constants A and C. It may be inferred that A must equal zero from the finite flux condition ($0 \le \phi(r) < \infty$), which called for only acceptable flux values. A physically valid solution for $r > 0$ cannot include the term $I_0(r/L)$, which goes to ∞ as $r \to \infty$. Then, $r > 0$ physically feasible solution has to be

$$\phi(r) = CK_0(r/L) \qquad \text{Eq. 3.156}$$

where C is a constant that can be determined from the source condition at $r \to 0$.

If S_0 is the source strength, then the number of neutrons crossing a cylinder outward in the positive r-direction must tend to S_0 as $r \to 0$.

So that the solution may be written as

$$\phi(r) = \frac{S_0}{2\pi D}K_0\left(\frac{r}{L}\right) \qquad \text{Eq. 3.157}$$

3.22.2 Solution in Two Different Diffusion Media

Different zones with varying compositions can make up a diffusion environment. As a result, there will be variations in the neutron flux distribution per zone as well as in the diffusion coefficient (see Section 3.2.2 of this chapter) and absorption macroscopic cross-section (see Section 1.9 of this book for more information). The diffusion equations in zones 1 and 2 must be resolved in order to determine the flux distribution in different zones.

$$D_1\nabla^2\phi_1 - \Sigma_{a1}\phi_1 = 0 \qquad |x| \le \frac{a}{2}$$
$$D_2\nabla^2\phi_2 - \Sigma_{a2}\phi_2 = 0 \qquad \frac{a}{2} \le |x| \le \frac{b}{2} \qquad \text{Eq. 3.158}$$

where b is the outside dimension of the diffusion environment, including the extrapolated distance, and a is the actual width of zone 1. When two distinct diffusion mediums are involved, the interface boundary requirements are essential and need to be met:

Physically, the neutron flux and the normal component of the neutron current must be continuous at interfaces between two distinct diffusion mediums (for example, between the reactor core and the neutron reflector). Stated differently, φ and J are not permitted to exhibit a jump.

1. **Interface Conditions ϕ and J**

$$\phi_A\big|_{\text{interface}} = \phi_B\big|_{\text{interface}}$$
$$J_{An}\big|_{\text{interface}} = J_{Bn}\big|_{\text{interface}} \qquad \text{Eq. 3.159}$$

It must be stated that if the diffusion coefficients in the two mediums are different, the flux gradient will exhibit a jump since J must be continuous. We must employ two more boundary conditions since these two diffusion equations require four boundary conditions to be solved.

2. **Finite Flux Condition**

   With the possible exception of contrived singular points in a source distribution, the solution must be finite in those areas where the equation is true. The mathematical expression for this boundary condition is

$$0 \le \phi(\vec{r}, t) < \infty \qquad \text{Eq. 3.160}$$

3. **Source Condition**

   Given that all neutrons entering through the source's enclosing region must originate from the neutron source, its existence may be employed as a boundary condition. The source geometry determines this boundary condition, which for a planar source may be expressed mathematically as

Planar Source

$$\lim_{x\to 0} J(x) = \frac{S}{2} \qquad \text{Eq. 3.161}$$

For $x > 0$, these diffusion equations have the following appropriate solutions.

$$\phi_1(x) = A_1\exp(x/L_1) + C_1\exp(-x/L_1) \qquad \text{Eq. 3.162 (a)}$$

and

$$\phi_2(x) = A_2\exp(x/L_2) + C_2\exp(-x/L_2) \qquad \text{Eq. 3.162 (b)}$$

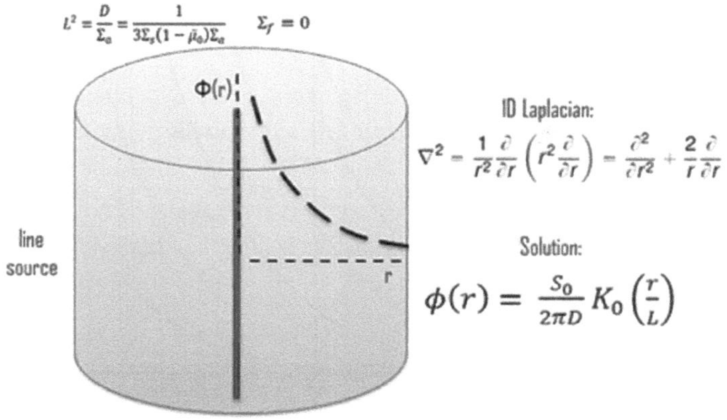

FIGURE 3.23 An infinite cylindrical geometry configuration [5].

where the four boundary conditions must be applied in order to find the four constants. Figure 3.23 depicts the usual neutron flux pattern in a straightforward two-region diffusion issue.

In Figure 3.23, the solution of the diffusion equation in two different non-multiplying diffusion media are depicted.

3.23 DIFFUSION EQUATION – FINITE CYLINDRICAL REACTOR

The diffusion equation solution for the case of multiplying systems is presented in this section.

The environment was thought to be non-multiplying in the preceding section. A neutron source located in the center of a coordinate system emits neutrons, which subsequently freely flow through media in a non-multiplying environment.

Now that we have fissionable nuclei in the multiplying system, we can examine neutron diffusion (i.e., $\sum_f \neq 0$).

We will likewise examine the spatial distribution of neutrons in this multiplying system, but unlike in the non-multiplying environment, these neutrons have the ability to initiate nuclear fission processes. Refer to Figure 3.24.

Also, in this section, we will solve the following diffusion equations of the following form as

$$D\Delta\phi(\vec{r}) - \sum{}_a\phi(\vec{r}) + \nu\sum{}_f\phi(\vec{r}) = 0 \qquad \text{Eq. 3.163}$$

in various geometries that satisfy the boundary conditions. In this equation, ν is number of neutrons emitted in fission and $\sum_f$ is macroscopic cross-section of fission reaction. ϕ denotes a reaction rate. For example, a fission of ^{235}U by thermal neutron yields 2.43 neutrons.

It should be mentioned that we will not use any outside sources to solve the diffusion equation. Since such an equation is a linear homogeneous equation in the flow, this is crucial. As a result, every multiple is a solution if we discover one for the equation. This indicates that it is impossible to infer the neutron flux's absolute value from the diffusion equation. This is completely distinct from issues with external sources, which establish the neutron flux's absolute value.

3.23.1 Multiplication

In a multiplication system (a nuclear reactor), precisely every fission must trigger another fission in order for the chain reaction to be stable and self-sustaining. At least one neutron that triggers the fission of another nucleus must be produced on average by each fissioning nucleus. Additionally, the system's reaction rate – the number of fissions that occur per unit of time – must remain constant.

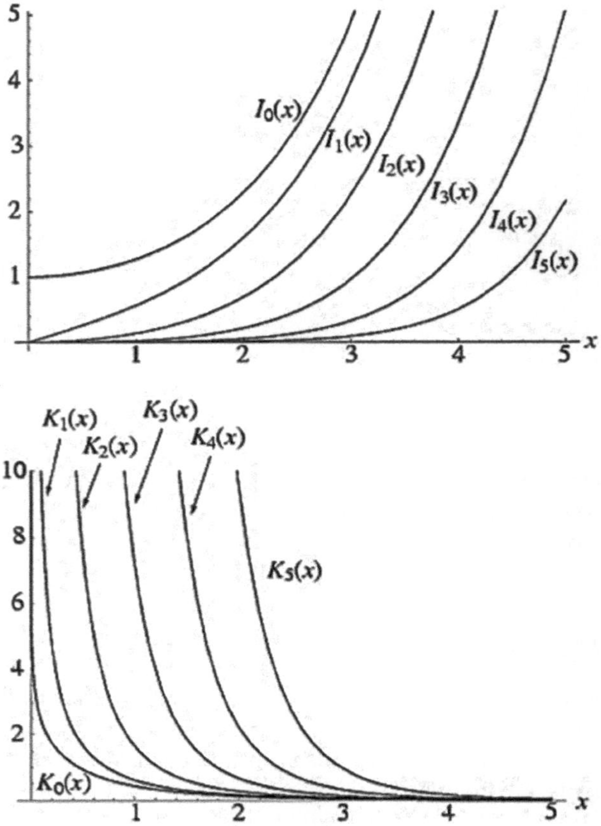

FIGURE 3.24 Bessel's function solution plots.

(Source: www.wikipedia.org)

The multiplication factor provides an easy way to represent this requirement. The ratio of neutrons lost by absorption in the previous generation to the number of neutrons created by fission in one generation is known as the infinite multiplication factor. The following is a mathematical expression for this.

$$k_\infty = \frac{\text{Neutron production from fission in one neutron generation}}{\text{Neutron absorption in the preceding neutron generation}}$$

In a multiplying system, the change in the fission neutron population from one generation to the next is measured by the infinite multiplication factor.

- $\boldsymbol{k_\infty < 1}$: Assume a multiplying system's multiplication factor is smaller than 1.0. In that scenario, the chain reaction will never be self-sustaining as the amount of neutrons declines with time (with the mean generation time). The subcritical state is the name given to this situation.
- $\boldsymbol{k_\infty = 1}$: When a multiplying system's multiplication factor is 1.0, the chain reaction will be self-sustaining and the neutron population will not vary over time. The critical state is the name given to this situation.

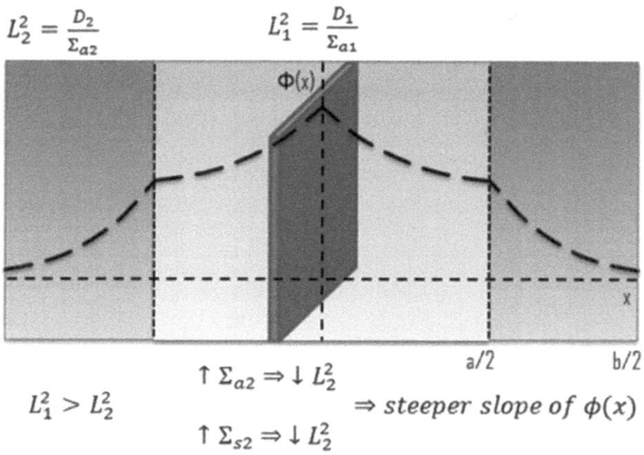

FIGURE 3.25 Two-region diffusion problem schema [10].

- $k_\infty > 1$: In the event that a multiplying system's multiplication factor exceeds 1.0, it generates more neutrons than it requires to maintain itself. As the mean generation time increases, the quantity of neutrons increases exponentially with time. The supercritical state is the name given to this phenomenon.

Figure 3.25 shows a visual representation of all three of the aforementioned infinite multiplication factors in terms of reactor criticality.

3.24 SOLUTION FOR DIFFUSION EQUATION – FINITE CYLINDRICAL REACTOR

As shown in Figure 3.26, let us suppose a uniform reactor (multiplying system) is shaped like a cylinder with a physical radius of R and a height of H. This finite cylindrical reactor is located at the origin of coordinates in cylindrical geometry. We must substitute the cylindrical form of the Laplacian in order to solve the diffusion equation.

3D Laplacian

$$\nabla^2 \equiv \frac{1}{r}\frac{\partial}{\partial r}\left(r\frac{\partial}{\partial r}\right)+\frac{1}{r^2}\frac{\partial^2}{\partial \theta^2}+\frac{\partial^2}{\partial z^2}$$

2D Laplacian

$$\nabla^2 \equiv \frac{\partial^2}{\partial r^2}+\frac{1}{r}\frac{\partial}{\partial r}+\frac{\partial^2}{\partial z^2}$$

Since there is no dependence on angle Θ, we can replace the 3D Laplacian with its 2D form and solve the problem in radial and axial directions. Since the flux is a function of radius $-r$ and height $-z$ only ($\phi(r,z)$), the diffusion equation (i.e., Figure 3.26) can be written as

$$D\left[\frac{d^2\phi(r,z)}{dr^2}+\frac{1}{r}\frac{d\phi(r,z)}{dr}+\frac{d^2\phi(r,z)}{dz^2}\right]-\Sigma_a \phi(r,z)+\nu\Sigma_f \phi(r,z)=0 \qquad \text{Eq. 3.164 (a)}$$

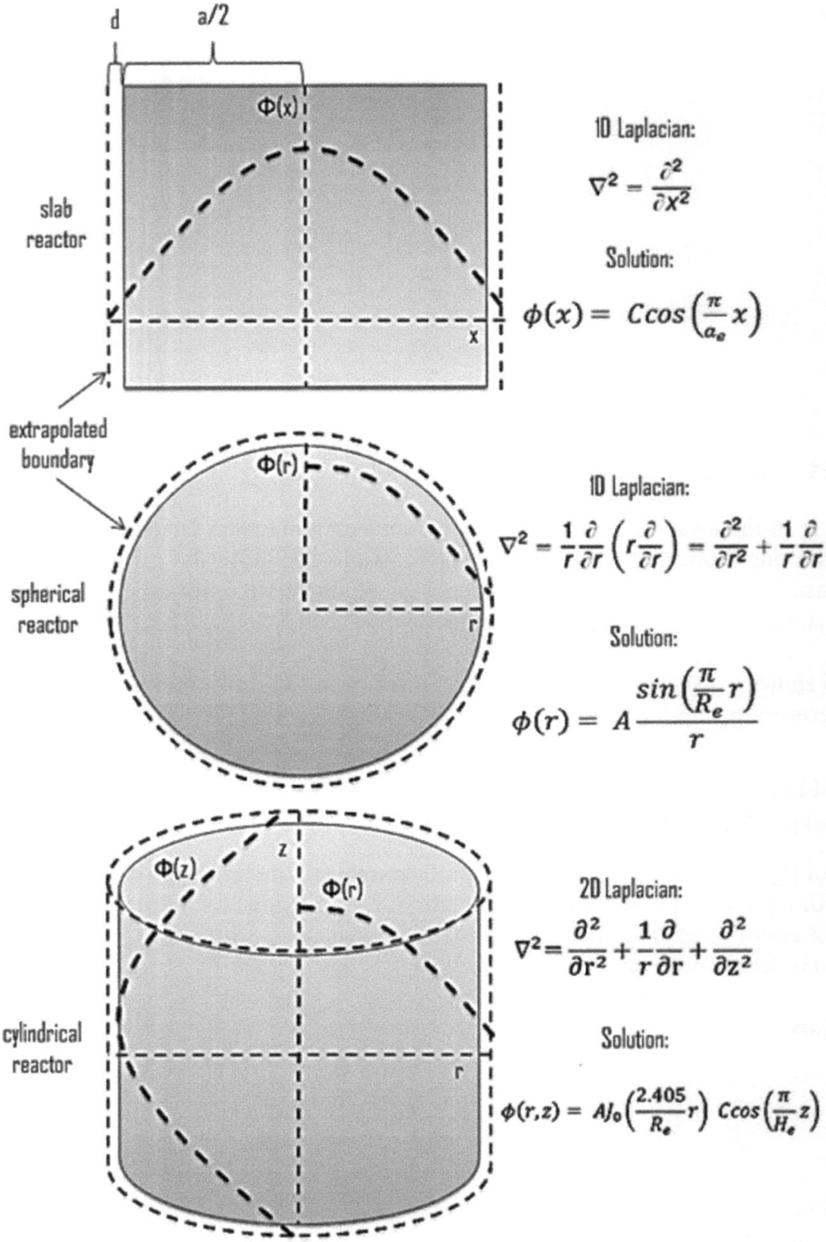

FIGURE 3.26 Simple reactor core geometry configurations.

(Source: www.wikipedia.org)

or

$$\frac{d^2\phi(r,z)}{dr^2} + \frac{1}{r}\frac{d\phi(r,z)}{dr} + \frac{d^2\phi(r,z)}{dz^2} + B_g^2\phi(r,z) = 0 \qquad \text{Eq. 3.164 (b)}$$

where

$$B_g^2 = \frac{\nu \Sigma_f - \Sigma_a}{D} \qquad \text{Eq. 3.164 (c)}$$

The solution of this diffusion equation is based on the use of the separation-of-variables technique, therefore,

$$\phi(r,z) = R(r)Z(z) \qquad \text{Eq. 3.165}$$

where the functions *R(r)* and *Z(z)* need to be found. When we divide this by *R(r)Z(z)* and enter it into the diffusion equation, we get

$$\frac{1}{R(r)}\frac{1}{r}\frac{d}{dr}\left[r\frac{dR(r)}{dr}\right] + \frac{1}{Z(z)}\frac{d^2Z(z)}{dz^2} = -v^2 - \kappa^2 = -B_g^2 \qquad \text{Eq. 3.166}$$

For the equation to have a solution, both terms must be constants as the first term depends only on r and the second only on z. Assume that v^2 and κ^2 are the constants. All of these variables added together must equal $B_g^2 = v^2 + \kappa^2$. Variables may now be separated, and the neutron flow needs to meet the differential equations shown below.

Radial Direction

$$\frac{d^2R(r)}{dr^2} + \frac{1}{r}\frac{dR(r)}{dr} + v^2R(r) = 0 \qquad \text{Eq. 3.167}$$

Axial Director

$$\frac{d^2Z(z)}{dz^2} + \kappa^2 Z(z) = 0 \qquad \text{Eq. 3.168}$$

Solution for the Radial Direction

Mathematicians are familiar with Bessel's equation, which is the differential equation for radial direction. In this instance, the solutions to the Bessel's equation are referred to as the first and second class Bessel functions, or $J_a(X)$ and $Y_a(X)$, respectively. The answer is shown graphically in Figure 3.27.

For $r > 0$, this differential equation has two possible solutions, $J_0(\text{v}r)$ and $Y_0(\text{v}r)$, the Bessel functions of order zero, which give a general solution.

$$R(r) = AJ_0(\text{v}r) + CY_0(\text{v}r) \qquad \text{Eq. 3.169}$$

From finite flux condition ($0 \le \phi(r) < \infty$), which required only reasonable values for the flux, it can be derived that C must be equal to zero. The term $Y_0(\text{v}r)$ goes to $-\infty$ as $r \to \infty$ and therefore cannot be part of a physically acceptable solution. The physically acceptable solution must then be

$$R(r) = AJ_0(\text{v}r)$$

According to the vacuum boundary condition, the relative neutron flux close to the border must have a slope of $-1/d$, meaning that at a distance d beyond the barrier, the flux would extrapolate linearly to 0. This zero-flux border condition is simpler and may be expressed analytically as follows:

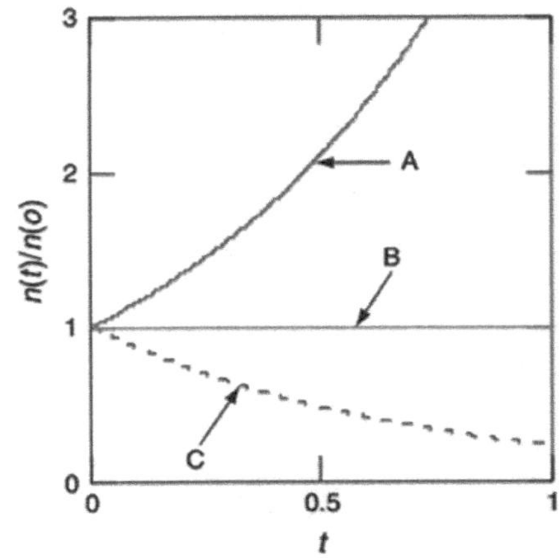

FIGURE 3.27 Reactor criticality. (a) A supercritical state; (b) a critical state; (c) a subcritical state [10].

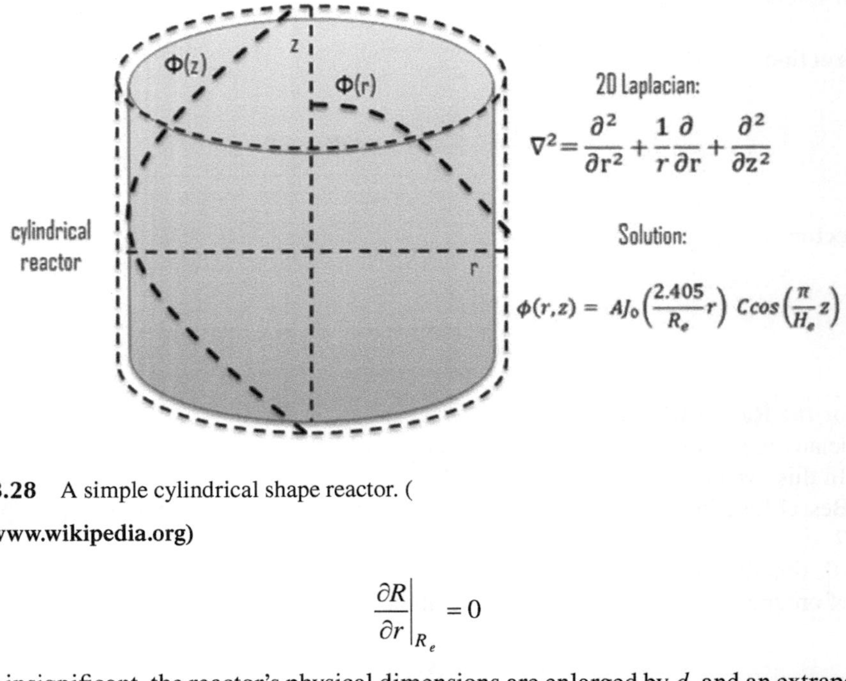

FIGURE 3.28 A simple cylindrical shape reactor. (

(Source: www.wikipedia.org)

$$\left.\frac{\partial R}{\partial r}\right|_{R_e} = 0 \qquad \text{Eq. 3.170}$$

If d is not insignificant, the reactor's physical dimensions are enlarged by d, and an extrapolated barrier of dimension $R_e = R + d$ is created; this condition may be expressed as $\phi(R+d) = \phi(R_e) = 0$.

Therefore, the solution must be $R(R_e) = AJ_0(\mathrm{v}r) = 0$. The function of $J_0(r)$ has several zeroes. The first is at $r_1 = 2.405$, and the second at $r_2 = 5.6$. However, because the neutron flux cannot have regions of negative values, the only physically acceptable value for v is $2.405/R_e$. The solution of the diffusion equation is

$$R(r) = A\,J_0\left(\frac{2.405}{R_e}r\right) \qquad \text{Eq. 3.171}$$

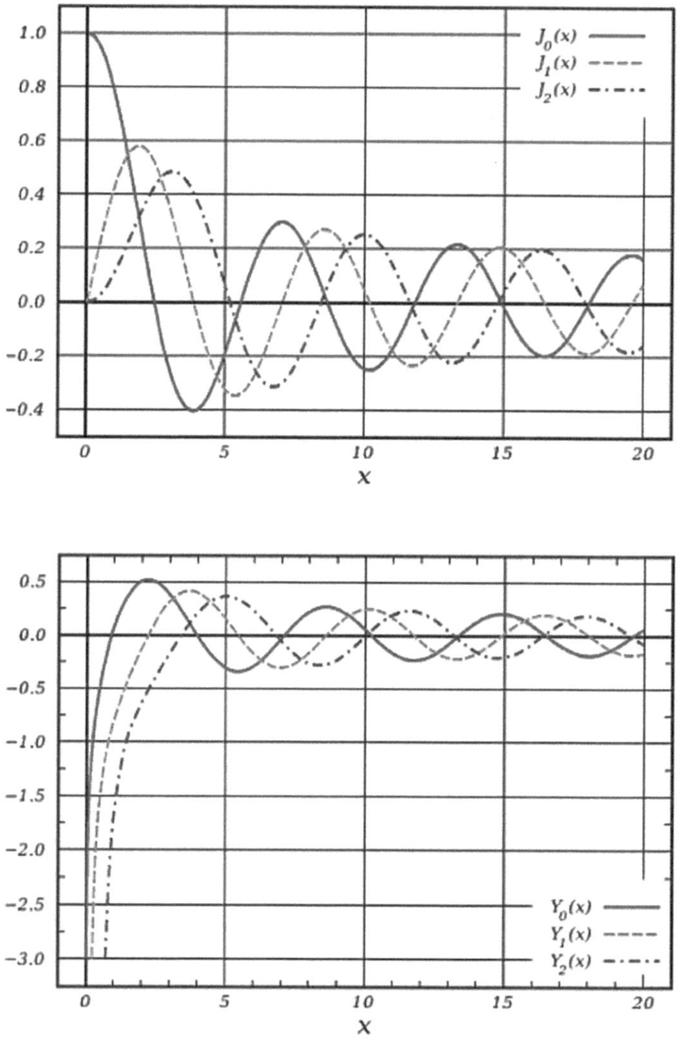

FIGURE 3.29 Graphical Bessel functions solutions.

(Source: www.wikipedia.org)

Solution for the Axial Direction

Since the axial direction solution (infinite slab reactor) has already been solved, it has the same answer. In an axial direction, the answer is

$$Z(z) = C\cos\left(\frac{\pi}{H_e}z\right) \quad \text{Eq. 3.172}$$

Solution for the Radial and the Axial Directions

Thus, the following is the complete solution for the neutron flux distribution in the finite cylindrical reactor.

$$\begin{cases} \phi(r,z) = A J_0\left(\frac{2.405}{R_e} r\right) C\cos\left(\frac{\pi}{H_e} z\right) \\ B_g^2 = \nu^2 + \kappa^2 = \left(\frac{2.405}{R_e}\right) + \left(\frac{\pi}{H_e}\right) \end{cases} \quad \text{Eq. 3.273}$$

where B_g^2 is the total geometrical buckling.

Since the diffusion equation cannot provide the constants A and C, which provide the neutron flux's absolute value, they must be added. The critical reactor can run at any power level, and the neutron flux can be any amount. Note that this flux shape only applies in a fictitious scenario in a homogeneous, uniform cylindrical reactor operating at low power levels (also known as "zero power criticality").

For instance, the neutron flow in a power reactor core can be as high as 3.11×10^{13} neutrons•cm^{-2}•s^{-1}, however, these figures are very dependent on a number of factors (fuel kind, fuel burnup, fuel enrichment, fuel pattern position, etc.).

A power reactor's criticality (k_{eff}) is unaffected by power level unless thermal reactivity feedbacks are in place; in the absence of reactivity feedbacks, a power reactor operates between 10E-8% and 1% of rated power.

REFERENCES

1. H. Van Dam, T. H. J. J. van der Hagen, J. E. Hoogenboom, *Nuclear Reactor Physics*, 2005. Retrieved from http://www.janleenkloosterman.nl/reports/ap3341.pdf.
2. B. Zohuri, *Neutronic Analysis for Nuclear Reactor Systems*, 2nd edition, Springer Publishing Company, 2019.
3. *Reactor Physics Constants*, 2nd Edition, Argonne National Laboratory Report ANL-5800, Section 3.3, 1963.
4. J. R. Lamarsh, A. J. Baratta, *Introduction to Nuclear Engineering*, 4th edition, Pearson, 2018.
5. J. Duderstadt, L. Hamilton, *Nuclear Reactor Analysis*, John Wiley Publishing Company, 1976.
6. A. Henry, *Nuclear Reactor Analysis*, The MIT Press, 1975.
7. Argonne National Laboratory, Report, *Reactor Physics Constants*, ANL-5800, 1979.
8. J. R. Lamarsh. *Introduction to Nuclear Reactor Theory*, 2nd ed., Addison-Wesley, 1983.
9. W. M. Stacey, *Nuclear Reactor Physics*, John Wiley & Sons, 2001.
10. S. Glasstone. *Nuclear Reactor Engineering: Reactor Systems Engineering*, Springer; 4th edition, 1994.
11. W. S. C. Williams. *Nuclear and Particle Physics.* Clarendon Press; 1st edition, 1991.
12. G. R. Keepin. *Physics of Nuclear Kinetics.* Addison-Wesley Pub. Co; 1st edition, 1965.
13. R. R. Burn. *Introduction to Nuclear Reactor Operation*, 1988.
14. U.S. Department of Energy, Nuclear Physics and Reactor Theory, *DOE Fundamentals Handbook*, Volume 1 and 2, 1993.
15. K. O. Ott, W. A. Bezella, *Introductory Nuclear Reactor Statics*, Revised edition, American Nuclear Society, 1989.
16. K. O. Ott, R. J. Neuhold, *Introductory Nuclear Reactor Dynamics*, American Nuclear Society, 1985.
17. D. L. Hetrick. *Dynamics of Nuclear Reactors*, American Nuclear Society, 1993.
18. E. E. Lewis, W. F. Miller, *Computational Methods of Neutron Transport*, American Nuclear Society, 1993.
19. A. Z. Paydar, S. K. M. Balgehshiri, B. Zohuri, *Advanced Reactor Concept (ARC), A new Nuclear Power Plant Perspective Producing Energy*, 1st edition, Elsevier, 2023.

4 Nuclear Reactor Kinetics

4.1 INTRODUCTION

The study of the neutron flux's time dependence for assumed changes in the macroscopic cross-sections is known as reactor kinetics. Reactor kinetics without feedback is another name for this concept.

The point kinetics equations are typically used by engineers to investigate the kinetic behavior of the system. They serve a vital role in reactor management and are important from the perspective of reactor kinetics and reactor safety, even though the quantity of delayed neutrons per fission neutron is relatively modest (usually less than 1%) and does not contribute much to power output.

Among other things, we shall examine how nuclear reactors behave over time. The most crucial aspect of nuclear reactor safety is comprehending how the neutron population in a reactor behaves over time in response to either deliberate changes in the reactor's reactivity or to unforeseen and unusual circumstances.

Transient neutron flux fluctuations brought on by a departure from the critical state or some reactivity insertion are the subject of nuclear reactor kinetics. These circumstances can occur as a result of unintentional disruptions in the reactor's steady state functioning, environmental changes like a shift in the boron concentration, or operational changes like the motion of the control rods. Refer to Figure 4.1.

In general, we can state the following.

- **Reactor kinetics:** The study of the neutron flux's time dependence for assumed changes in the macroscopic cross-sections is known as reactor kinetics. Reactor kinetics **without feedback** is another name for this concept.
- **Reactor dynamics:** The study of reactor dynamics examines how the neutron flux changes over time when the macroscopic cross-sections are permitted to rely on the neutron flux level. Reactor kinetics **with feedback** and spatial effects are substitutes for it.

However, before moving on to the next phase, we must characterize the time-dependent behavior of nuclear reactors using the following time scale classification.

- **Short-term kinetics** characterizes incidents that take place during periods of time less than a few seconds. This includes how a reactor reacts to unanticipated and unusual circumstances as well as to a planned change in reactivity. We shall introduce point kinetics equations, in particular, in this section.
- **Medium-term kinetics** characterizes events that take place over a period of hours to days. This includes, in particular, how neutron poisons (such as xenon poisoning or spatial oscillations) affect reactivity.
- **Long-term kinetics** characterizes events that take place across several months or even years. This includes all long-term fuel composition changes brought on by fuel burnup.

The point kinetics equations and short-term kinetics are the main topics of this chapter. Since prompt and delayed neutrons are crucial to short-term reactor dynamics, we must first begin with an introduction to them. They serve a vital role in reactor management and are important from the

DOI: 10.1201/9781003590194-4

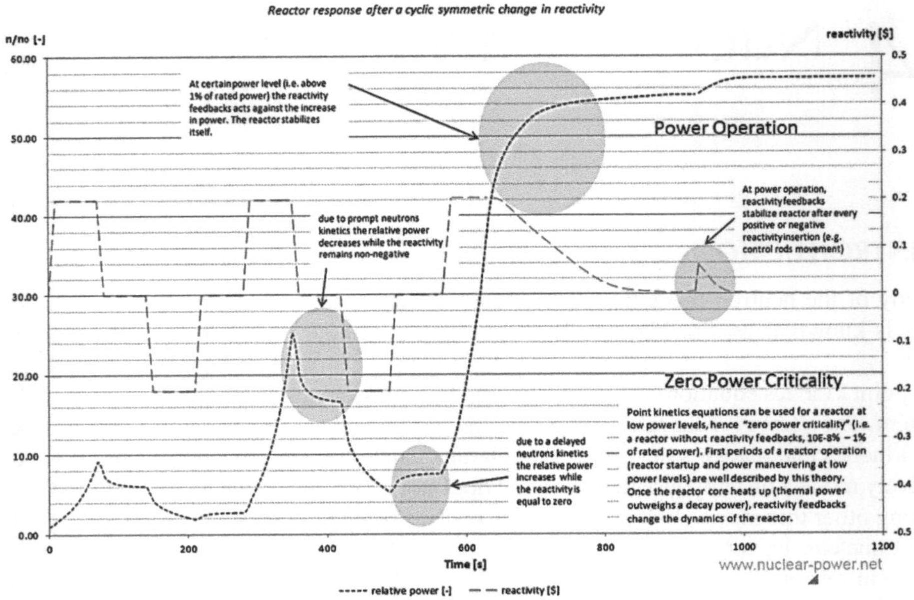

FIGURE 4.1 Reactor response after a cyclic change in reactivity.

(Source: Nuclear-Power.net).

perspective of reactor kinetics and reactor safety, even though the quantity of delayed neutrons per fission neutron is relatively modest (usually less than 1%) and does not contribute much to power output. When they are present, a reactor's dynamic temporal response to a change in reactivity is totally altered, making it controllable by control devices like moving drums or control rods in some types of gas-cooled reactors.

A reactor can be operated in a quick subcritical, delayed critical condition, thanks to delayed neutrons. Every power reactor has safety mechanisms to keep it from ever reaching prompt criticality and is built to function in a delayed critical condition.

The significant difficulty in solving the whole energy-dependent diffusion equation for a point source is reflected in the fundamental neutron balance equation, which is finally expressed as [1–2]

$$\frac{1}{\upsilon}\frac{\partial\phi(\vec{r},E,t)}{\partial t} = \chi(E)\int_0^\infty dE'\nu\Sigma_f(E')\phi(\vec{r},E',t) + Q(\vec{r},E,t)$$
$$+\int_0^\infty dE'\Sigma_s(E'\to E)\phi(\vec{r},E',t) - \Sigma_t(E)\phi(\vec{r},E,t) + \nabla\cdot D(E)\nabla\phi(\vec{r},E,t) \qquad \text{Eq. 4.1}$$

where ν is the average number of neutrons released per fission, $\chi(E)$ is the fission spectrum, and v is the neutron speed. The remaining variables are defined as before in Chapter 2 of this book and in [1–2].

Additionally, "point kinetics" is also used to characterize short-term issues commonly encountered in nuclear reactor kinetics when an action results in a perturbation of reactor system parameters. These disturbances or alterations showed up as a change in the temperature of the main system, which in turn caused a change in the neutron moderation rate and, ultimately, a change in the multiplication factor. Point kinetics gets its name from the approach most frequently employed to address such short-term issues is predicated on the idea that the reactor flux's shape remains constant and that the reactor would behave like a point. Many evaluations of reactor systems are based on this method, with the exception of those in which flux forms are known to change with time. In order to account for flux shape change, many studies are being performed using more intricate approximations, such as complete three-dimensional kinetics [3].

Furthermore, many approximations, including the *point kinetics approximation*, are taken into account in the nuclear reactor kinetics approach.

This matter is the subject of Section 4.2.

4.2 SIMPLE POINT KINETIC EQUATION

Since the number of neutrons is multiplied by a factor k_{eff} from one generation to the next, as we have seen in earlier chapters, the multiplication environment (i.e., nuclear reactor) behaves like an exponential system, meaning that the power increase is exponential rather than linear.

It is clear that the change in the fission neutron concentration from one generation to the next is measured by the effective multiplication factor in a multiplying system.

- $k_{eff} < 1$. The chain reaction will never be self-sustaining if the multiplication factor for a multiplying system is less than 1.0, as the amount of neutrons decreases over time (with the mean generation time). The subcritical state is the name given to this situation.
- $k_{eff} = 1$. The chain reaction will be self-sustaining and the neutron population will not vary over time if the multiplying system's multiplication factor is equal to 1.0. The critical state is the name given to this situation.
- $k_{eff} > 1$. In the situation where a multiplying system's multiplication factor exceeds 1.0, it generates more neutrons than it requires to maintain itself. As the mean generation time increases, the quantity of neutrons increases exponentially with time. This state is referred to as the supercritical state.

However, the length of a neutron generation – that is, the number of times in a second that we must multiply the neutron population by a factor k_{eff} – has not yet been covered. The exponential growth rate is determined by this time. However, as previously said, there are two kinds of neutrons: prompt and delayed, which completely alter the system's kinetic behavior. Thus, such a conversation will not be insignificant. Refer to Figure 4.2.

The point kinetics equations are typically used by engineers to investigate the kinetic behavior of the system. Since the shape of the neutron flux and the neutron density distribution are disregarded in this simplified formalism, the term "point kinetics" is employed. As a result, the reactor is lowered to a point. We will begin with point kinetics in its most basic form, as seen in Section 4.2.1, and then go on to the next section to introduce point kinetics.

4.2.1 Simple Point Kinetic Equation Derivation

Here, the reactor kinetic equation is calculated in its most basic form. Without specifically taking into consideration the temporal development of the flux's geographical distribution, it is possible to show how the core average or total reactor power changes over time. We do need to depict the creation of neutrons with certain time delays, even in the simplest kinetic formulation. A simple form of the neutron source term, as shown in Equation 4.1, is obtained at the instantaneous release of these fissionable neutrons. In this case, however, we assume that all fission neutrons are released instantly, ignoring the presence of delayed neutrons. Given this situation, we can duly recognize that a certain fraction β of neutrons in the fission process is typically produced with some time delays through a series of radioactive decays of fission products.

As an example of the mentioned process, we can look at how radioactivity impacts ${}^{87}_{35}Br$, which undergoes a β-decay into a metastable state of ${}^{87}_{36}Kr^{*}$, which decays into the ground state of ${}^{87}_{36}Kr$ with the emission of a neutron. Figure 4.3 presents ${}^{87}_{36}Kr^{*}$ where decay takes place with a half-life $t_{1/2} = 0.1$ millisecond, which is negligible short compared with $t_{1/2} = 55.9$ seconds, associated with the β-decay of ${}^{87}_{35}Br$ itself [2].

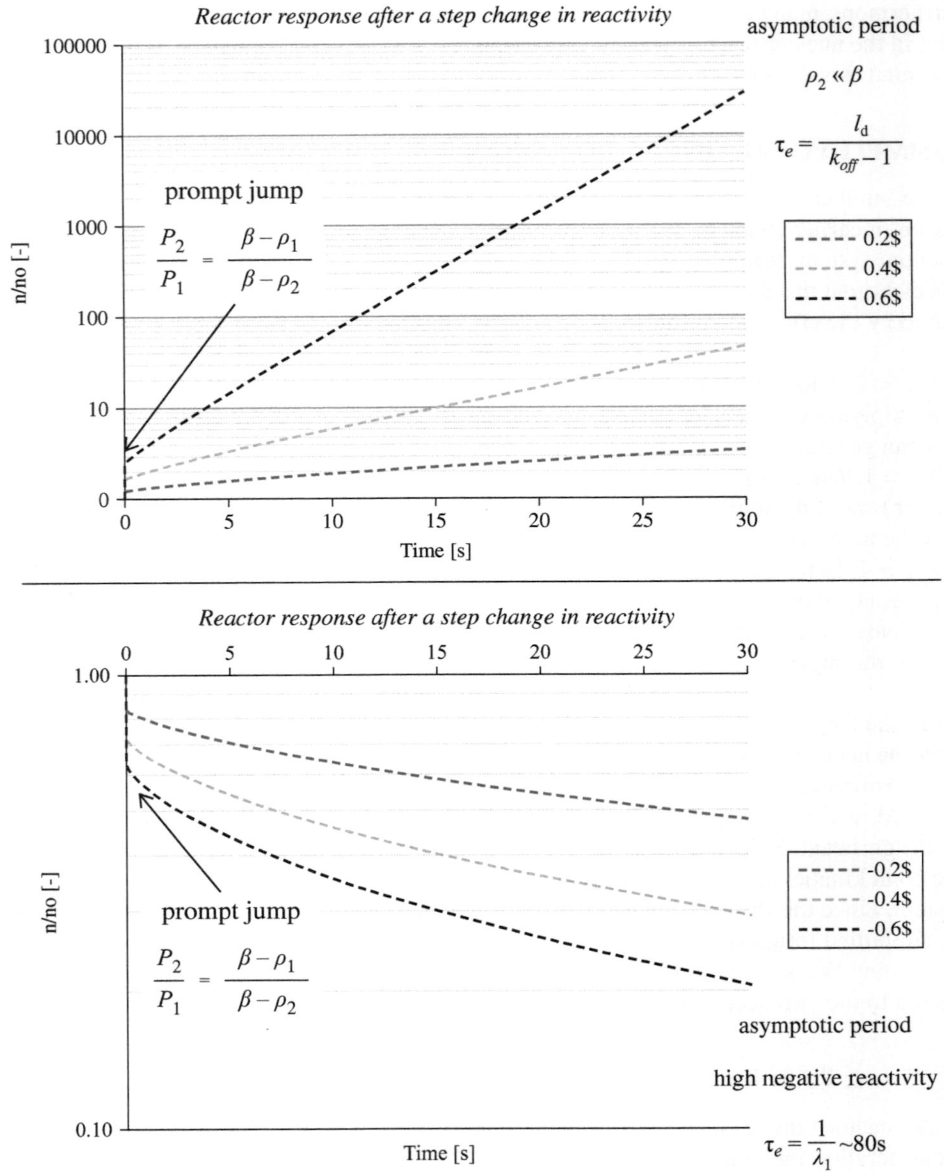

FIGURE 4.2 Reactor responses after a step change in reactivity.

Thus, we will easily see the effective half-life of the decay chain resulting in the neutron release characterized by $t_{1/2} = 55.9$ seconds and the yield of the fission product $^{87}_{35}Br$ ranges from the value of 2.5 x 10^{-5} or 0.025% for fission of ^{235}U induced by thermal neutrons to 8.0 x 10^{-5} for thermal fission of ^{239}Pu.

The decay chain contributing to the delayed emission of neutrons is summarized as below.

$$\frac{87}{35}Br \xrightarrow{t_{1/2}=55.9s} \frac{87}{36}Kr* \xrightarrow{t_{1/2}=0.1ms} \frac{86}{36}Kr + \frac{1}{0}n$$

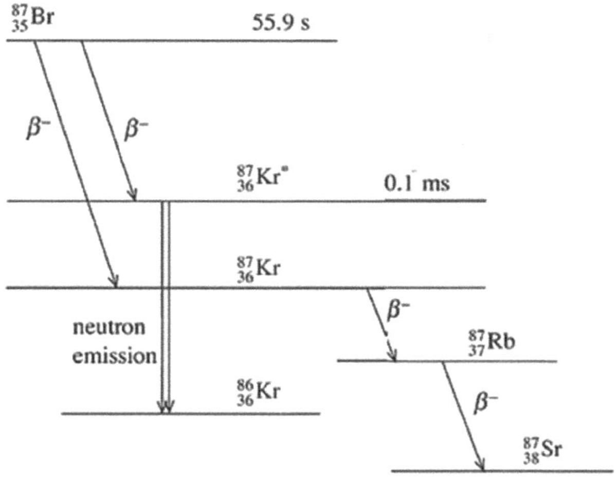

FIGURE 4.3 Schematics of decay chains for fission $^{87}_{35}$**Br**

As per [2], the neutrons produced through this decay chain will contribute to the overall neutron balance essentially with a half-life $t_{1/2} = 55.9$ seconds associated with the β-decay of $^{87}_{35}Br$. Thus, $^{87}_{35}Br$ is known as a precursor for this group of delayed neutrons. Dozens of delayed neutron precursors have been observed [4] to date, and for convenience, they are usually broken up into six equivalent groups [5].

However, suggestions [6] have been made in recent years that different groupings for delayed neutrons should be used going forward. For thermal fission of ^{235}U, the total delayed neutron yield is $\beta = 0.0065$, while for ^{239}Pu, the yield is only 0.0023. For more details or a summary of the decay, please refer to [2] and Table 2.2.

We use $n(t)$to represent the number of neutrons as a function of time, t to represent the prompt neutron lifetime, and ℓ to represent the average time from prompt neutron emission to either its absorption (fission or radiative capture) or its escape from the system (i.e., reactor) in order to continue with our derivation of the simple point kinetics equation.

In a unit time interval dt, the average number of neutrons that vanish is $n(dt/L)$. However, an average of k generation new neutrons are produced for every neutron that vanishes. Lastly, the variation in neutron counts over a unit time interval dt is as follows:

$$\begin{cases} dn = -n\left(\dfrac{dt}{\ell}\right) + kn\left(\dfrac{dt}{\ell}\right) \\ \dfrac{dn(t)}{dt} = -\left(\dfrac{1}{\ell}\right)n(t) + \left(\dfrac{k}{\ell}\right)n(t) \\ \downarrow \\ \dfrac{dn}{dt} = \left(\dfrac{k-1}{\ell}\right)n \end{cases} \quad \text{Eq. 4.2}$$

Equation 4.2 integrated with regard to unit time t will yield

$$n(t) = n(0)\exp\left(\frac{k-1}{\ell}\right)t = n(0)\exp\left(\frac{t}{\tau_e}\right) \quad \text{Eq. 4.3}$$

where

$n(t)$ = Transient reactor power
$n(0)$ = Initial reactor power
$\tau_e = \dfrac{\ell}{k-1}$ = Reactor period
ℓ = Prompt neutron lifetime
t = Time

The time needed for the neutron density to vary by a factor of $e = 2.718$ is known as the reactor period, τ_e, or e-folding time. Typically, the reactor period is stated in seconds or minutes. The reactor power changes more quickly when τ_e is smaller. Both positive and negative reactor periods are possible.

Bear in mind to reach the form of Equation 4.3, we consider that if there are neutrons in the system at $t = 0$, that is, if $n(0) > 0$ (i.e., some means of fissionable source in the system), then in the absence of an external source, we can deduce Equation 4.3. See Figure 4.4. If a time-independent chain reaction occurs in this situation without a source, the system is considered critical. When the multiplication factor k equals 1, as shown in Equation 4.4, the system is obviously critical. The infinite media is comparable to this, hence we may say that [7]:

$$k \begin{cases} >1 & \textit{Supercritical} \\ =1 & \textit{Critical} \\ <1 & \textit{Subcritical} \end{cases} \qquad \text{Eq. 4.4}$$

The behavior of multiplication is demonstrated in Figure 4.4 and Figure 4.4(a).

If the system has a time-independent source S_0, we must solve Equation 4.2 in a different way, as follows:

$$\frac{d}{dt}n(t) = S(t) + \left(\frac{k-1}{\ell}\right)n(t) \qquad \text{Eq. 4.5}$$

where $S(t)$ is the source at time period t.

We must solve Equation 4.5 with respect to $S(t) \to S_0$. Applying the integrating factor technique for first-order differential equation as presented in Appendix A, Reference [7] with *initial condition* (IC) $n(0) = 0$, we obtain the following solution:

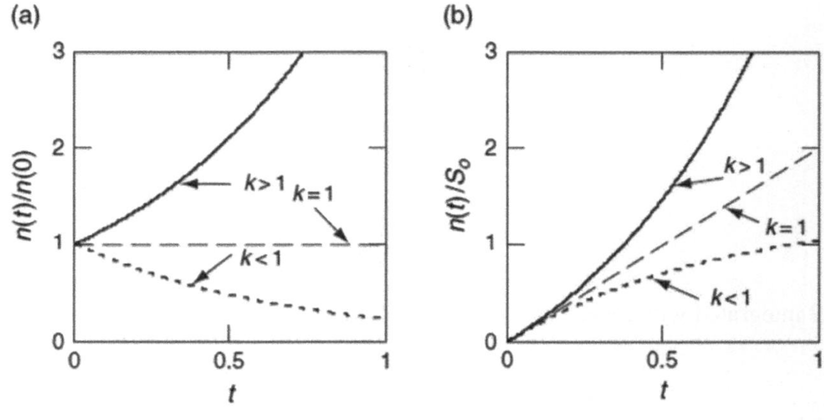

FIGURE 4.4 Neutron populations in subcritical, critical, and supercritical systems. (a) n(0) = 1, no source present, (b) n(0) = 0, source present. [7]

$$n(t) = \frac{\ell S_0}{(k-1)}\left\{\exp\left[\left(\frac{(k-1)}{\ell}\right)t\right] - 1\right\} \quad \text{Eq. 4.6}$$

In this instance, multiplication systems with a time-independent source present behave as shown in Figure 4.4(b). The neutron population in a supercritical system, $k > 1$, increases at an accelerating pace and eventually reaches an exponential level. If we first change the negative signs as seen in Equation 4.2 to rewrite the solution of Equation 4.6 in the new form as follows, the behavior of the subcritical system, $k < 1$, becomes more transparent.

$$n(t) = \frac{\ell S_0}{(1-k)}\left\{1 - \exp\left[\left(\frac{(k-1)}{\ell}\right)t\right]\right\} \quad \text{Eq. 4.7}$$

Therefore, for $k < 1$ in the subcritical situation, the population stabilizes over time to a time-independent solution: After the first rise in neutrons, the exponential component decays away.

$$n(t \to \infty) = \frac{\ell S_0}{(1-k)} \quad \text{Eq. 4.8}$$

In the case of a critical system, for which $k = 1$, the behavior of Equation 4.6 or 4.7 is more predictable since both the denominator and the term in parentheses equal zero. However, as shown above, we may expand the exponential portion of both equations as a power series if we consider the limit to be $k \to 1$.

$$\exp(x) = 1 + x + \frac{1}{2}x^2 + \cdots \quad \text{Eq. 4.9}$$

Using Equation 4.9 in conjunction with Equation 4.7, we obtain

$$n(t) = \frac{\ell S_0}{(k-1)}\left\{1 + \left(\frac{(k-1)}{\ell}\right)t + \frac{1}{2}\left(\frac{(k-1)^2}{\ell^2}\right)t^2 + \cdots - 1\right\} \quad \text{Eq. 4.10}$$

Manipulating Equation 4.10 by cancelling terms, and taking the limit of multiplication factor $k \to 1$, we induce the following equation as

$$n(t) = S_0 t \quad \text{Eq. 4.11}$$

Thus, the neutron population in a critical system will increase linearly with time in the presence of a time-independent source.

Figure 4.4 summarizes the findings of the aforementioned assertions and considerations. A time-independent population of neutrons can only be generated in two circumstances, which are as follows:

1. An essential system without an external supply (i.e., $k = 1$ and $S = 0$)
2. A subcritical system where a neutron source is present ($k < 1$ and $S > 0$).

Keep in mind that the definition of criticality is independent of the source and only dependent on the multiplication factor. However, the presence or absence of a source has a significant impact on the system's behavior [7].

4.3 SIMPLE POINT KINETIC EQUATION WITHOUT DELAYED NEUTRONS

The point kinetics equation (in a specific version) governs the neutron kinetics of the system without a source and delayed neutrons. According to this formula, the excess neutron production (by fission) less neutron loss (by absorption) in a single prompt neutron lifespan equals the temporal change of the neutron population. It is clear that prompt neutron lifetime plays a part. Simply put, multiplying systems respond more quickly when their lives are shorter.

The solution of this equation yields the most basic version of the point kinetics equation (without source and without delayed neutrons) if there are neutrons in the system at $t = 0$, that is, if $n(0) > 0$.

$$\begin{cases} \dfrac{dn(t)}{dt} = \left[\dfrac{(k-1)}{\ell}\right] n(t) \\ n(t) = n(0)\exp\left[\dfrac{(k-1)}{\ell}\right] t \end{cases} \quad \text{Eq. 4.12}$$

Reactivity and prompt generation time, Λ, are frequently used to formulate this straightforward point kinetics equation as follows:

$$\begin{cases} \dfrac{dn(t)}{dt} = \dfrac{\rho}{\Lambda} n(t) \\ n(t) = n(0)\exp\left(\dfrac{\rho}{\Lambda}\right) t \end{cases} \quad \text{Eq. 4.13}$$

where

$\rho = \dfrac{(k-1)}{k}$ = Reactivity, which describes the deviation of an effective multiplication factor from unity.

$\Lambda = \dfrac{1}{k_{eff}}$ = Average time between a prompt neutron emission and an absorption that alone produces fission, or prompt neutron generation time.

The point kinetics equation can be expressed in any way. For computations, the formula that uses Λ, prompt neutron generation time, is often superior. This is true because variations in the absorption cross-section, not the fission cross-section, are what produce the majority of reactivity transients. While the prompt generation time is constant throughout these transients, the prompt neutron lifespan varies.

Example 4.1: Assuming that the prompt neutron lifespan is $\ell \sim 2 \times 10^{-5}$ seconds, we should step-raise k (k_∞, neutron multiplication factor) by just 0.01%, or 10 pcm or ~1.5 cents, so that $k_\infty = 1.0000$ will become $k_\infty = 1.0001$.

It should be mentioned that in the case of Light Water Reactors (LWRs), this reactive insertion (10 pcm) is extremely little. For LWRs, the reactivity insertions of the order of 1 pcm are almost impossible. The reactor time in this instance will be

$$T = \frac{\ell}{(k_\infty - 1)} = \frac{2\times10^{-5}}{(1.0001-1)} = 0.2 \text{ Seconds}$$

This is a fairly brief time frame. The reactor's neutron flux (and power) would grow by a factor of $e^5 = (2.718)^5$ in 1 second, and by $e^{50} = (2.718)^{50}$ in 10 seconds, after the reactor had gone through 50 cycles.

Furthermore, the reaction of such a minor reactivity insertion would be much more unthinkable in the case of fast reactors, where prompt neutron durations are of the order of 10^{-7} seconds. If it is 10^{-7}, the time frame will be

$$T = \frac{\ell}{(k_\infty - 1)} = \frac{10^{-7}}{(1.0001-1)} = 0.001\,\text{s}$$

Controlling reactors with such kinetics would be extremely challenging. Fortunately, no multiplication system exhibits this behavior. Since actual reactor times are far longer than those calculated above, it is easier to manage the nuclear chain reaction. The existence of delayed neutrons is what causes the extended durations.

Please take note that we did discuss delayed neutrons to some extent at the beginning of this chapter. We will go into deeper detail regarding neutronics and its behavior in Chapter 5 of this book titled, "Introduction to Nuclear Energy from Neutrons to Fast Spectrum".

4.3.1 Percent Mille – Unit of Reactivity

The following definition applies for the following statements as

Units of Reactivity: Reactivity is a dimensionless number in mathematics; however, it may be expressed in a variety of units. Since they precisely reflect a departure from prompt criticality conditions, units normalized to the delayed neutron fraction (such as cents and dollars) are the most often used units for research reactors.

The most widely used units for power reactors are $\%\Delta\ K/K$ or pcm. The explanation is straightforward. Because the normalization factor, or effectively delayed neutron fraction, varies greatly with fuel burnup, using money is challenging. With fuel burnup, the delayed neutron percentage in LWRs falls (for example, from $\beta_{eff} = 0.007$ at the start of the cycle to $\beta_{eff} = 0.005$ at the end).

The fuel's isotopic alterations are to blame for this. It is easy. The fissile substance in fresh uranium fuel is simply ^{235}U. In the meantime, ^{239}Pu fission becomes more significant (up to 50% in some situations) as fuel burns off. A multiplication system's core delayed neutron fraction (the weighted average of the constituent delayed neutron fractions) falls as a consequence of ^{239}Pu producing substantially fewer delayed neutrons (0.0021 for thermal fission).

$$\beta_{\text{core}} = \Sigma P_i \cdot \beta_i$$

Section 4.2.6 and Equation 4.18 provide a more thorough explanation of the aforementioned connection.

- **Percent Mille:** One thousandth of a percent or $\%\Delta\ K/K$, is the reactivity unit (equivalent to $10^{-2} \times 10^{-3} = 10^{-5}$ of k_{eff}). Since reactivity insertion values are often rather modest and units of pcm enable reactivity to be expressed in whole numbers, many LWRs employ this unit. Reactivity insertion of the order of units of percent mille (PCM), every one step is often caused by operational changes such the movement of control rods. There is a significant impact when the effective delayed neutron percentage varies with fuel burnup. A reactor's reaction to the same reactivity insertion (in units of pcm) at the beginning of the cycle (BOC) and the end of the cycle (EOC) differs because of the change in β_{eff}.

One control rod step, for instance, results in a larger reaction at EOC than at BOC. The value of the control rod rises with fuel burnup, thus even if we assume that one step results in the same reactivity insertion (for example, +10 pcm) in both situations, this assumption is not necessarily accurate.

(10 pcm = 1.43 cents for β_{eff} = 0.007; 10 pcm = 2.00 cents for β_{eff} = 0.005)
$k_{eff} = 0.99 \ \rho = (k_{eff} - 1)/(k_{eff}) \ \ k_{eff} = -0.01 \ \rho = -0.01 * 10^5 = -1000$ pcm

4.3.2 Prompt Neutron Lifetime

The average period between a prompt neutron emission and its escape from the system or its absorption (either by fission or radiative capture) is known as the prompt neutron lifespan or ℓ. Both multiplying and non-multiplying systems define this parameter. The prompt neutron lifetimes in both systems are highly dependent on

- Material composition of the system
  - multiplying–non-multiplying system
  - system with or without thermalization
  - isotopic composition of the system
- Geometric configuration of the system
  - Homogeneous or heterogeneous system
  - Shape of the entire system
  - Size of the system

The prompt neutron lifespan in an infinite reactor (without escape) is equal to the sum of the diffusion time t_d and the slowing down time t_s, as seen here.

$$\ell = t_s + t_d \qquad \text{Eq. 4.14}$$

Both $t_s \lll t_d$ and $\ell \approx t_d$ are present in an infinite thermal reactor. In thermal reactors, the usual prompt neutron lifetime is around 10^{-4} seconds. Longer neutron lifetimes typically occur in systems where thermalization of the neutrons is required for absorption.

Prompt neutron lifetimes are significantly shorter in systems where the majority of neutrons are absorbed at higher energy and neutron thermalization is controlled, such as in fast reactors.

In fast reactors, the prompt neutron lifetime is typically about 10^{-7} seconds. Refer to Table 4.1 (Slowing Down and Diffusion Times for Thermal Neutrons in an Infinite Medium Reactor), for further details.

4.3.3 Prompt Generation Time – Mean Generation Time

I is the average duration between two generations of prompt neutrons at k_{eff} = 1 in multiplying systems, where the absorption of a prompt fission neutron can start a fission reaction.

The prompt neutron generation time is the name given to this period. Figure 4.5 shows how the asymptotic time period depends on the reactivity needed for various reactor designs with varying prompt neutron lifetimes.

The average time between a prompt neutron emission and a capture that alone produces fission is known as the prompt neutron generation time (or mean generation time), or Λ, denoted by the following equation:

$$\Lambda = \frac{1}{k_{eff}} p \qquad \text{Eq. 4.15}$$

The quick generation time of power reactors varies as the fuel burns up, it rises when fuel burns up in LWRs. The answer is straightforward: A significant amount of fissile material (about 4% of ^{235}U) is present in new uranium fuel. This results in a considerable excess of reactivity, which needs

TABLE 4.1
Slowing Down and Diffusion Times for Thermal Neutrons in an Infinite Medium Reactor

	Slowing Down Time (s)	Diffusion Time (s)
Water	5.3×10^{-6}	2.4×10^{-4}
Heavy Water	5.0×10^{-5}	6.0×10^{-2}
Beryllium	5.7×10^{-5}	4.2×10^{-3}
Graphite	1.4×10^{-4}	1.6×10^{-2}

(Source: Robert Reed Burn, Introduction to Nuclear Reactor Operation, 1988)

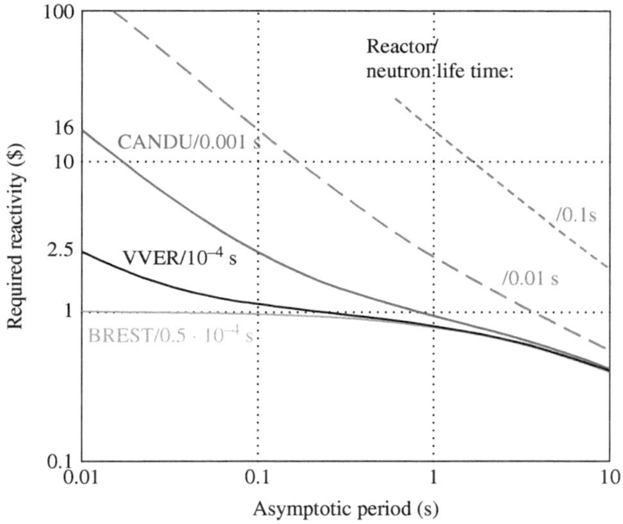

FIGURE 4.5 Dependencies of asymptotic time period on the reactivity.

to be made up for either by burnable absorbers or chemical shim (in the case of Pressurized Water Reactors (PWRs)). Please take note that we have discussed this topic in the next subsection. By chemical shim, we mean that boric acid is dissolved in the coolant/moderator.

The high chance of absorption in the fuel and the high probability of absorption in the moderator cause the prompt neutron lifetime to be low and significantly shorter. Fuel burnup reduces the quantity of fissile material and moderator absorption, allowing the prompt neutron to "live" for a significantly longer period of time.

4.3.3.1 Boric Acid – Chemical Shim

By "chemical shim", we indicate that the coolant/moderator includes dissolved boric acid. The white powder known as boric acid (molecular formula: H_3BO_3) dissolves in water. Chemical shim, or boric acid, is employed in PWRs to offset excessive reactor core reactivity during fuel burnup (long-term reactivity control). The concentration of boric acid is greatest at the start of the particular fuel cycle (see Figure 4.6). A reactor has to be refueled at the conclusion of this cycle since the concentration of boric acid is nearly zero.

In certain conditions, the chemical shim can also regulate subtle power variations. Boric acid must be diluted in order to remove ^{10}B from the reactor core and lessen its poisoning impact if power is to be increased. Because boric acid is uniformly dissolved in the coolant throughout the reactor

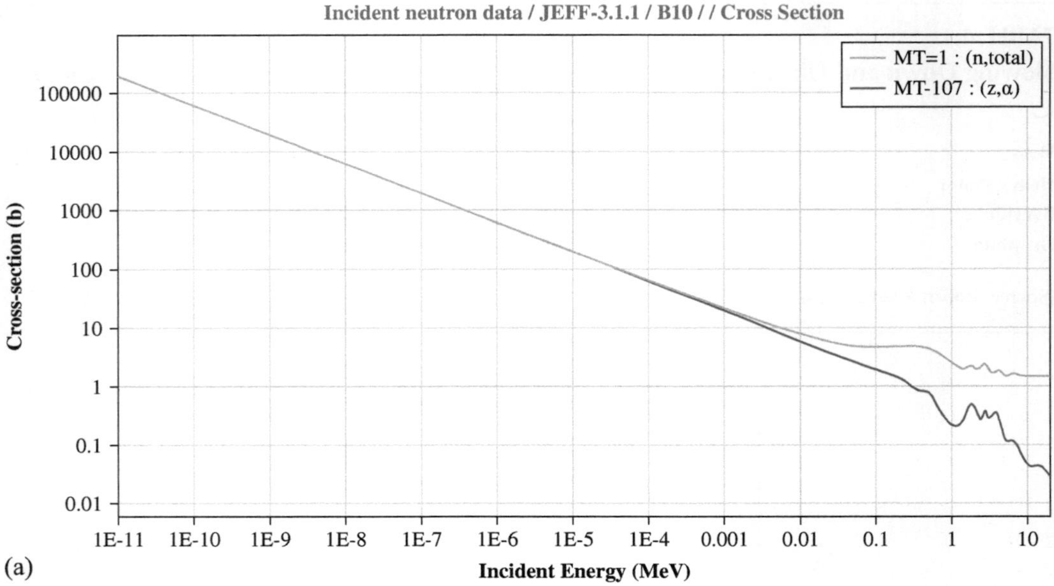

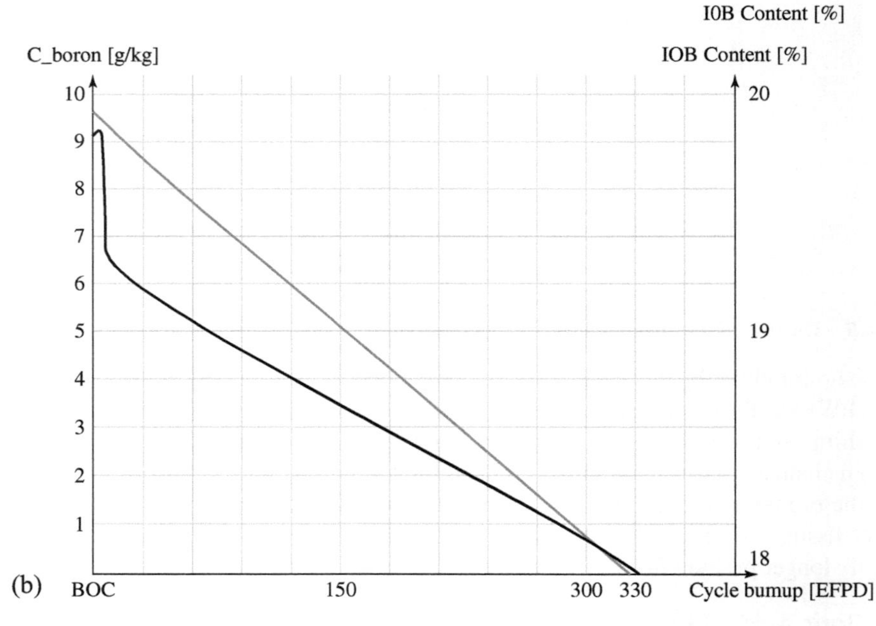

FIGURE 4.6 (a, b) Chemical shim of boric acid as applications of Boron in nuclear reactor.

(Source: JANIS (Java-based Nuclear Data Information Software); The JEFF-3.1.1 Nuclear Data Library)

core, it avoids the uneven neutron flux density in the core as compared to burnable absorbers (long-term reactivity control) or control rods (quick reactivity control).

However, it is undesirable for high boric acid concentrations to result in a positive moderator temperature coefficient. More burnable absorbers need to be employed in this situation. Refer to Figure 4.6(a, b) for examples.

Furthermore, this approach controls reactivity slowly. Changing the concentration of boric acid (dilute or borate) in the main loop typically takes a few minutes. It is necessary to employ control rods for quick adjustments in responsiveness.

4.3.4 Effect of Prompt Neutron Lifetime on Nuclear Safety

The prompt neutron lifetime (PNL) corresponds to essential neutron physical parameters of the reactor core. Its value is mostly determined by the moderator type and the energy of the fission-causing neutrons. Its significance for nuclear reactor safety has long been recognized.

Reactor kinetic response may be significantly enhanced by longer PNLs (the longer PNL simply yields slower power growth).

For instance, reactors should be able to tolerate a jump-like insertion of reasonably high (~1$ or even more) positive reactivity under reactivity-initiated accident (RIA) circumstances. In this case, the PNL is crucial. As a result, a reload safety evaluation procedure should include PNL verification.

Reactor cores can occasionally be altered to raise the PNL and enhance nuclear safety, particularly in specific fast reactors.

4.4 SIMPLE POINT KINETIC EQUATION WITH DELAYED NEUTRONS

The simple point kinetics equation with delayed neutrons is the most straightforward equation regulating the neutron kinetics of the system with delayed neutrons. According to this equation, the excess neutron production (by fission) less the neutron loss (by absorption) in one mean generation time with delayed neutrons (ℓ_d) equals the time change of the neutron population. Hence,

$$\ell_d = (1-\beta)\cdot\ell_p + \Sigma\ell_i\cdot\beta_i \Rightarrow \ell_d = (1-\beta)\cdot\ell_p + \Sigma\tau_i\cdot\beta_i \qquad \text{Eq. 4.16}$$

where

$(1-\beta)$ = Fraction of all neutrons emitted as prompt neutrons
ℓ_p = Prompt neutron lifetime
τ_i = Mean precursor lifetime, the inverse value of the decay constant $\tau_i = 1/\lambda_i$
λ_i = Weighted decay constant

Since $\tau = \Sigma\tau_i\cdot(\beta_i/\beta) = 13.05$ seconds is the weighted delayed generation time, $\lambda_i = (1/\tau_i) \approx 0.08$ s^{-1} is the weighted decay constant.

While delayed neutrons make up a very small portion of total neutrons in the core, the quantity, 0.08 s^{-1}, is comparatively substantial and dominates the reactor time response. The computation of a weighted mean generation time with delayed neutrons provides the clearest example of this.

$$\ell_d = (1-\beta)\cdot\ell_p + \Sigma\tau_i\cdot\beta_i = (1-0.0065)\cdot 2\times10^{-5} + 0.085$$
$$= 0.00001987 + 0.085 \approx 0.085$$

To put it briefly, the mean generation time with delayed neutrons is around ~0.1 s, as opposed to ~10^{-5} as seen in Section 4.3.2 "Prompt Neutron Lifetime", where the delayed neutrons were excluded.

The role of ℓ_d is clear. Simply put, multiplying systems respond more slowly over longer lives. It is also clear that reactivity (k_{eff} – 1) plays a part. The multiplication system just responds more when the reactivity is higher.

Once more, the simplest point kinetics equation with delayed neutrons (which is comparable to the case without delayed neutrons) is obtained by solving this equation if there are neutrons in the system at $t = 0$, that is, if $n(0) > 0$.

$$\begin{cases} \dfrac{dn(t)}{dt} = \left[\dfrac{(k-1)}{\ell_d}\right] n(t) \\ n(t) = n(0)\exp\left[\dfrac{(k-1)}{\ell_d}\right] t \end{cases} \quad \text{Eq. 4.17}$$

Example 4.2: Assuming that the average generation time with delayed neutrons is approximately 0.085 and that k (k_∞, neutron multiplication factor) is only 0.01% (i.e., 10 pcm or approximately 1.5 cents), $k_\infty = 1.0000$ will rise to $k_\infty = 1.0001$.

It should be mentioned that in the case of LWRs (e.g., one step by control rods), such reactive insertion (10 pcm) is extremely modest. For LWRs, reactivity insertions on the order of 1 pcm are almost impossible. The reactor time, in this instance, will be

$$T = \frac{\ell_d}{(k_\infty - 1)} = \frac{0.085}{(1.0001-1)} = 850 \text{ Sec}$$

This is a very long time. In addition, the reactor's neutron flux (and power) would increase by a factor of e = 2.718 in about 14 minutes, which is a completely different dimension of the response on reactivity insertion than in the case where delayed neutrons were not present, where the reactor period was 1 second.

4.4.1 Six Groups of Delayed Neutrons

Precursors in the case of delayed neutron groups can be categorized based on their half-lives. These kinds of groupings are known as delayed neutron groupings or precursor groups. Although six delayed neutron groups are often utilized, other groups may be used instead. Six delayed neutron groups typically offer adequate precision for analysis of periods of the order of 5 seconds; a larger number of groups may be required for longer durations. For every precursor group k, partial delayed neutron yields ν_{dk} are defined. The average number of group k precursors created per fission is represented by the partial delayed neutron group yield. Partial delayed neutron fractions can, therefore, be described as follows:

$$\beta_k = \frac{\nu_{dk}}{\nu} \quad \text{Eq. 4.18}$$

Obviously, we have

$$\sum_{k=1}^{k_{max}} \beta_k = \beta \quad \text{Eq. 4.19}$$

Table 4.2 displays the delayed group constant values for ^{235}U based on Rose's [8] data.

Although reactor kinetic calculations that take into account several beginning circumstances would be accurate, they would also be extremely complex, as shown in Figure 4.7. G.R. Keepin [5] and his colleagues proposed classifying the precursors according to their half-lives. Consequently, six delayed neutron groups are commonly used to represent delayed neutrons. Non-linear least-squares fits to experimental results are used to determine the yields and decay constants (λ) of these groups. The following list of drawbacks applies to this model:

- Every constant for every set of antecedents fits the data empirically.
- The decay constants of certain precursors cannot be matched with them.
- Each fissionable nuclide has a separate set of these constants.
- The neutron energy spectrum also affects these constants.

TABLE 4.2
Delayed Neutron Data for Thermal Fission in ^{235}U [8].

Group	Decay Constant, λ_k (1/s)	Delayed Yield, ν_{dk} (n/Fission)	Delayed Fraction, β_k
1	0.01334	0.00585	0.000240
2	0.03274	0.003018	0.001238
3	0.1208	0.002881	0.001182
4	0.3028	0.002659	0.002651
5	0.8495	0.002648	0.001087
6	2.853	0.001109	0.000455
Total		0.016700	0.006854

A finer energy resolution in the delayed neutron spectra and a higher accuracy of the delayed neutron yields are sought, even if this six-group parameterization still meets the needs of commercial enterprises.

The asymptotic die-away time constants linked to the three dominating precursors with the longest half-lives, ^{87}Br, ^{137}I, and ^{88}Br, were found to be inaccurately reproduced by the half-lives in the six-group structure.

Since almost all observations of delayed neutron activity have been made for fast or thermal neutron-induced fission, this model might not be adequate, particularly in the case of epithermal reactors. The precision and energy resolution may be crucial in the case of fast reactors, where the nuclear fission of six fissionable isotopes of uranium and plutonium is significant.

4.4.2 Photoneutrons

Gamma radiation in nuclear reactors is important for both subcriticality control and reactor dynamics. Particularly in nuclear reactors that use Be reflectors (certain experimental reactors) or D_2O moderators (CANDU reactors). Neutrons are often called photoneutrons since they may also be created in (γ, n) reactions.

Under some conditions, a neutron can be expelled from a nucleus by a high-energy photon, sometimes known as a gamma ray. It happens when its energy surpasses the neutron's binding energy in the nucleus. The binding energies of most nuclei are greater than 6 MeV, which is higher than the energy of the majority of fission gamma rays. However, there are not many nuclei with low enough binding energies for practical use. These are ^{13}C, ^{6}Li, ^{7}Li, ^{9}Be, and ^{2}D. The chart shows that the lowest thresholds are ^{2}D at 2.226 MeV and ^{9}Be at 1.666 MeV. Refer to Table 4.3.

When it comes to deuterium, gamma rays with a minimum energy of 2.22 MeV can interact with it to create neutrons.

$$\frac{2}{1}D + \gamma \rightarrow \frac{1}{1}H + n$$

In the kinetic calculations, photoneutrons are often considered the same as ordinary delayed neutrons since fission products with specific delays can produce gamma rays, and the mechanism is very similar to that via which a "true" delayed neutron is emitted. Like "real" precursors, photoneutron precursors can also be categorized by their decay constant. By displaying the composition of the whole power flow. Table 4.4 illustrates the relative significance of source neutrons in CANDU reactors.

U^{235}			
Group	**Decay constant λ_i (Sec^{-1})**	**Yield (neutrons per fission)**	**Fraction β_i**
1	0.0124	0.00052	0.000215
2	0.0305	0.00346	0.001424
3	0.111	0.00310	0.001274
4	0.301	0.00624	0.002568
5	1.14	0.00182	0.000748
6	3.01	0.00066	0.000273
Total yield: 0.0158			
Total delayed fraction (β):			0.0065
Pu^{239}			
Group	**Decay constant λ_i (Sec^{-1})**	**Yield (neutrons per fission)**	**Fraction β_i**
1	0.0128	0.00021	0.000073
2	0.0301	0.00182	0.000626
3	0.0124	0.00129	0.000443
4	0.0325	0.00199	0.000685
5	1.12	0.00052	0.000181
6	2.69	0.00027	0.000092
Total yield: 0.0061			
Total delayed fraction (β):			0.0021
U^{233}			
Group	**Decay constant λ_i (Sec^{-1})**	**Yield (neutrons per fission)**	**Fraction β_i**
1	0.0126	0.00057	0.000224
2	0.0337	0.00197	0.000777
3	0.139	0.00166	0.000655
4	0.325	0.00184	0.000723
5	1.13	0.00034	0.000133
6	2.50	0.00022	0.000088
Total yield: 0.0066			
Total delayed fraction (β):			0.0026

FIGURE 4.7 Six groups of delayed neutron.

TABLE 4.3
Nuclide with Low Photodisintegration Threshold Energies

Nuclide	Threshold (MeV)	Reaction
^2D	2.225	$^2H(\gamma, n)^1H$
^6Li	3.697	$^6Li(\gamma, n+p)^4He$
^6Li	5.67	$^6Li(\gamma, n)^5Li$
^7Li	7.251	$^7Li(\gamma, n)^6Li$
^9Be	1.667	$^9Be(\gamma, n)^8Be$
^{13}C	4.9	$^{13}C(\gamma, n)^{12}C$

TABLE 4.4
Relative Important of Source Neutrons IN CANDU

Neutrons	Contribution
Prompt Neutrons	99,470,000,000,000
Delayed Neutrons	500,000,000,000
Photoneutrons	30,000,000,000
Spontaneous Fission	1
Total (10^{14} n $cm^{-2} \cdot s^{-1}$)	100,000,000,000,000

Deuterium nuclei are always present (~0.0156%) in the light water of LWRs, even though photoneutrons are crucial, particularly in CANDU reactors. Additionally, tiny quantities of D_2O are produced when neutrons are captured in the hydrogen nucleus of the water molecules in the moderator. The concentration of heavy water is increased as a result. Consequently, photoneutrons from D_2O are likewise considered extra groups of delayed neutrons with distinctive decay constants λ_j and effective group percentages in LWR kinetic calculations.

Gamma rays from the fission products will accumulate significantly after a nuclear reactor has been running at maximum power for a while. Following shutdown, the strong gamma flux from short-lived fission products will quickly drop. The decay of long-lived fission products that emit delayed high-energy gamma rays causes the photoneutron source to gradually diminish over time, reducing by a small amount every day. With a half-life of 12.75 days, ^{140}Ba is the fission product with the longest half-life and gamma ray energy above the threshold.

The length of time the reactor was in operation before the shutdown and the power level at which it was operating before shutdown determine how much fission product is present in the fuel components (i.e., Figure 4.8). When a reactor is subcritical during a long-term shutdown, photoneutrons – which are often a main source in reactors – ensure that there is enough neutron flux on source range detectors.

While delayed neutrons provide reactor management, photoneutrons are crucial for low power operation, and fission neutrons enable a self-sustaining chain reaction. The nuclear chain reaction, which happens when one nuclear event averages out one or more subsequent nuclear reactions, is seen in Figure 4.9. Furthermore, two, three, or more free neutrons that can trigger further fissions may be produced throughout the fission process. In nuclear reactor physics, this series of fission events is referred to as the fission chain reaction.

Only under the right conditions and in the right multiplication environment can the chain reaction occur. Let us say a single neutron triggers two further fissions. In such a scenario, both the reactor power (reaction rate) and the amount of neutrons in the multiplication system will rise over time.

FIGURE 4.8 Typical fuel elements.

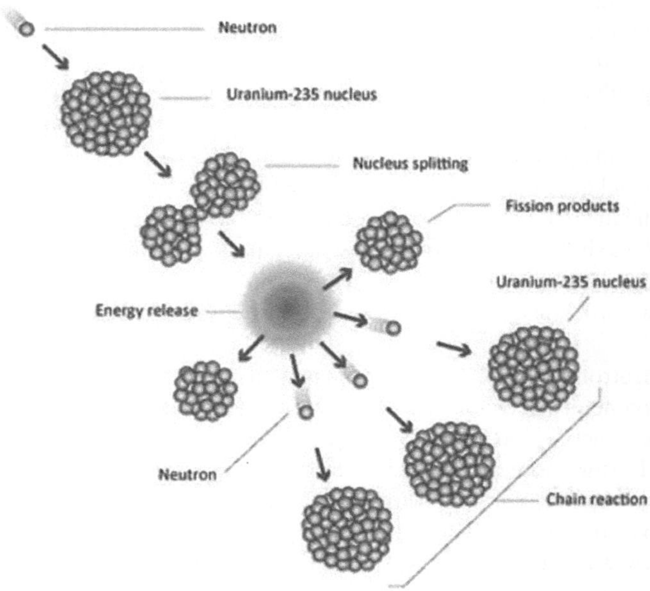

FIGURE 4.9 Fission chain reaction.

In order to stabilize such a multiplication environment, the system's non-fission neutron absorption must be increased (for example, by inserting control rods). Furthermore, the nuclear reactor, which is the multiplication environment, acts like an exponential system, meaning that the power gain is exponential rather than linear.

4.4.3 Effective Delayed Neutron Fraction – β_{EFF}

This section begins by examining the overall yield of delayed neutrons per fission V_d, which is dependent on the following:

- An isotope that has undergone fission, as indicated in Table 4.5.
- As Figure 4.10 illustrates, fission is induced by a neutron's energy.

However, keep in mind that relative units, often known as the delayed neutron fraction (DNF), are handy to utilize in reactor kinetic calculations.

The DNF is equal to the precursor yield fraction β in the steady state condition of criticality with $k_{eff} = 1$ and is defined as follows:

$$DNF = \frac{delayed\ neutrons}{prompt\ neutrons + delayed\ neutrons}$$

$$\beta = \frac{precursor\ atoms}{prompt\ neutrons + precursor\ atoms} = precursor\ yield\ fraction \qquad \text{Eq. 4.20}$$

$$\beta = \sum_i \beta_i$$

where β_i is likewise defined as the proportion of neutrons that emerge as delayed neutrons in the *i*th group, and Equation 4.20 is comparable to Equation 4.19 for grouping purposes.

The delayed neutrons in each group show up with a more or less definite energy spectrum, but the quick neutrons release a continuous spectrum. Compared to the most prompt neutrons, the delayed neutrons are often released with significantly less energy. It is clear how these two parameters differ from one another. The multiplying system's specific reactivity determines the DNF. Conversely, β is independent of reactivity. In the event that the reactor's neutron count changes quickly, these two factors – DNF and β – are not the same.

TABLE 4.5
Six Groups of Precursors

	Possible Precursor Nuclide	Mean Energy (MeV)	Average Half-Life of the Group (s)			Delayed neutron fraction (%)		
i			^{235}U	^{239}Pu	^{233}U	^{235}U	^{239}Pu	^{233}U
1	^{87}Br, ^{142}Cs	0.25	55.72	54.28	55.0	0.021	0.0072	0.0226
2	^{137}I, ^{88}Br	0.56	22.72	23.4	20.57	0.140	0.0625	0.0786
3	^{138}I, ^{89}Br, $^{(93,94)}Rb$	0.43	6.22	5.60	5.00	0.126	0.0444	0.0658
4	^{139}I, $^{(93,94)}Kr$, ^{143}Xe, $^{(90,92)}Br$	0.62	2.3	2.13	2.13	0.252	0.685	0.730
5	^{140}I, ^{145}Cs	0.42	0.61	0.618	0.615	0.074	0.018	0.0135
6	(Br, Rb, As, etc.)	–	0.23	0.257	0.277	0.027	0.0093	0.0087
			Total			0.64	0.21	0.26

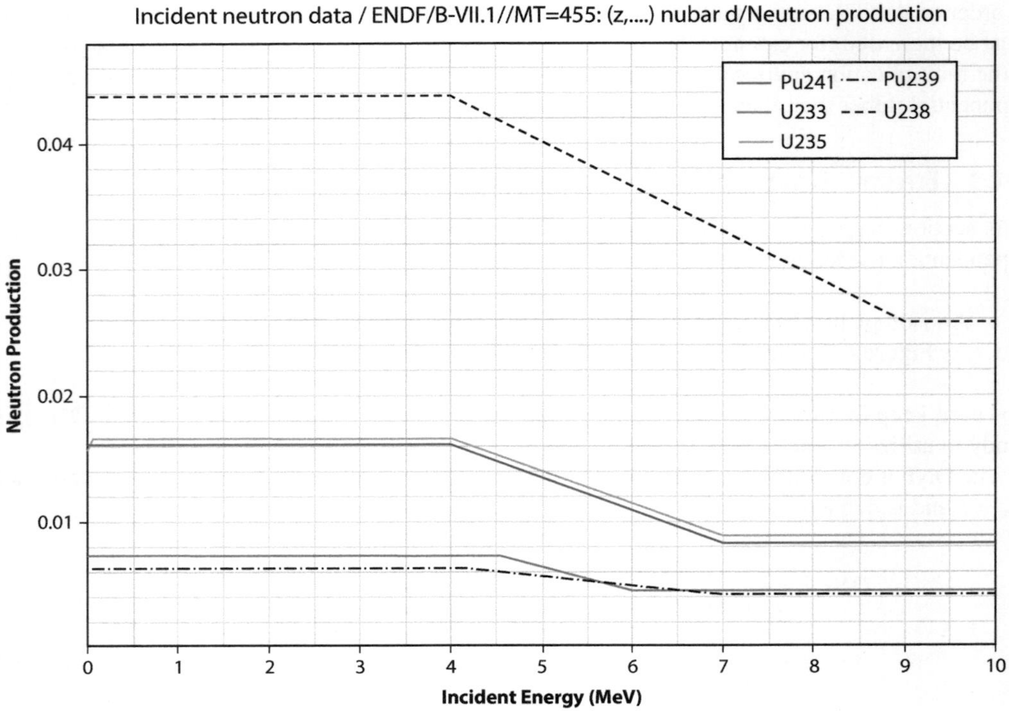

FIGURE 4.10 Delayed neutron production – MT-455.

As fuel burns up in LWRs, the delayed neutron percentage falls. The fuel's isotopic alterations are to blame for this. The fact that ^{235}U is the sole fissile element found in new uranium fuel is clear and straightforward.

However, the significance of ^{239}Pu fission rises (in some cases up to 50%) during fuel burnup. A multiplication system's core delayed neutron fraction (the weighted average of the constituent DNFs) falls as a result of ^{239}Pu producing substantially fewer delayed neutrons. This also explains why fuel burnup makes the neutron spectrum in the core harsher, as shown here.

$$\beta_{core} = \sum P_i \cdot \beta_i \qquad \text{Eq. 4.21}$$

where

P_i = Fraction of power generated by isotope i.

Example 4.3: Assume the reactor is at the start of its cycle and that ^{235}U fission generates around P_i= 98% of its power, while ^{238}U fission contributes P_i= 2% due to rapid fission. Utilizing Equation 4.21, we determine the core delayed neutron percent.

$$\begin{aligned}\beta_{core} = \sum P_i \cdot \beta_i &= [0.98 \times \beta_{235} + 0.02 \times \beta_{238}] \\ &= [0.98 \times 0.0065 + 0.02 \times 0.02] \\ &= 0.0064 + 0.0083 \\ &= 0.0067\end{aligned}$$

Now that this section has been introduced, we may go on to the DNF. Equation 4.17 shows that β is the proportion of delayed neutrons in the core during formation, or at high energy.

However, in the case of thermal reactors, the fission can be triggered primarily by thermal neutrons, which are useful for studying the behavior of thermal reactors. At thermal energy, the same proportion is known as the effective delayed neutron proportion, or β_{eff}.

The reactor's capacity to thermalize and use each generated neutron is indicated by the effective delayed neutron percentage. Since delayed neutrons differ from prompt neutrons produced straight from fission in their characteristics, the DNF β and the effective delayed neutron fraction (EDNF) β_{eff} are not the same. The energy of delayed neutrons is often smaller than that of prompt neutrons. The initial energy of prompt neutrons ranges from 1 MeV to 10 MeV, with an average of 2 MeV. The initial energy of delayed neutrons ranges from 0.3 MeV to 0.9 MeV, with an average of 0.4 MeV.

As a result, compared to the 2 MeV prompt neutron, a delayed neutron in thermal reactors travels a lower energy range to become thermal and is also less likely to be lost through leakage or parasitic absorption. However, because their average energy is lower than the minimum needed for fast fission to occur, delayed neutrons are likewise less likely to induce fast fission.

The importance factor is a concept that refers to the tendency of these two effects – a lower fast fission factor and a greater fast non-leakage probability for delayed neutrons – to cancel each other out. The effective delayed neutron fraction and the average delayed neutron fraction are related by the significance factor. Thus, the average DNF multiplied by the importance factor yields the EDNF, which is shown as follows:

$$\beta_{eff} = \beta \cdot I \qquad \text{Eq. 4.22}$$

The efficiency of the prompt and delayed neutrons in triggering a subsequent fission event varies. The efficacy of the various groups of delayed neutrons will also vary since the energy distribution of the delayed neutrons varies from group to group as well. Furthermore, a variety of fissionable isotopes can be found in a nuclear reactor. As a result, in certain situations, the significance factor alone is not enough, and an important factor needs to be established.

For example, when highly enriched fuel is used in a small thermal reactor, the fast non-leakage probability will rise more than the fast fission factor will fall, and the importance factor will be larger than 1.

The important factor will be smaller than 1 (about 0.97 for a commercial PWR) in a big thermal reactor with low enriched fuel, since the rise in the fast non-leakage probability will be dominated by the drop in the fast fission factor.

The rise in the fast non-leakage probability will be dominated by the decline in the fast fission factor in big fast reactors, and β_{eff} is around 10% smaller than β.

The power gain will rely on the delayed neutron half-lives, which can be managed technically, as long as the criticality at reactor initiation stays within the range $1.0 < k_{eff} < (1.0 + \beta)$. The PNL, which is around 10^{-3} seconds for CANDU and graphite-moderated gas cooled reactors, 10^{-4} to 10^{-5} seconds for LWRs, and $\sim 10^{-7}$ seconds for fast reactors, will determine the power gain for $k_{eff} > (1.0 + \beta)$. As with the Chernobyl catastrophe, this is uncontrollable. ^{235}U has a better DNF than the majority of actinide isotopes.

For detailed values of these characteristics for various reactor designs and infrastructures, see Table 4.6.

4.4.4 Mean Generation Time with Delayed Neutrons

The weighted average of the prompt and delayed neutron generation periods is known as mean generation time with delayed neutrons, or ℓ_d. The weighted average of the mean precursor lives of the six (or more) groups of delayed neutron precursors is the delayed neutron generation time, or τ_i.

It should be mentioned that the mean lifespan of their antecedents ($t_s + t_d << \tau_i$) is much longer than the genuine lifetime of delayed neutrons (the slowing down and diffusion times). Consequently,

TABLE 4.6
Table of Main Kinetic Parameters

	LWR	CANDU	Fast Reactor
$\Lambda\,[s]$	2 x 10^{-5}	1 x 10^{-3}	4 x 10^{-7}
β_{eff}	0.006	0.005	0.0035
$\lambda\,[s^{-1}]$	0.08	0.08	0.08

the mean lifespan of a neutron from the ith group, or $\tau_i = \ell_i$, is likewise equal to τ_i, and the following is the equation for mean generation time with delayed neutrons:

$$\ell_d = (1-\beta)\cdot\ell_p + \Sigma\ell_i\cdot\beta_i$$
$$\ell_d = (1-\beta)\cdot\ell_p + \Sigma\tau_i\cdot\beta_i \quad \text{Eq. 4.23}$$

where

- $(1-\beta)$ = Fraction of all neutrons emitted as prompt neutrons
- ℓ_p = Prompt neutron lifetime
- τ_i = Mean precursor lifetime, the inverse value of the decay constant $\tau_i = 1/\lambda_i$
- The weight-delayed generation time is given by $\tau = \Sigma\tau_i\cdot(\beta_i/\beta) = 13.05$ sec
- Therefore, the weighted decay constant $\lambda = (1/\tau) \approx 0.08\ \text{sec}^{-1}$

Despite the fact that delayed neutrons make up a very small portion of total neutrons in the core, the quantity, 0.08 second^{-1}, is very high and dominates the reactor time response. The computation of a weighted mean generation time with delayed neutrons provides the clearest example of this as

$$\ell_d = (1-\beta)\cdot\ell_p + \Sigma\tau_i\cdot\beta_i = (1 - 0.0065) \times 2 \times 10^{-5} + 0.085 = 0.00001987 + 0.085 \approx 0.085$$

Essentially, the average generation time with delayed neutrons is around 0.1 second, as opposed to ~10^{-5} as in Section 4.3.2 "Prompt Neutron Lifetime", where the delayed neutrons were not included.

4.4.5 Effect of Delayed Neutrons on Reactor Control

As we previously mentioned, delayed neutrons are vital for reactor control and are necessary from the perspective of reactor kinetics and reactor safety, even though their number per fission neutron is relatively small (usually less than 1%) and does not greatly contribute to power generation. When they are present, a reactor's dynamic temporal response to a change in reactivity is totally altered, allowing control systems like control rods to regulate it.

As previously mentioned, delayed neutrons enable the operation of a reactor in a delayed critical and quick subcritical condition. Every power reactor has safety mechanisms to keep it from ever reaching prompt criticality and is built to function in a delayed critical condition.

The quick criticality for most PWRs happens once β_{eff} is inserted with positive reactivity (i.e., $k_{eff} \approx 1.006$ or $\rho = +600$ pcm). It is nearly hard to insert such a reactivity insertion in power reactors (both in normal and abnormal operation), particularly when the reactor is in power operation mode and a reactivity insertion would cause the reactor core to heat up.

The negative reactivity from the moderator and fuel temperature coefficients balances the positive reactivity insertion because of the presence of reactivity feedback. From this perspective, the existence of delayed neutrons is also significant as it gives reactivity feedback time to respond to undesired reactivity insertion.

4.4.6 Interactive Chart – Infinite Multiplying System without Source and Delayed Neutrons

Examine the reactor's reaction in comparison to the scenario of an infinite multiplying system without a source of delayed neutrons (or set the $\beta = 0$). Refer to Figure 4.11.

4.5 POINT KINETIC EQUATION

Due to their numerous simplifications, the two earlier basic point kinetics equations are only approximations. For higher reactivity insertions, when there is a notable difference between the production of prompt and delayed neutrons, the straightforward point kinetics equation with delayed neutrons totally breaks down. Consequently, a more precise model is needed. Without using any assumptions, the precise point kinetics equations that may be obtained from the general neutron balance equations are

$$\begin{cases} \dfrac{dn}{dt} = \left[\dfrac{k(1-\beta)-1}{\ell}\right] n + \displaystyle\sum_i \lambda_i C_i \\ \dfrac{dC_i}{dt} = \left[\dfrac{k\beta_i}{\ell}\right] n - \lambda_i C_i \end{cases} \qquad \text{Eq. 4.24}$$

The production of prompt neutrons in the current generation, $k(1-\beta)n/\ell$, less the total number of neutrons in the previous generation, $-(n/\ell)$, is the first term on the right-hand side (RHS) of the neutron equation. The current generation's production of delayed neutrons is the second term. As can be shown, the neutron absorption rate is identical to that of the basic model $-(n/\ell)$. However, there is a difference between the delayed channel that results from radioactive disintegration of precursor nuclei ($\lambda_i C_i$) and the direct route for quick neutron generation $(1-\beta)$.

The generation of the ith group's precursors and their decay following the decay constant λ_i are balanced in the precursor equation. As can be observed, the rate of production is proportional to the

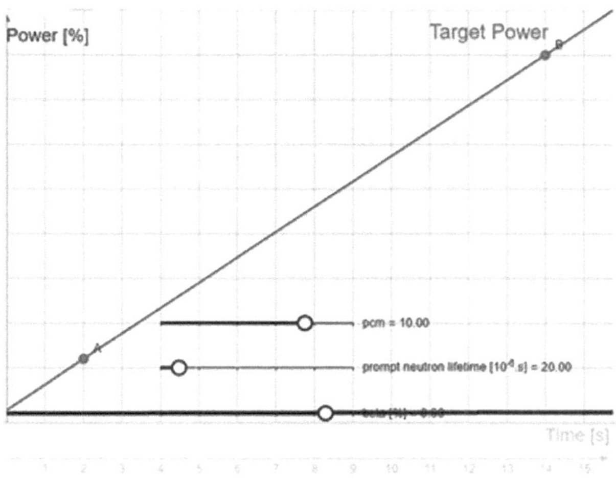

FIGURE 4.11 Interactive chart for the power of the reactor.

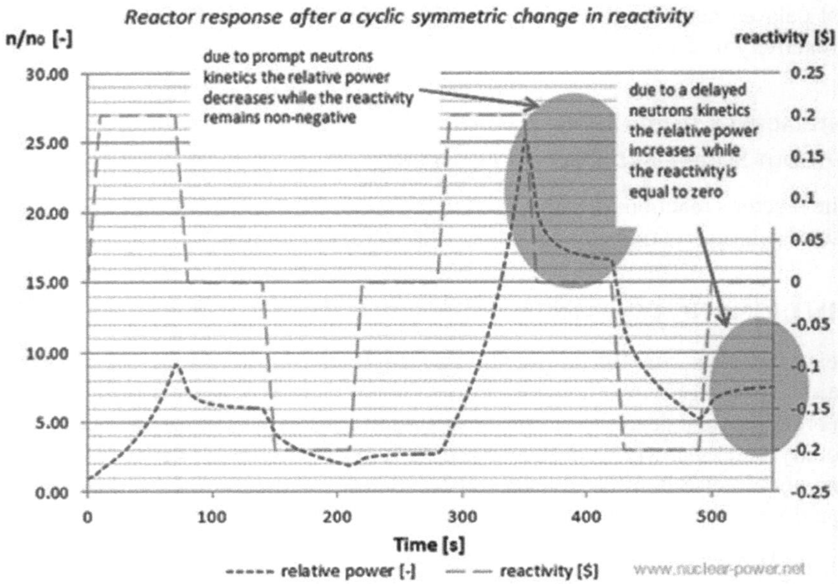

FIGURE 4.12 Solution of point kinetic equations for a cyclic change in reactivity.

(Source: www. Nuclear-power.net)

number of neutrons times β_i, which is the fraction of neutrons that emerge as delayed neutrons in the ith group, and the rate of precursor decay equals the radioactivity rate ($\lambda_i C_i$).

As can be shown, the point kinetics equations consist of two differential equations: one for the concentration of precursors $C(t)$and another for the neutron density $n(t)$. Refer to Figure 4.12.

Once more, reactivity $\left[\rho = (k-1)/k\right]$ and prompt generation time Λ are frequently used to define the point kinetics equations as follows:

$$\begin{cases} \dfrac{dn}{dt} = \left(\dfrac{\rho - \beta}{\Lambda}\right) n + \sum_i \lambda_i C_i \\ \dfrac{dC_i}{dt} = \left(\dfrac{\beta_i}{\Lambda}\right) n - \lambda_i C_i \end{cases} p \qquad \text{Eq. 4.25}$$

The point kinetics equation can be expressed in any way. For computations, the formula that uses Λ, or prompt neutron generation time, is often superior. This is true because variations in the absorption cross-section, not the fission cross-section, are what produce the majority of reactivity transients. While the prompt generation time is constant throughout these transients, the prompt neutron lifespan varies.

The reactivity of a reactor is defined by the previous equation, which characterizes the effective multiplication factor's departure from unity. The reactivity is zero at critical circumstances.

The reactor is farther away from criticality the higher the absolute magnitude of reactivity in the reactor core. Reactivity can actually be used to gauge how far a reactor has deviated from criticality. In Sections 4.5.1–4.5.3, we may categorize the various reactor states and the associated outcomes based on reactivity as follows:

4.5.1 Prompt Critical State

The prompt critical state is defined as $k_{eff} > 1;\ \rho \geq \beta_{eff}$, where the EDNF is less than the reactor's reactivity. In this instance, the neutron population may be increased and neutron losses balance by the creation of prompt neutrons alone. As quickly as the prompt neutron production lifespan (~10^{-5} s), the quantity of neutrons is growing exponentially with time.

4.5.2 Delayed Supercritical State

The following describes the delayed supercritical and prompt subcritical states: $k_{eff} > 1; 0 < \rho < \beta_{eff}$, when a reactor's reactivity is less than the EDNF but more than zero. In this instance, delayed neutrons are required to maintain the chain reaction since quick neutron production is inadequate to offset neutron losses. While the mean generation lifetime with delayed neutrons is around 0.1 second, the neutron population grows significantly more slowly.

4.5.3 Delayed Critical State

The prompt subcritical and delayed critical state is defined as $k_{eff} = 1;\ \rho = 0$. In this scenario, a reactor's reactivity is 0. In this instance, delayed neutrons are required to maintain the chain reaction since quick neutron production is inadequate to offset neutron losses. The chain reaction will be self-sustaining and the neutron population will not fluctuate over time. This condition is identical to the basic classification's crucial state.

4.5.4 Delayed Subcritical State

The prompt subcritical and delayed subcritical state is defined as $k_{eff} < 1;\ \rho < 0$, when a reactor's reactivity is less than zero. In this instance, the chain reaction is not self-sustaining as the generation of all neutrons is inadequate to offset neutron losses. The reactor is in what is often known as the subcritical multiplication state if the reactor core has internal or external neutron sources.

4.6 INHOUR EQUATION

The set (1 + 6) of linear ordinary differential equations with constant coefficients that can be solved analytically is part of the point kinetics equation model if the reactivity is constant. Reactivity and reactor period are related when six-group point kinetics equations are solved using Laplace transformation.

When employed as a unit of reactivity that correlated to an e-fold change in neutron density for one hour, this relationship – known as the Inhour equation – was derived from the inverse hour.

Furthermore, in order to elaborate on the "Inhour formula", we assume that if six groups of delayed neutrons are accounted for, the solutions of point kinetics equations for all sums of seven exponentials as $\sum_{i=0}^{6} \alpha_i \exp(\omega_i t)$, where ω_i for $i = 0, 1, \cdots, 6$ are the solutions of a seventh-order algebraic equation and cannot be found from a formula. The solutions must be determined numerically, delayed neutron fraction β_i in the ith group can no longer be obtained, and acquiring insight into the physical nature of any given transient is more difficult. Thus, in form of the seven equations set in the form $n(t) = A\exp(\omega t)$ can be established [9].

To better understand the Inhour formula and the origin of its naming convention: in the early days of reactor technology, the reactor period ω was expressed in "inverse hours". The value of reactor reactivity (ρ) was defined such that ω = 1 h^{-1} represented a reactor period of 1 hour.

Looking at Section 4.5 and Point Kinetics Equation without and Neutron Source $S(t)$ as well as looking at reactivity ρ formulation, a reactor's time-dependent behavior is extremely sensitive to even slight variations in the multiplication of on. Writing the kinetics equations in terms of the reactivity ρ, which is defined as follows, highlights this behavior.

$$\rho = \frac{k-1}{k} \qquad \text{Eq. 4.26}$$

Hence,

$$\rho \begin{cases} >0 & \textit{Supercritical} \\ =0 & \textit{Critical} \\ <0 & \textit{Subcritical} \end{cases} \qquad \text{Eq. 4.27}$$

Designating the prompt neutron generation time or mean generation time Λ as

$$\Lambda = \frac{1}{k} \qquad \text{Eq. 4.28}$$

Moreover, if we take the neutron kinetics equation with the source neutron $S(t)$ as below

$$\frac{d}{dt}n(t) = S(t) + \frac{1}{\ell}\left[(1-\beta)k - 1\right]n(t) + \sum_{i=1}^{6}\lambda_i C_i(t) \qquad \text{Eq. 4.29}$$

and

$$\frac{d}{dt}C_i(t) = \beta_i\left(\frac{k}{\ell}\right)n(t) - \lambda_i C_i(t) \quad i = 1, 2, \cdots, 6 \qquad \text{Eq. 4.30}$$

With Equations 4.29 and 4.30 as well as the definition of ρ and Λ, we may simplify the kinetics equations so that Equations 4.29 and 4.30 can be induced as

$$\frac{d}{dt}n(t) = S(t) + \left[\frac{(\rho-\beta)}{\Lambda}\right]n(t) + \sum_{i=1}^{6}\lambda_i C_i(t) \qquad \text{Eq. 4.31}$$

and

$$\frac{d}{dt}C_i(t) = \left(\frac{\beta_i}{\Lambda}\right)n(t) - \lambda_i C_i(t) \quad i = 1, 2, \cdots, 6 \qquad \text{Eq. 4.32}$$

Of course, we may approximate $\Lambda \approx 1$ without significantly altering equation solutions in many situations when the reactivity is small, then, $k \approx 1$. For specific details, see Lewis [7].

Additionally, the reactor period can be determined by solving the kinetics equations with the source term set to zero ($S(t) = 0$). Under this constraint, we seek a solution to the system of seven equations described above, expressed as a sum of seven exponentials $\sum_{i=0}^{6}\alpha_i \exp(\omega_i t)$, where ω_i for $i = 0, 1, \cdots, 6$ is set for the reactor period and also can be expressed as the precursor concentration of the ith group $C_i(t) = B_i \exp(\omega t)$, and the equation for the neutron flux as $n(t) = A\exp(\omega t)$ where A, B_i, and ω are constants. Inserting these two recent expressions into Equations 4.32 and 4.31, respectively, we will obtain the following results:

$$\omega A = \frac{(\rho-\beta)}{\Lambda}A + \sum_{i=1}^{6}\lambda_i B_i \qquad \text{Eq. 4.33}$$

and

$$\omega B_i = \frac{\beta}{\Lambda} A - \lambda_i B_i \qquad \text{Eq. 4.34}$$

Solving the second equation for B_i in terms of A and inserting the result into Equation 4.33, we obtain, after cancelling out the A values,

$$\omega = \frac{(\rho - \beta)}{\Lambda} + \frac{1}{\Lambda} \sum_{i=1}^{6} \frac{\beta_i \lambda_i}{\omega + \lambda_i} \qquad \text{Eq. 4.35}$$

The terms on the right are then combined using Equation 4.25, allowing us to write

$$\omega = \frac{\rho}{\Lambda} - \frac{1}{\Lambda} \sum_{i=1}^{6} \left(\frac{\beta_i}{\omega + \lambda_i} \right) \omega \qquad \text{Eq. 4.36}$$

Manipulating Equation 4.36 and solving for reactor reactivity ρ, the result obtained is

$$\rho = \left(\Lambda + \sum_{i=1}^{6} \frac{\beta_i}{\omega + \lambda_i} \right) \omega \qquad \text{Eq. 4.37}$$

where $\omega_0 = 1/\tau_e$, which is known as the Inhour equation, since the units of ω are commonly taken as inverse hours.

Coming back to the RHS graph against, as shown in Figure 4.13, allows one to evaluate the solution of Equation 4.37.

Regardless of whether the reactivity is positive or negative, we can see that there are seven solutions which form $\omega_1 > \omega_2 > \cdots > \omega_7$ by drawing a horizontal line for a given value of ρ. Additionally, based on the aforementioned, the neutron population of $n(t) = A\exp(\omega t)$ is as follows:

$$n(t) = \sum_{n=1}^{7} A_i \exp(\omega_i t) = \sum_{i=0}^{6} A_i \exp(\omega_i t) \qquad \text{Eq. 4.38}$$

Once more, Figure 4.13 shows that only ω_1 is the positive case for positive reactivity; as a result, the other components quickly vanish, producing an asymptotic solution that looks like this

$$n(t) \simeq A_1 \exp\left(\frac{t}{T} \right) \qquad \text{Eq. 4.39}$$

with $T = 1/\omega_1$ serving as the reactor period indicator. A negative period results from a negative reactivity, as we already mentioned in Figure 4.13. As a result, while $T = 1/\omega_1$ will fade away more slowly than the others, all of the ω_i are negative. Equation 4.37 is, therefore, applicable to both positive and negative reactivities of the reactor of interest.

The plots in Figure 4.14 are for reactivities of $\rho = \pm 0.1\beta$, using parameters for Uranium-235 and for prompt neutron generation time or mean generation time of $\Lambda = 50 \times 10^{-6}$ seconds.

The prompt jump magnitude at the beginning of the curves is approximately $|A_1 - n(0)|$. To determine the reactivity required to produce a given period or vice versa – a plot of ρ versus T must be constructed using the delayed neutrons data for a particular fissionable isotope or isotopes, and for a given prompt neutron lifetime.

Because the half-lives of the neutron-emitting fission products are the main factor influencing the neutron population's increase or decay after the prompt jump [7], it should be noted that the exponential behavior illustrated in Figure 4.14 happens very slowly.

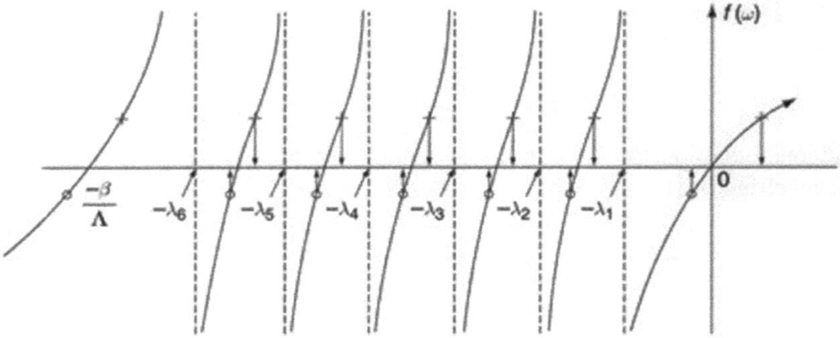

FIGURE 4.13 A plot of the Inhour Formula. [9]

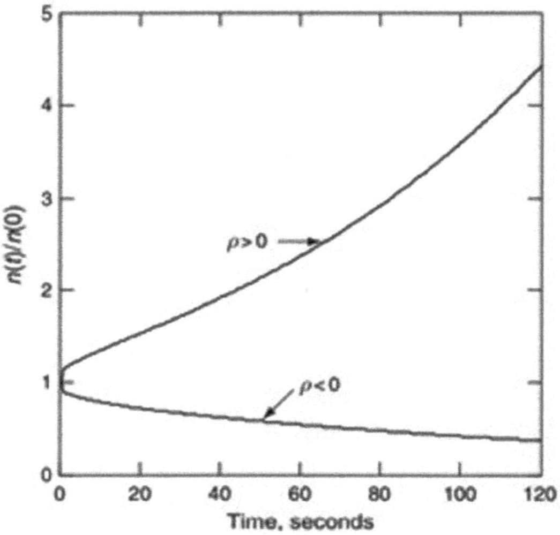

FIGURE 4.14 Neutron populations following reactivity insertion of ±0.01 β. [7]

As we indicated before, this asymptotic behavior may be represented in the form of $n(t) \propto \exp(t/T)$, where T was again defined as the reactor period. Depending on whether the reactor is supercritical or subcritical, the period is often either positive or negative. It is perhaps the most significant number that can be derived from the kinetics equations and is the amount of time needed for the reactor power to rise or decrease by a factor of e.

Notably, in some textbooks of neutronic analysis, the final Equations 4.32 and 4.36 are presented as

$$\begin{cases} \left.\begin{matrix} \dfrac{dC_i(t)}{dt} = \dfrac{\beta_i}{\Lambda} n(t) - \lambda_i C_i(t) \\ C_i(t) = C_{i,0} \exp(s,t) \end{matrix}\right\} \to C_{i,0} = \dfrac{1}{\Lambda} \dfrac{\beta_i}{s + \lambda_i} n_0 \\ \\ \rho_0 = s\left(\Lambda + \displaystyle\sum_{i=0}^{6} \dfrac{\beta_i}{s + \lambda_i}\right) \\ s_0 = \dfrac{1}{\tau_0} \end{cases} \quad \text{Eq. 4.40}$$

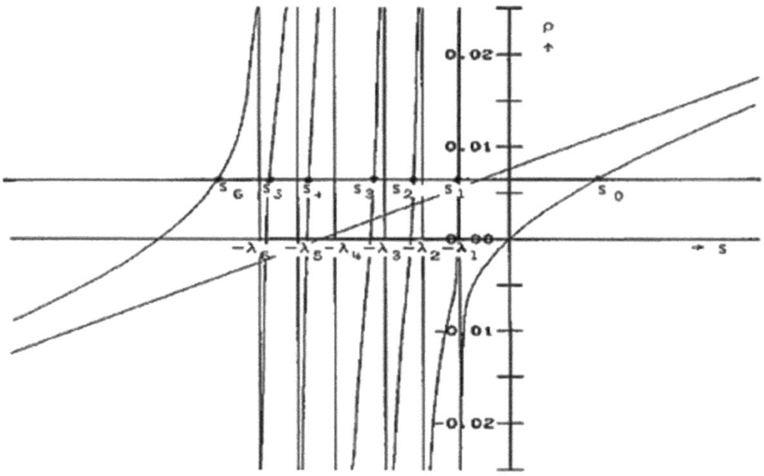

FIGURE 4.15 A graphical demonstration of the roots $s_0 - s_6$

which in case of the Equation 4.40, the Inhour equation in terms of ρ_o and s_0 is presented since the constants of s_{0-6} were originally determined in inverse hours. For a given value of the reactivity ρ the associated values of s_{0-6} are determined with this equation.

Figure 4.15 shows the relation between ρ and roots s graphically and from this figure it can be seen that for a given value of ρ seven solutions exist for s.

Only S_0 is positive for positive reactivity, as shown in Figure 4.15, and the other terms quickly vanish as previously mentioned, producing an asymptotic solution that looks like this:

$$\begin{cases} n(t) = \sum_{i=0}^{6} A_i \exp(s_i t) \\ \downarrow \\ n(t) = A_0 \exp(s_0 t) \end{cases} \quad \text{Eq. 4.41}$$

where $s_0 = 1/\tau_0$ is the stable reactor period or asymptotic period of the reactor. This root, s_0, is positive for $\rho > 0$ and negative for $\rho < 0$; therefore this root describes the reactor response, which is lasting after the transition phenomena have died out. Additionally, the graphic illustrates how a negative phase results from a negative reactivity: All of the s_i are negative, but the root s_0 will die away more slowly than the others. Consequently, both positive and negative reactivity insertions can use solution $n(t) = A_0 \exp(s_0 t)$.

A plot of ρ versus τ_e must be created using the delayed neutron data for a certain fissionable isotope or mix of isotopes and for a specified prompt generation time in order to calculate the reactivity needed to generate a given period. It is convenient to use either of the aforementioned forms of the Inhour equation to get the stable reactor period that arises from a particular reactivity insertion.

Additionally, Uranium-235 (^{235}U) data are used in this type of graphic in Figure 4.16. Keep in mind that as ρ surpasses β, the time rapidly decreases. The second component in Equation 4.31 changes from negative to positive, indicating that the chain reaction is sustained without delayed neutrons, which is what the condition $\rho = \beta$ means by quick critical.

The difference between the reactor period and the prompt leap disappears once prompt criticality is reached; at this point, the rate of exponential rise is mostly determined by the prompt neutron

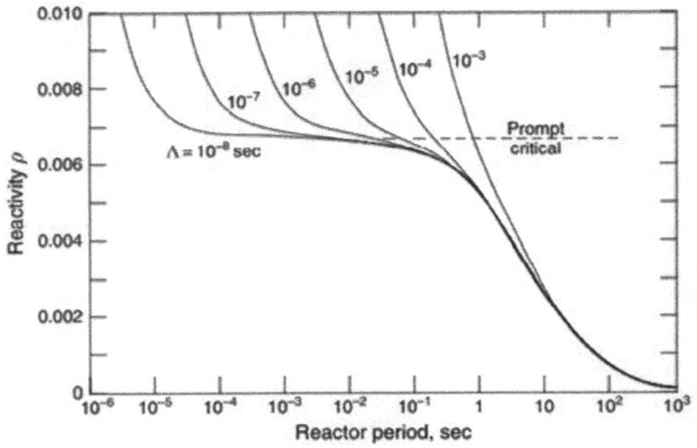

FIGURE 4.16 Reactivity versus reactor period for various prompt neutron generation times.

lifetime rather than the delayed neutron half-lives. In fact, the time gets so short as the quick critical approaches that it becomes extremely difficult, if not impossible, to regulate the reactor mechanically using tools like spinning drums or control rod movement. Reactivity is sometimes expressed in dollars, $\$ = \rho/\beta$, or cents, since it is so crucial to avoid nearing prompt critical.

The rate at which the neutron population may be reduced for negative reactivities has an asymptotic limit. As you can see from Figure 4.13, the greatest half-life of delayed neutrons, or $T = (1/\lambda_1) = 0.6933t_{1/2}\big|_{max}$, which is around 56 seconds for a uranium reactor, determines the lowest negative period that may occur, or the quickest that a reactor's output can be reduced following the initial prompt drop.

Figure 4.13 shows that the maximum value of omega, ω_1, is also modest when the reactivity is minimal. Let us say that we take $\omega_1 \ll \lambda_i$ for all λ_i for a tiny reactivity. Next, we may remove ω_1 from Equation 4.37's denominator, which results in

$$\rho = \left(\Lambda + \sum_{i=0}^{6} \frac{\beta_i}{\lambda_{i\theta}} \right) \omega_1 \qquad \text{Eq. 4.42}$$

The equation of average decay constant in its general form can be defined as

$$\frac{1}{\lambda} = \frac{1}{\beta} \sum_{i=0}^{6} \beta_i \frac{1}{\lambda_i} \qquad \text{Eq. 4.43}$$

Then, eliminating the summation, and solving for ω_1, we obtain the result as

$$\omega_1 = \frac{\rho}{\left(\Lambda + \dfrac{\beta}{\lambda} \right)} \qquad \text{Eq. 4.44}$$

Generally speaking, $\Lambda \ll \beta/\lambda$, yields the following:

$$T \approx \frac{\beta}{\rho\lambda} \qquad \text{Eq. 4.45}$$

Therefore, the delayed neutron characteristics β and λ almost entirely control the reactor time for modest reactivities, whether positive or negative. Given that delayed neutrons make up such a minor portion of fission neutron production, this may appear unexpected. However, a duration far longer than the prompt neutron lifespan is obtained when this tiny percentage is multiplied by the average half-life, $(\beta/\lambda) = 0.693\beta t_{1/2}$.

Since the time period is very short above the prompt critical, ω_1 is quite massive. In this situation, we may take $\omega_1 \gg \lambda_i$ for all λ_i in Equation 4.36, reducing it to $\rho = A\omega_1 + \beta$, or equivalently $\omega_1 = (\rho - \beta)/\Lambda$.

Because it is proportional to the prompt neutron generation time and independent of the delayed neutron half-lives, the reactor duration for $\rho > \beta$ is, therefore, extremely brief.

$$T \approx \frac{\Lambda}{(\rho - \beta)} \qquad \text{Eq. 4.46}$$

4.6.1 Reactor Period Form of Inhour Equation

The reactor period term τ_e was defined in Section 4.6 and is presented as Equation 4.46 below.

$$\tau_e = \frac{\ell}{\rho} + \frac{\beta_{eff} - \rho}{\lambda_{eff} \cdot \rho + \dot{\rho}} \qquad \text{Eq. 4.47}$$

where

ℓ = Prompt neutron lifetime
β_{eff} = Effective delayed neutron fraction
λ_{eff} = Effective delayed neutron precursor decay constant
τ_e = Reactor period
ρ = Reactivity

This formula's first term, the prompt term, causes an instantaneous power boost known as the prompt leap to occur just after the positive reactivity insertion. The reason for this power increase is that as soon as the reactivity is introduced, the rate at which prompt neutrons are produced changes. Following the prompt leap, the rate of power change cannot accelerate beyond the time delay that the antecedent half-lives permit.

There are a lot of fascinating events when delayed neutrons are present. The typical reactor reaction to continual reactivity injection is depicted in Figure 4.17.

Additionally, the second term in this formula is known as the delayed term or prompt delay, according to Equation 4.47. Control rods or another reactivity control device, such as spinning drums in various types of reactor infrastructures, can be used to regulate the reactor since the presence of delayed neutrons makes the power rise manageable.

4.6.2 Reactor Period and Startup Rate

Since reactivity cannot be measured directly, it is not included in most power reactor operations and is not limited in most technical specifications. Rather, they define what is known as a starting rate, which is the limiting rate of neutron power growth (as determined by Excore detectors), particularly for PWRs.

The number of factors of 10 that power changes in a minute is known as the reactor start-up rate, or SUR. As a result, SUR is measured in decades per minute (DPM), or powers of 10 per minute. The following formula provides the link between reactor power and starting rate:

$$n(t) = n(0) \cdot 10^{\text{SUR} \cdot t} \qquad \text{Eq. 4.48}$$

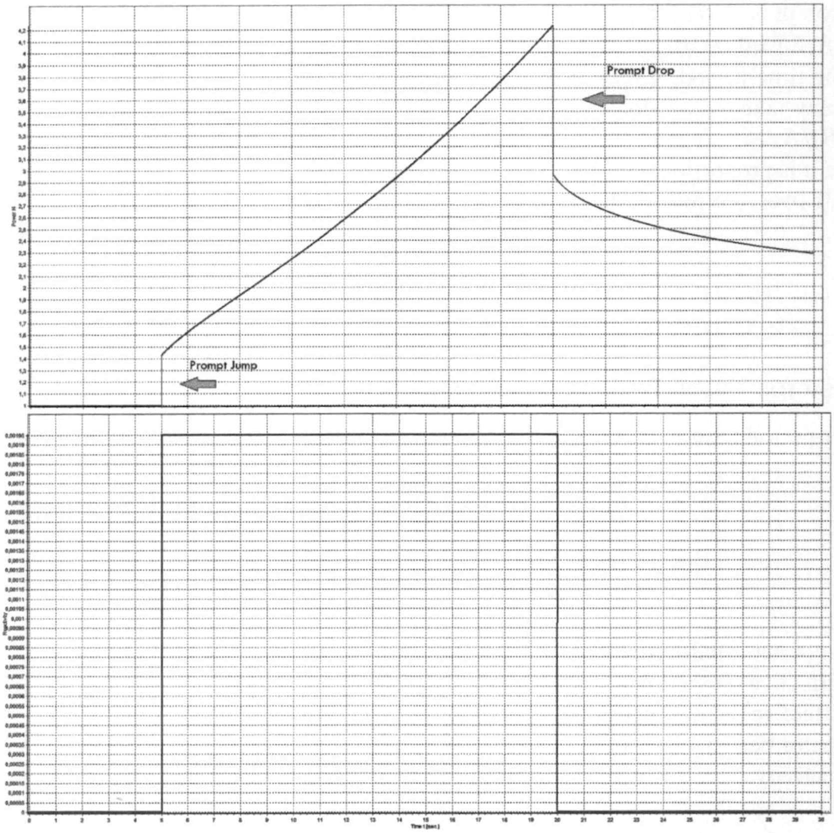

FIGURE 4.17 Reactor response on constant reactivity insertion.

where, in this equation, the following parameters are defined as

SUR = Reactor startup rate (DMP)
t = Time during reactor transient (minutes)

The reactor power changes more quickly the greater the SUR value. Either a positive or negative beginning rate is possible. Reactor power is rising if SUR is positive. Reactor power is dropping if SUR is negative. The following formulas show how the reactor period and starting rate are related.

$$\frac{n(t)}{n(0)}\exp\left(\frac{t(\text{sec})}{\tau_e}\right) = 10^{sur\cdot t(\text{min})}$$

$$\exp\left(\frac{t(\text{sec})}{\tau_e}\right) = \exp 2.303\cdot \text{sur}\cdot t(\text{min}) \qquad \text{Eq. 4.49}$$

$$\frac{60}{\tau_e} = 2.303\cdot \text{SUR}$$

$$\text{SUR} = \frac{26.06}{\tau_e}$$

Example: Assume that reactor $k_{eff} = 1.0005$ has a generation time $\ell_d = 0.01$ s. Determine the reactor period (τ_e), doubling time (DT), and SUR for this condition.

$\rho = 1.0005 - 1/1.0005 = 50$ pcm
$\tau_e = \ell_d/(k-1) = 0.1/0.0005 = 200$ s
$\text{DT} = \tau_e \cdot \ln 2 = 139$ s

4.7 SPECIAL CASES OF INHOUR EQUATION

The following special cases are discussed.

4.7.1 $\rho << \beta$ Low Reactivity

Presume that ρ is very small. As a result, the reactivity's first root, s_0, is likewise rather little. Since $s_0 << \lambda_i$ is present in this instance, we may disregard the term s_0 in the Inhour equation's denominator. Next, the Inhour equation is formed.

$$\rho_0 = s\left(\Lambda + \sum_{i=0}^{6} \frac{\beta_i}{s+\lambda_i}\right) \xrightarrow{s<<\lambda_i} s\left(\Lambda + \sum_{i=0}^{6} \frac{\beta_i}{\lambda_i}\right) \to \tau_e\left(\Lambda + \sum_{i=0}^{6} \tau_i \beta_i\right)$$

$$= \frac{\ell_d}{k_{eff} - 1} \qquad \text{Eq. 4.50}$$

thus

$$\tau_e = \frac{\ell_d}{k_{eff} - 1}$$

This specific example, which also uses the mean generation time with delayed neutrons (ℓ_d), yields the same period as the simple point kinetics equation, as can be seen.

$$\frac{d}{dt} n(t) = \frac{(k-1)}{\ell_d} n(t)$$

$$n(t) = n(0) \exp\left[\frac{(k-1)}{\ell_d} t\right] \qquad \text{Eq. 4.51}$$

The delayed neutron characteristics also control the reactor period nearly entirely for modest reactivities, whether positive or negative. The period in seconds (τ_i) is crucial, even if the quantity of delayed neutrons is just a hundredth of a percent of the total.

It is assumed that $s_0 << \lambda_i$. Since $\tau_i = 1/\lambda_i$ has a maximum value of 80 seconds, this method works for periods τ_e that are longer than 80 seconds. For reactivities up to roughly $\rho = 0.0005 = 50$ pcm, however, this formula is practical and precise enough for the majority of uses.

4.7.2 $\rho \geq \beta$ High Reactivity – Prompt Criticality

Suppose that ρ is greater than β. In this instance, the neutron population may be increased and neutron losses are balanced by the creation of prompt neutrons alone. Prompt neutron production lifespan (~10^{-5} s) is very small but quantity of neutrons is growing exponentially with time.

As a result, the reactivity's first root, s_0, grows larger than each λ_i. Since $\lambda_i << s_0$ is present in this instance, we may disregard the term λ_i in the Inhour equation's denominator. Next, the Inhour equation is formed.

$$\rho_0 = s\left(\Lambda + \sum_{i=0}^{6} \frac{\beta_i}{s}\right) \xrightarrow{\lambda_1 << s} \left(\Lambda + \sum_{i=0}^{6} \frac{\beta_i}{s}\right) \qquad \text{Eq. 4.53}$$

$$\rightarrow \rho_0 = s\Lambda + \beta \rightarrow \tau_e = \frac{\ell}{k_{eff} - 1}$$

This particular case produces the same time as the case of the simple point kinetics equation without delayed neutrons, as shown in Figure 4.18.

Since a reactor's reaction is obviously altered when it reaches the reactivity of β (650 pcm, for example), this circumstance can only be coincidental. The difference between the reactor period and the prompt leap disappears once prompt criticality is attained; at this point, the rate of exponential rise is mostly determined by the prompt neutron lifetime rather than the delayed neutron half-lives.

It would be extremely challenging to regulate reactors with such kinetics mechanically, for example, by moving control rods. Reactors must operate with enough margin to maintain their reactivity far below the quick criticality criterion.

Such dynamics may arise in design basis accidents (DBAs). Reactors must, for instance, be able to tolerate a jump-like insertion of quite high (~1$ or even more) positive reactivity under RIA circumstances, such as in the event of control rod ejection. In these situations, the PNL is crucial. Reactor kinetic response may be significantly enhanced by longer PNLs (the longer PNL simply yields slower power growth).

Consequently, a reload safety evaluation (RSE) procedure should be used to validate the PNL. Reactivity feedbacks, particularly those based on the Doppler temperature coefficient (DTC), provide the foundation for managing this accident. In terms of reactor stability, this coefficient is the most crucial. In general, the moderator temperature coefficient (MTC) is seen to be even less significant than the DTC. The first and most crucial feedback that will offset the added positive reactivity is the DTC, particularly in the event of a control rod ejection.

While the fuel temperature coefficient takes effect nearly immediately, the time it takes for heat to be transmitted to the moderator is typically measured in seconds. Because it responds instantly to changes in fuel temperature, this coefficient is also known as the quick temperature coefficient.

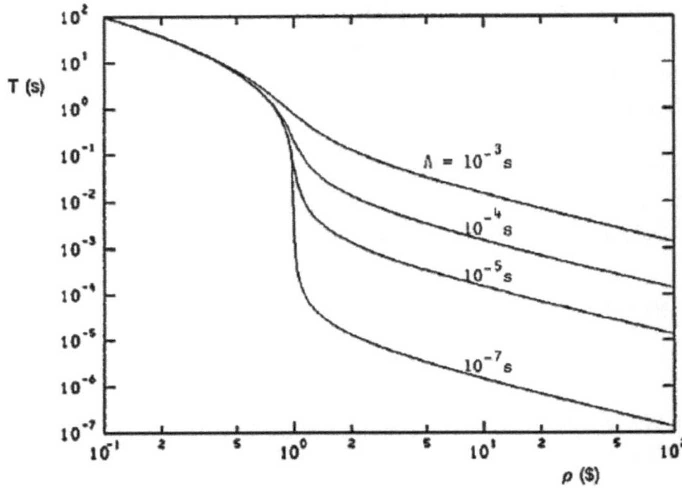

FIGURE 4.18 Reactor period versus reactivity.

4.7.3 $\rho < \beta$ Smallest Negative Period Possible

Assume that $\rho \ll \beta$ and ρ are negative. This situation is common for reactor emergency shutdowns. Another name for this occurrence is a SCRAM or reactor trip. The quick insertion or descent of the control rods into the core to halt the fission chain reaction is known as a reactor "SCRAM" (or "trip"). All of the control rods in PWRs are typically installed in 2 to 4 seconds. In this instance, the multiplication system's responsiveness rapidly turns negative and surpasses β (in absolute numbers).

However, the neutron flow would not instantly drop to zero even if an infinite negative reactivity could be added. Almost instantly, prompt neutrons will be absorbed. It adheres to the prompt drop pattern.

Consequently, the neutron flux that results will be

$$\frac{n_2}{n_1} = \frac{\beta - \rho_1}{\beta - \rho_2} \xrightarrow{\rho_2 \to \infty} n_2 = \beta \cdot n_1 \qquad \text{Eq. 4.54}$$

It is clear that the neutron flow cannot go below βn_1. The actual numbers are far greater.

The integral worth of all control and emergency rods (PWRs) is for example –9,000 pcm. It is equal to $\rho = -9{,}000/600 = -15\beta = -0.09$ ($\beta = 600$ pcm $= 0.006$).

The cue decrease for this negative reaction is equivalent to

$$\frac{n_2}{n_1} = 0.006/(0.006 + 0.09) = 0.063$$

This is around 10 times more than what would occur with an injection of infinite negative reactivity.

After that, the neutron flux keeps decreasing in accordance with the steady period. The decay constant of the long-lived precursor's group, $s_0 = -\lambda_1$, is where the first root of the reactivity equation is found. Then, $\tau_e = -1/\lambda_1 = -80$ seconds is the shortest negative stable period. There is no way to decrease the neutron flow faster than this time frame. However, in a matter of tens of seconds, the thermal power originating from nuclear fission falls below the thermal power originating from decay heat due to the rapid drop, which produces an instantaneous reduction to around 6% of rated power.

4.8 PROMPT JUMP APPROXIMATION

The rod drop and source jerk approaches can be used to experimentally determine reactor parameters by leveraging the prompt leap that follows incremental changes in reactivity.

The change may be seen in Figure 4.19, where the first portion of the flux transients that occur in Figure 4.14 have been enlarged. The rapid increase in neutron populations after reactivity insertion of $\pm 0.10\beta$ is seen in Figure 4.19.

It should be noted that the neutron population changes significantly more slowly after first jumping quickly.

During the time it takes for the neutrons to make the first run up or down, the precursor concentration changes very little, acting much more slowly than the neutron population. Precursor behavior defined by Equation 4.32 is slow even when reacting to a quick change in neutron population because the decay constant multiplying C_i is modest. This behavior is caused by the long precursor half-lives in comparison to the prompt generation time. On the other hand, because of the short value of prompt neutron production time Λ in the denominator of Equation 4.31, the neutron population initially reacts very quickly. However, the neutron reaction slows down after a short period of time because the value of the last component in Equation 4.32 changes more slowly.

In the two scenarios that follow, we may utilize this behavior to estimate the prompt jump magnitudes.

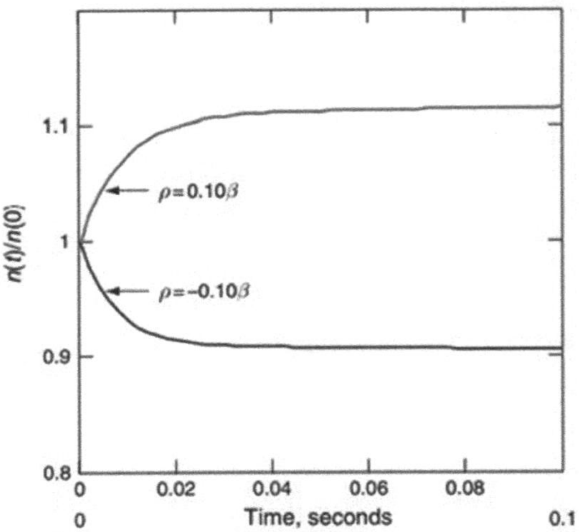

FIGURE 4.19 Prompt jump in neutron population.

4.8.1 Rod Drop

Assume that a source-free critical reactor is in a steady state of operation. Equation 4.32 states that the precursor concentrations and neutron population are connected by

$$C_{io} = \left(\frac{\beta_i}{\lambda_1 \Lambda}\right) n_0 \qquad \text{Eq. 4.55}$$

The precursor concentration C_{i0} will virtually stay the same for as long as $\lambda_i t \square$ after the rod is inserted. Consequently, Equation 4.31's source-free version is as follows:

$$\frac{d}{dt} n(t) = -\frac{(|\rho| + \beta)}{\Lambda} n(t) + \frac{\beta}{\Lambda} n_0 \qquad t << \left(\frac{1}{\lambda_i}\right) \qquad \text{Eq. 4.56}$$

When a negative reactivity is indicated by $\rho = -|\rho|$, Equation 4.20 is used. To find a solution, we apply the integrating factor method. After some simplification, we obtain, with an initial condition of $n(0) = n_0$ and an integrating factor of $\exp\left[\left(|\rho + \beta|\right)/\Lambda\right]$:

$$n(t) = \frac{\beta}{(|\rho| + \beta)} n_0 + \frac{|\rho|}{(|\rho| + \beta)} n_0 \exp\left\{-\frac{1}{\Lambda}(|\rho| + \beta)t\right\} \qquad \text{Eq. 4.57}$$

When the second term decays, which happens extremely quickly, we have periods that are significantly longer than the prompt generation time but less than the half-lives of the delayed neutrons (i.e., $\Lambda/(|\rho| + \beta) << t << 1/\lambda_i$ that is about neutron population) as follows:

$$n_1 = \frac{\beta}{(|\rho| + \beta)} n_0 \qquad \text{Eq. 4.58}$$

Therefore, the ratio of neutrons immediately before to immediately after the installation of the control rod may be used to calculate the reactivity drop in dollars.

$$\frac{|\rho|}{\beta} = \frac{n_0}{n_1} - 1 \qquad \text{Eq. 4.59}$$

4.8.2 Source Jerk

A subcritical reactor ($\rho = -|\rho|$) with a neutron source (S_0) is used for the source jerk. Since the system is in a steady state, the derivatives under these circumstances are as follows: Kinetic equations hold a solution for which n and C_i are time-independent.

$$0 = S_0 + \frac{1}{\ell}\left[(1-\beta)k\right]n + \sum_{i=1}^{6}\lambda_i C_i \qquad i = 1,2,\cdots,6 \qquad \text{Eq. 4.60}$$

Therefore, $C_{io} = (\beta_i/\lambda_i\Lambda)n_0$ is the experiment's beginning condition. A rapid negative jump in the neutron population occurs after the source is abruptly removed (i.e., jerked).

When the source term is set to zero and $C_i(t)$ swapped out for C_{i0} period $t << 1/\lambda_i$, we can once more use Equation 4.31 to represent the transient. The connection between reactivity, DNF, and decline in neutron population is described by Equations 4.57 and 4.58. Equation 4.55 again yields a solution, which is provided by Equation 4.56. As a result, we see that the neutron population decreases noticeably from n_0 to n_1 when a negative reactivity $\rho = -|\rho|$ is inserted into a critical reactor (the rod drop experiment) or when the source is removed from a subcritical system with reactivity $\rho = -|\rho|$ (the source jerk experiment).

4.8.3 Rod Oscillator

Equations 4.31 and 4.32 may be used to examine a third experimental approach. The equations might not be as easily solved in the rod oscillator experiment, though. The equations must be solved using Laplace transforms or similar methods when a sinusoidal reactivity of the type $\rho(t) = \rho_0 \sin(\omega t)$ is present. Here, we only express the outcome: After transients have subsided, the neutron population will fluctuate as follows if n_0 is the initial neutron population. Thus,

$$n(t) = n_0\left[1 + A(\omega)\sin(\omega t) + \frac{\omega\rho_0}{\Lambda}\right] \qquad \text{Eq. 4.61}$$

where $A(\omega)$ is a function of the frequency. If $\sin(\omega t)$ is averaged over time the sinusoidal term vanishes leaving $n = n_0(1 + \omega\rho_0/\Lambda)$. Thus, plotting n/n_0 versus frequency determines ρ_0/Λ, the ratio of reactivity to prompt neutron generation time.

4.8.4 Prologue to Reactor Dynamics

Thermal feedback effects have not yet been taken into consideration in our discussion of the time-dependent behavior of chain reactions. Because of this, these therapies are commonly called zero-power kinetics. However, the material densities will alter if the energy produced by fission is sufficient to raise the system's temperature.

Macroscopic cross-sections will also alter since they are related to densities. The resonance cross-sections, which broaden and flatten with rising temperature due to the Doppler broadening, will provide additional feedback effects in addition to the changes in material density.

The Maxwell–Boltzmann distribution's temperature dependency causes the spectrum to solidify in thermal reactors as well. When combined, these feedback effects will change the kinetic

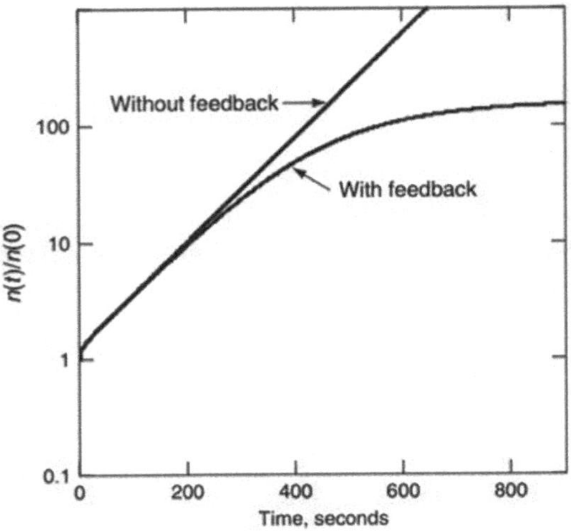

FIGURE 4.20 Effect of negative temperature feedback on neutron population.

equations' values. The reactivity is by far the most affected. Reactor design must thus ensure that the feedback for temperature rise is negative under all operating situations.

For instance, Figure 4.14's curves are affected by negative feedback in the manner described below. If negative feedback is present, the curve will flatten when the neutron population grows to a size where a temperature rise may occur. Over time, it will settle and may even drop.

The positive reactivity curve of Figure 4.14 has been redrawn on a logarithmic scale in Figure 4.20. Additionally, a curve for the same reactivity insertion has been included, but this time the effects of negative temperature feedback are included after a reactivity insertion of $+ 0.10\beta$.

The straight-line behavior on the logarithmic plot indicates that both curves initially follow the same period. However, the feedback curve becomes concave downward and stabilizes at a constant power as the power increases. The original reactivity insertion has now been fully offset by the negative feedback.

REFERENCES

1. B. Zohuri, *Neutronic Analysis for Nuclear Reactor Systems*, 2nd edition, Springer Publishing Company, 2019.
2. J. C. Lee, *Nuclear Reactor Physics and Engineering*, 1st edition, John Wiley & Sons Publishing Company, 2020.
3. J. R. Lamarsh, A. J. Baratta, *Introduction to Nuclear Engineering*, 4th edition, Pearson, 2018.
4. M. C. Brady, T. R. England, Delayed Neutron Data and Group Parameters for 43 Fissioning Systems, *Nucl. Sci. Eng.* 103, 129 (1989).
5. G. R. Keepin, *Physics of Nuclear Kinetics*, Addison-Wesley, 1965.
6. T. A. Parish, W. S. Charlton, N. Shinohara, M. Andoh, M. C. Brady, S. Raman, Status of Six-Group Delayed Neutron Data and Relationships Between Delayed Neutron Parameters from the Macroscopic and Microscopic Approaches, *Nucl. Sci. Eng.* 131, 208 (1999).
7. E. E. Lewis, *Fundamental of Nuclear Reactor Physics*, Academic Press, 2008.
8. P. F. Rose (ed.), ENDF/B-VI Summary Documentation. Cross Section Evaluation Working Group, Report BNL-NCS-17541 (ENDF-201), 1991.
9. A. Henry, *Nuclear Reactor Analysis*, The MIT Press, 1975.

Nuclear and Reactor Physics

1. J. R. Lamarsh, *Introduction to Nuclear Reactor Theory*, 2nd edition, Addison-Wesley, 1983.
2. J. R. Lamarsh, A. J. Baratta, *Introduction to Nuclear Engineering*, 3rd edition, Prentice-Hall, 2001.
3. W. M. Stacey, *Nuclear Reactor Physics*, John Wiley & Sons, 2001.
4. S. Glasstone, *Nuclear Reactor Engineering: Reactor Systems Engineering*, 4th edition, Springer, 1994.
5. W. S. C. Williams, *Nuclear and Particle Physics*, 1st edition, Clarendon Press, 1991.
6. G. R. Keepin, *Physics of Nuclear Kinetics*, 1st edition, Addison-Wesley Pub. Co., 1965.
7. R. R. Burn, *Introduction to Nuclear Reactor Operation*, 1988.
8. U.S. Department of Energy, Nuclear Physics and Reactor Theory, *DOE Fundamentals Handbook*, Volume 1 and 2, January 1993.

Advanced Reactor Physics

1. K. O. Ott, W. A. Bezella, *Introductory Nuclear Reactor Statics*, Revised edition, American Nuclear Society, 1989.
2. K. O. Ott, R. J. Neuhold, *Introductory Nuclear Reactor Dynamics*, American Nuclear Society, 1985.
3. D. L. Hetrick, *Dynamics of Nuclear Reactors*, American Nuclear Society, 1993.
4. E. E. Lewis, W. F. Miller, *Computational Methods of Neutron Transport*, American Nuclear Society, 1993.

5 Introduction to Nuclear Energy from Neutrons to Fast Spectrum

5.1 INTRODUCTION

In Canada, fissionable nuclear energy from uranium has produced approximately 4 million kWh of non-carbon electricity from 60,000 tons of naturally occurring uranium mined during the past 50 years*. Since the Manhattan Project in Los Alamos, New Mexico, and the subsequent launch of the nuclear-powered submarine, Nautilus, in the United States, fissionable nuclear reactors have made nuclear energy more accessible. Today, many Generation III (GEN-III) reactors are commercially producing energy for the production of electricity power not only in the United States and Canada, but also worldwide. These reactors are free from carbon sources and have no impact on the effects of global warming.

To generate the same amount of electricity, 58,000 times as much coal is needed, and 3.6 billion tons of carbon dioxide (CO_2) is released. Around 40 times as much nuclear energy comes from uranium sources globally, and this number is rising thanks to Generation IV (GEN-IV) Small Modular Reactor (SMR) technology. The need for nuclear energy as a production source is increasing due to the world's population expansion, and this trend is consistent [1–2].

Surprisinglyfew % of the current uranium ore is used to generate nuclear energy in commercial nuclear reactors; the remainder is now regarded as long-term radiotoxic yet radioactive waste in the form of "nuclear waste". However, each unused uranium atom still has the same amount of nuclear energy as each atom in the very limited amount of uranium that was used.

We must re-examine the current waste management concerns in light of the Advanced Reactor Concept (ARC) technology, which uses high energy neutrons in fast-spectrum reactors as part of SMRs. Furthermore, although this idea is not new, it is no longer covered in the most recent nuclear physics and engineering curriculum.

Nuclear power choices that are inexpensive and safe can be provided by technologically advanced SMRs. The sophisticated SMRs that are presently being developed in the United States alone come in a range of sizes, technology possibilities, and development scenarios from numerous businesses.

These cutting-edge reactors can be used for power generation, process heat, desalination, and other industrial applications. They are expected to range in size from a few megawatts in the form of microreactors [3] to hundreds of megawatts, like the General Electric Hitachi (GEH) Boiling Water Reactor Experiment (BWRX-300). Advanced SMRs have several benefits, including comparatively small size, lower capital costs, flexibility in location and footprint, and the ability to add power incrementally. Additionally, as combined cycle technology advances, the thermal output efficiency of these reactors increases [4].

In addition to providing unique security, non-proliferation, and safety benefits, SMRs also thoroughly analyze and support strategies for GEN-IV, the next generation of nuclear power plants, to be used safely as a productive source of renewable energy [5].

* Canadian Nuclear Association

DOI: 10.1201/9781003590194-5

However, for several millennia of climate mitigation, nuclear energy must develop into a plentiful, essential source of non-carbon energy. Selecting the appropriate neutronic configuration is essential to bringing this about. Less than 1% of the energy stored in uranium fuel is being extracted by slow neutrons, mostly using the uncommon, readily split Uranium-235 atoms. Natural uranium contains only 0.7% of these atoms. The millions of tons of depleted uranium that remain after enrichment, the leftover nuclear fuel "waste", and the remaining fuel can all contain more than 100 times as much energy as can be extracted by fast neutrons. Today's nuclear reactor fleets, however, do not use fast neutrons to harvest energy.

In addition to explaining why, this chapter progressively reintroduces those fast neutrons into the framework of our current understanding. It offers a thorough and understandable non-specialist explanation of the main neutron interactions in various reactors with the use of graphs and diagrams. This includes the crucial inelastic scatter, which is now completely disregarded. The outcome is the understanding that current nuclear fuel waste stockpiles can be cleaned of their million-year radiotoxicity in decades and provide enormous amounts of energy for many hundreds of years with the use of fast neutrons in reactor cores.

Such fast neutron reactors and cost-effective used fuel recycling technologies are currently in place. Here, the story turns you into a near-expert guiding you into a future with carbon-free energy while taking you on a quick journey into the inner workings of atoms as a neutron [6].

Nuclear reactor kinetics deals with transient neutron flux variations that arise from a departure from the critical state, from some reactivity insertion. We have already discussed reactor kinetics in detail in Chapter 3 of this book. These circumstances can occur as a result of unintentional disruptions in the reactor's steady state functioning, environmental changes like a shift in the boron concentration, or operational changes like the motion of the control rods. The challenge of examining the time-dependent behavior of nuclear reactors is examined in this chapter.

Now this chapter will cover more details and granular aspect of nuclear energy from neutrons to fast spectrum.

5.2 PROMPT NEUTRON LIFETIME

Although we mentioned this subject briefly in Section 3.3.2 of Chapter 3 of this book, we provide some additional information here. The prompt neutron lifespan is the average time interval between a prompt neutron emission and its exit from the system or absorption (either fission or radiative capture). This parameter is supplied in both multiplying and non-multiplying systems, as discussed in Chapters 2 and 4 of this book. In both systems, the prompt neutron lifetimes (PNLs) are strongly influenced by

- The material composition of the system
- Multiplying–non-multiplying system
- The system with or without thermalization
- Isotopic composition of the system
- The geometric configuration of the system
- Homogeneous or heterogeneous system
- The shape of the entire system
- Size of the system

The total of the neutron's slowing down time (t_s) and diffusion time (t_d) in an infinite reactor without escape is the prompt neutron lifespan, which may be expressed as follows:

$$\ell = t_s + t_d \qquad \text{Eq. 5.1}$$

$t_s < t_d$ and, consequently, $\ell \approx t_d$ in an infinite thermal reactor. In thermal reactors, the usual PNL is around 10^{-4} seconds. Longer neutron lifetimes are typically seen in systems where absorption of the neutrons requires thermalization.

Shorter PNLs are seen in systems where the majority of the neutrons are absorbed at higher energy and neutron thermalization is inhibited (such as in fast reactors). Fast reactors typically have a PNL of around 10^{-7} seconds.

Refer to Table 5.1, Slowing Down and Diffusion Times for Thermal Neutrons in an Infinite Medium Reactor, for further details.

The natural energy that neutrons achieve and maintain in the ambient surroundings of the components in the reactor core, whether gas, liquid, or solid, from 20°C to approximately 400°C to 500°C, is also shown in Figure 5.1 and the analysis of power reactors focused on neutron energies in the thermal region at around 0.025 eV. This suggests that at these energies, neutrons interact very strongly with the fissile nuclei of U-235 or U-233 as well as Pu-239 or Pu-241 to split them and release their nuclear energies.

However, as shown in Figure 5.1, other nuclei are split by these neutrons less frequently – roughly 50 million times less frequently for the plentiful U-238 – and as a result, around 99% of the nuclear energy potential in mined uranium is not retrieved in such a reactor. The question now would be whether using higher-energy neutrons in the reactor cores would allow for the extraction of additional energy from the fuel.

The answer will be revealed by looking at Figure 5.1, which shows that when neutron energy increases from thermal energies of –0.025 eV to roughly 2.5 eV, the relative ratios of the neutron interactions, as shown by the fission and absorption cross-sections for the uranium fuel atoms, stay the same. As a result, neither significant changes in the relative absorption by and conversion of U-238 to fissile Pu-239 nor in the relative fission of any uranium isotope are anticipated in this energy regime [6].

The cross-sections were corrected for the relative concentrations of U-235 (0.72%) and U-238 (99.28%) in natural uranium, as Figure 5.1 illustrates the fluctuation in the respective absorption and fission cross-sections of U-235 and U-238 for neutrons of different energy.

In summary, the resonance absorption region is found at neutron energies higher than 2.5 eV, where there is a dramatic swing in cores section values for little variations in neutron energy. Such fluctuations raise the degree of uncertainty in reactor core property estimates [6].

On average, however, cross-section stability is once more seen at neutron energy values over around 10,000 eV. Furthermore, the cross-section values in this area deviate from the steady relative ratios that defined their energy dependence in the vicinity of thermal energies. This shift in relative properties from roughly 10,000 eV to the energy of freshly formed fission neutrons, which have an average energy of roughly 2 MeV, allows for substantially higher fuel usage directly and full use of U-238 through recycling.

TABLE 5.1
Thermal Neutron Diffusion and Slowing Down in an Infinite Medium Reactor

	Slowing Down Time (s)	Diffusion Time (s)
Water	5.3×10^{-6}	2.4×10^{-4}
Heavy Water	5.0×10^{-5}	6.0×10^{-2}
Beryllium	5.7×10^{-5}	4.2×10^{-2}
Graphite	1.4×10^{-4}	1.5×10^{-2}

(Source: Robert Reed Burn, Introduction to Nuclear Reactor Operation, 1988)

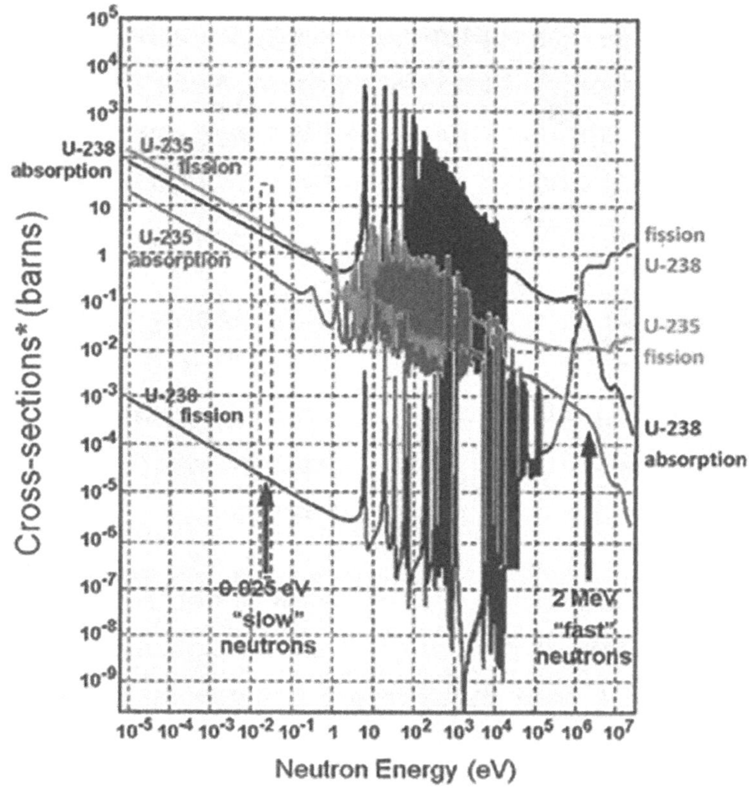

FIGURE 5.1 Variation relative absorption and fission cross-section of U-235 and U-238.

(Source: Brookhaven National Laboratory, ENF-III B Data)

This is exemplified most clearly in the Figure 5.2 and involves the interaction at a spectrum of higher energies of all four major neutron scattering events including the generally neglected inelastic scatter [6].

As indicated in Figure 5.2, the history of fuel levels of the major transuranic isotopes during the operation of a model metal-cooled reactor core over several fuel cycles and cycles of replacement with used fuel in the Canadian reactor, CANDU. The unlabeled green curves represent the build-up and partial extraction of fission products [6].

Additionally, Figure 5.2 shows that the concentration of Pu-239, the transmutation companion of U-238, initially rises with each fuel cycle. Any increasing in Pu-240 inventory in the fuel, the amount of isotope of Pu-239 is going to impurity, and lowering quality in chain reactions follows (see Figure 5.2(b)). Nevertheless, the accumulation of Pu-239 ceases when U-238 levels fall, and at that point, both U-238 and Pu-239 levels fall simultaneously, having reached a state that is nearly secular equilibrium.

Fuel must be refilled once the declining Pu-239 concentration reaches the same level as at the start of the cycle because the reactor shuts down and power levels cannot be maintained after this point. Every fuel cycle ends at this point. It is necessary to partially remove the fission products and replace them with more U-238 in order to replenish the fuel. Since the levels of fissile isotopes are the same as at the start of the previous cycle and are adequate to restart and continue the reactor's operation in a new U-238 cycle, no additional components are needed [6].

Keep in mind that CANDU, a Canadian reactor success story, is a nuclear reactor that runs on natural uranium as fuel and Pressurized Heavy Water (PHW) as a coolant and moderator. In 1954,

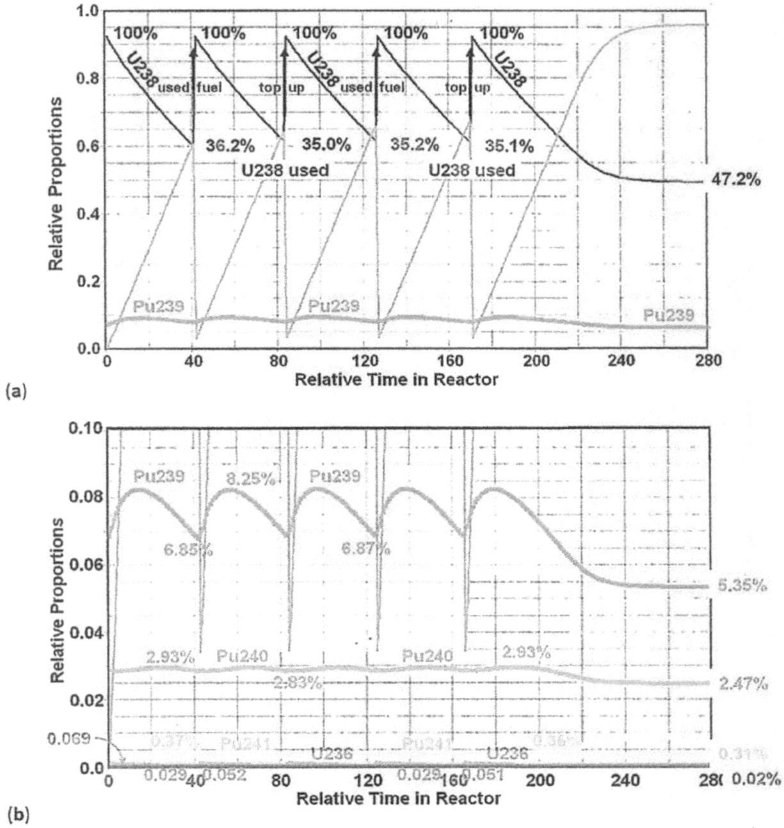

FIGURE 5.2 History of fuel levels within operational of a model metal-cooled reactor core [6].

Atomic Energy of Canada Ltd., Ontario Hydro, and the Canadian General Electric Company partnered to begin the development of the first CANDU reactor. See Figure 5.3.

As Figure 5.3 illustrates, CANDU is Pressurized Heavy Water Reactor (PHWR) nuclear power plant (NPP) and its general basic features are

1) Thermal–neutron spectrum;
2) Natural uranium dioxide (UO_2) fuel;
3) Fuel enrichment at about 0.7% U^{nat};
4) Indirect cycle using a steam generator, often known as a double-flow circuit (double loop), which also requires a pressurizer (not pictured);
5) Pressure-channel design: Figure 5.4 shows a Calandria vessel with horizontal fuel channels;
6) Reactor coolant and moderator separated, but both are heavy water;
7) Characteristics of the reactor coolant outlet: temperature around saturation (310°C) and pressure of about 9.9 MPa;
8) On-line refueling; and
9) Regenerative power cycle, subcritical pressure steam reheat in a Rankine steam–turbine cycle using light water as the working fluid and turbine steam inlet parameters of ~4.6 MPa and 259°C, respectively.

Figure 5.4 is schematic of a CANDU reactor fuel channel based on AECL design.

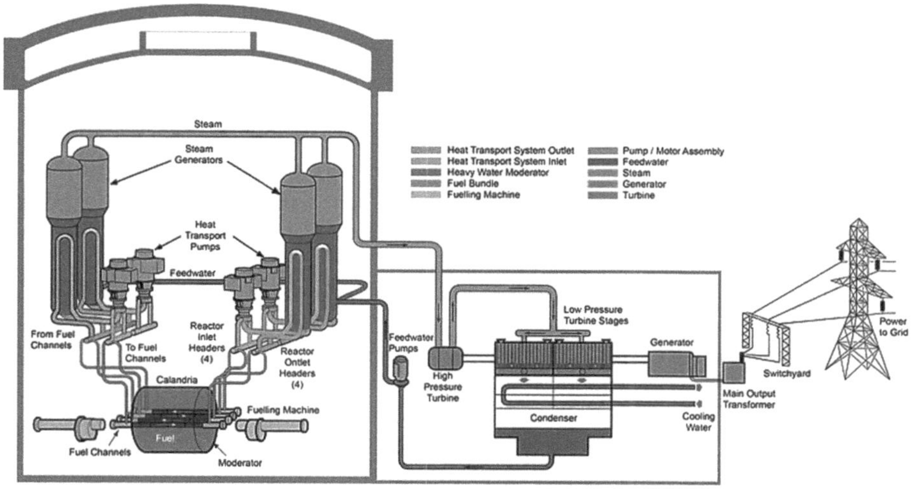

FIGURE 5.3 Schematic of CANDU-6 reactor PHWR.

(Source: Courtesy of Atomic Energy of Canada Limited (AECL))

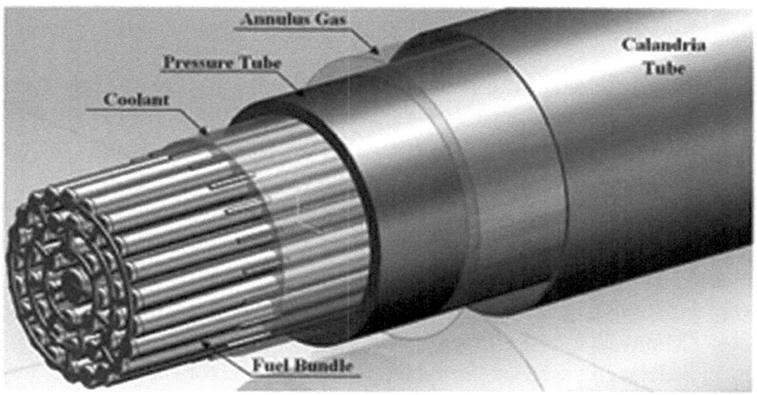

FIGURE 5.4 CANDU reactor fuel channel (based on AECL design).

(Source: Courtesy of KAERL)

As previously mentioned, the CANDU is a PWHR architecture that produces electricity in Canada. The abbreviation stands for its usage of (naturally occurring) uranium fuel and its deuterium oxide (heavy water) moderator.

5.3 PROMPT GENERATION TIME OR MEAN GENERATION TIME

A fission reaction can be started in multiplying systems by the absorption of a prompt fission neutron, which is equivalent to the average interval between two prompt neutron generations. The prompt neutron generation time is the name given to this period. The reader should consult Section 3.3.3 of Chapter 3 of this book for a detailed description of this topic. However, we are going into a little more detail about this topic in that part.

Numerous fission products from a nuclear reactor chain reaction may be regarded as possibly delayed neutron emitters. Brady and England [7] listed the 271 most significant precursors with delayed neutron emission capabilities. However, because of their extremely complex transmutation chain, low yield, and (or) short half-lives, it is very challenging to characterize all of these emitters experimentally. However, their aggregate behavior may be measured, and a few group models can be created where the abundances and decay constants are the mean values of different emitters with comparable decay constants [8].

The mean neutron generation time (Λ), which is the mean generation time between the neutron's birth and the following absorption-inducing fission, is the second significant kinetic parameter that describes the temporal behavior of the neutron population. In essence, $\Lambda$is simply the PNL (l_p) given $k \sim 1$. Numerous factors, including fuel enrichment, neutron fission cross-section, prompt neutron distribution function, average number of neutrons emitted per fission, neutron flux, and adjoint flux, affect the time it takes to generate a neutron.

A number of methods for calculating the effective delayed neutron fraction and neutron production time have been developed in the recent past. Iqbal et al. [9] used the perturbation calculations feature of CITATION to determine the effective delayed neutron fraction and prompt neutron production time of Pakistan Research Reactor-1 (PARR-1).

5.4 EFFECT OF PROMPT NEUTRON LIFETIME ON NUCLEAR SAFETY

More details on this subject can be found in Section 3.3.4 of Chapter 3 of this book, but we discuss them here as well.

Even though delayed neutrons make up a very small percentage (less than 1%) of all neutrons produced by fission, they are crucial for controlling fission chain reactions. Reactor operation would be difficult due to abrupt fluctuations in reactor power if prompt neutrons were the only thing present.

Understanding the effective delayed neutron characteristics and their decay constants is necessary for nuclear reactor control and accident analysis, as well as for converting a reactor period into reactivity. Traditionally, one has mostly depended on calculations to calculate the effective delayed neutron fraction (β_{eff}) in a given context due to the challenges in measuring it.

Keepin [10] offered a theoretical foundation for these computations in 1965, and it has been in use ever since. These often call for the computation of both a flux and an adjoint function as they entail an adjoint and spectrum weighting of the delayed neutron production rate.

The creation of a Monte Carlo estimator for $\beta_{eff}$was made possible by the addition of delayed neutrons to MCNP (Mont Carlo N-Particle), which began with version 4C in 2000 [11]. After determining if a neutron is delayed based on the delayed neutron yield for each isotope, which varies with energy, MCNP-4C samples the energy of the neutron from the relevant spectrum. We can compute the effective delayed neutrons fraction using a Monte Carlo approach thanks to this two-step sampling procedure. Numerous techniques have been put forth to calculate β_{eff} using the Monte Carlo approach. The works by Meulekamp and Van der Marck [12], Nauchi and Kameyama [13], and Spriggs et al. [14] provide detailed descriptions. The majority of these techniques require just minor adjustments to the MCNP source code.

It is only possible to utilize the so-called prompt approach [12] without changing the normal MCNP code. Meulekamp and Van der Marck [12] showed that for a set of different benchmarks, the prompt and the "Meulekamp–Marck" technique yield comparably satisfactory findings of β_{eff}. However, because it needs to run around 40 times longer, the prompt technique takes a lot longer. With the advent of faster computer hardware, the usual MNCP input computation time is quickly reducing (it is halved approximately every 18 months), therefore, this is not a significant concern. With the exception of heterogeneous systems [12], the *k*-ratio technique [14] also produces satisfactory results.

Additionally, it has been demonstrated that the variations in β_{eff} values computed using various datasets (JEFF-3.0, ENDF/B-VI, and JENDL-3.3) can occasionally be greater than the variations between the three approaches.

5.5 POWER DISTRIBUTION IN CONVENTIONAL REACTORS

The flux shape obtained from the diffusion theory is only a speculative scenario in a homogeneous, uniform cylindrical reactor operating at low power levels (also known as "zero power criticality"). We have informally thought that some effective homogenous mixture may represent the core, which is made up of thousands of fuel and control components, coolant, and structure. Because it ignores a core's heterogeneity, this is a very strong presumption. Refer to the portions in this book's Chapter 2 that discuss the "Diffusion Equation-Finite Cylinder", which is the solution for the finite cylindrical homogeneous reactor shown in Figure 5.5.

5.5.1 Heterogeneous Core

A large number of PWRs run on uranium fuel, which is uranium dioxide. The dark, semiconducting, solid uranium dioxide has extremely poor heat conductivity. However, uranium dioxide is known to behave in a certain way and has a very high melting point. After being compressed into pellets, the UO_2 is sintered into the solid. Refer to Figure 5.6.

Then, because zirconium alloys have a very low absorption cross-section (unlike stainless steel), these pellets are loaded and encased within a fuel rod (or fuel pin). The surface that covers the pellets is called fuel cladding. Fuel assembly, the foundational component of the nuclear reactor core, is the grouping of fuel rods or components. Depending on the reactor type, there are around 157 fuel assemblies in the reactor core (PWR type). Western PWRs employ a square lattice configuration, and assemblies are distinguished by the number of rods they include; in modern designs, this is usually 17 × 17. Fuel rod enrichment is never consistent. Both axial and radial differentiation of the enrichment can be present. The power distribution is greatly impacted by this configuration.

The Russian Voda Voda Energo Reactor (VVER) type reactor (shown in Figure 5.7) is comparable in length and construction to other PWR fuel assemblies, but it employs fuel that is distinguished by its hexagonal configuration. "Water–water energy reactor" (also known as "water-cooled water-moderated energy reactor") is what the Russian acronym VVER stands for. It is a modified Pressurized Water Reactor (PWR) design. The main primary characteristics that set the VVER apart from other PWRs are

- Horizontal steam generators
- Hexagonal fuel assemblies

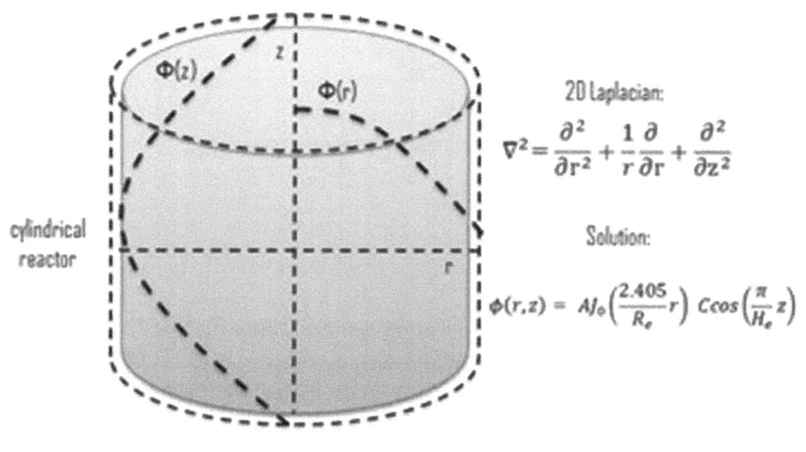

FIGURE 5.5 Solution for the finite cylindrical homogeneous reactor.

(Source: www.wikipedia.org)

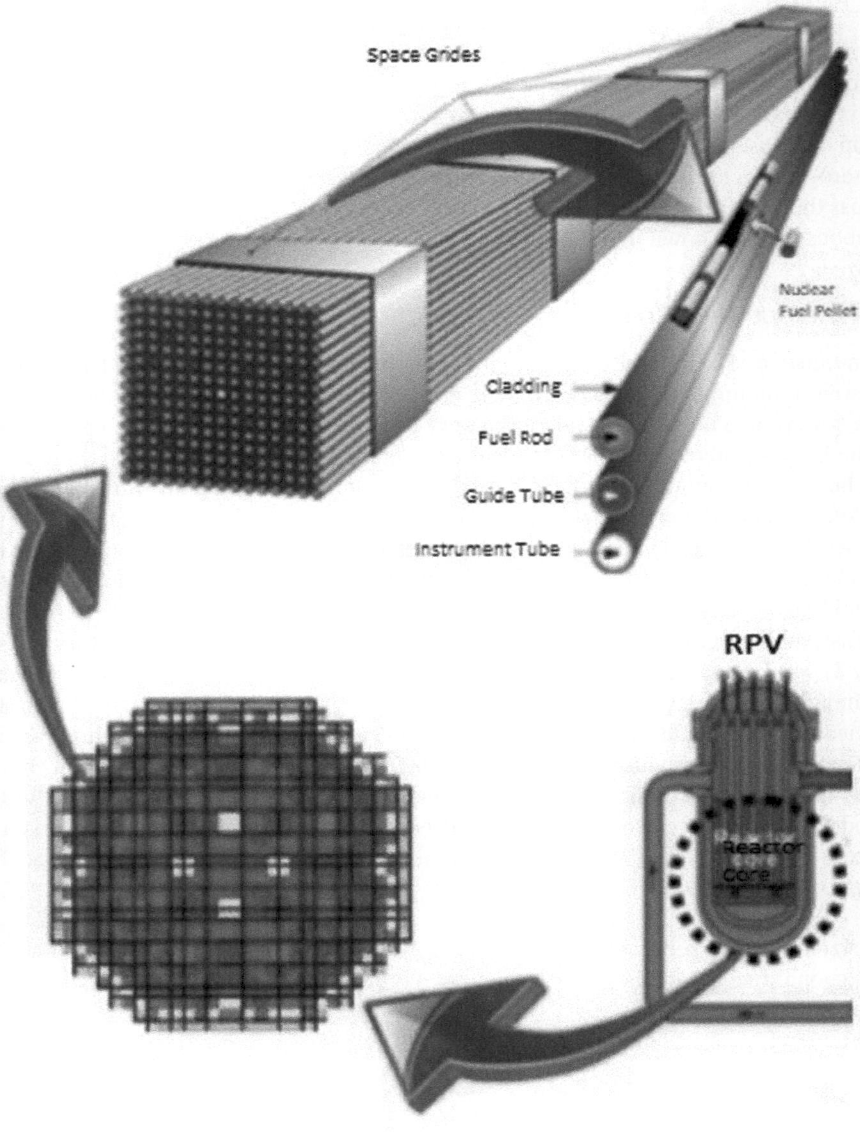

FIGURE 5.6 Uranium oxide (UO_2) fuel bundle assembly infrastructure.

(Source: www.nuclear-power.com)

- No bottom penetrations in the pressure vessel
- High-capacity pressurizers providing a large reactor coolant inventory

In order to prevent boiling at the typical operating temperatures of 220°C to over 320°C, reactor fuel rods are completely submerged in water maintained at pressures of 12.5, 15.7, and 16.2 MPa, respectively. An essential safety component of the reactor is the water's dual function as a moderator and coolant. A situation known as a negative void coefficient occurs when coolant circulation fails, reducing the neutron moderation impact of the water. This is because the increased heat produces steam bubbles that do not moderate neutrons, lowering reaction intensity and making up for the loss

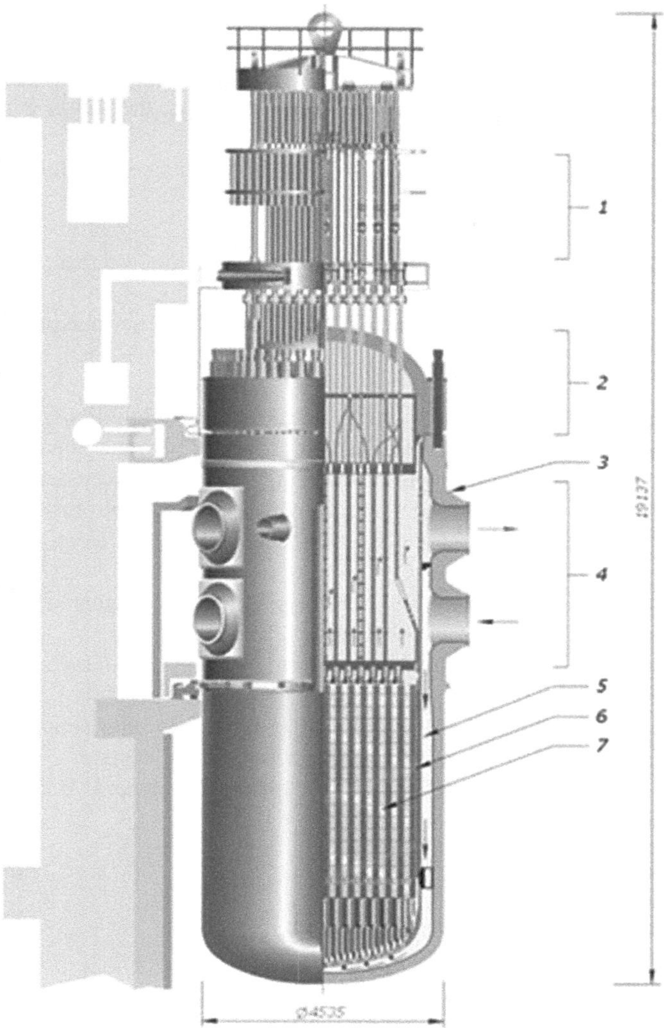

FIGURE 5.7 A WWER-1000 (or Voda Voda Energo Reactor (VVER)-1000).

(Source: www.wikipedia.org)

of cooling. Large steel reactor pressure containers surround later iterations of the reactors. Uranium dioxide (UO_2) or its equivalent, low-enriched (2.4–4.4%) U-235, is compressed into pellets and assembled into fuel rods.

Control rods, which may be placed into the reactor from above, regulate reactivity. Depending on how deeply they are inserted, these rods, which are composed of a substance that absorbs neutrons, might impede the chain reaction. The control rods can be fully inserted into the core to shut down the reactor in an emergency.

As previously mentioned, the water in the VVER's principal cooling circuits is maintained at a consistently high pressure to prevent boiling. The integrity of this circuit is vital as the water is exposed to radiation and transmits all of the heat from the core. One can differentiate four primary components.

1. **Reactor Vessel**: The nuclear chain reaction heats the fuel assemblies as water passes through them.

2. **Volume Compensator (Pressurizer)**: The volume compensator uses electrical heating and relief valves to adjust the balance between saturated steam and water, maintaining the water at a constant but regulated pressure.
3. **Steam Generator**: The water in the secondary circuit of the steam generator is boiled using the heat from the primary coolant water.
4. **Circulating Pump**: The pump makes sure that the water flows through the circuit in the right way.

Redundancy is built into the primary cooling system to ensure that the reactor core can continue to cool in an emergency.

The secondary circuit also comprises many subsystems from the perspective of the electrical output and secondary coolant loop.

1. **Steam Generator**: Heat from the primary circuit is used to boil the secondary water. The leftover water is extracted from the steam to make it dry before it enters the turbine.
2. **Turbine**: A turbine is powered by the expanding steam and is connected to an electrical generator. There are portions of the turbine with high and low pressure. Between these parts, steam is warmed to increase efficiency. One gigawatt of electricity is produced by VVER-1000 type reactors.
3. **Condenser**: Waste heat is released into a cooling loop when the steam cools and condenses.
4. **Deaerator**: It removes gases from the coolant.
5. **Pump**: Each of the circulation pumps is powered by a tiny steam turbine.

Before the deaerator and steam generator, coolant is reheated using steam from the turbine to improve process efficiency. This circuit's water should not be radioactive.

A WWER-1000 (or VVER-1000 as a direct transliteration of Russian В В Э Р-1000), a 1000 MWe Russian nuclear power reactor of PWR type.

1. Control rod drives
2. Reactor cover or vessel head
3. Reactor pressure vessel
4. Inlet and outlet nozzles
5. Reactor core barrel or core shroud
6. Reactor core
7. Fuel rods

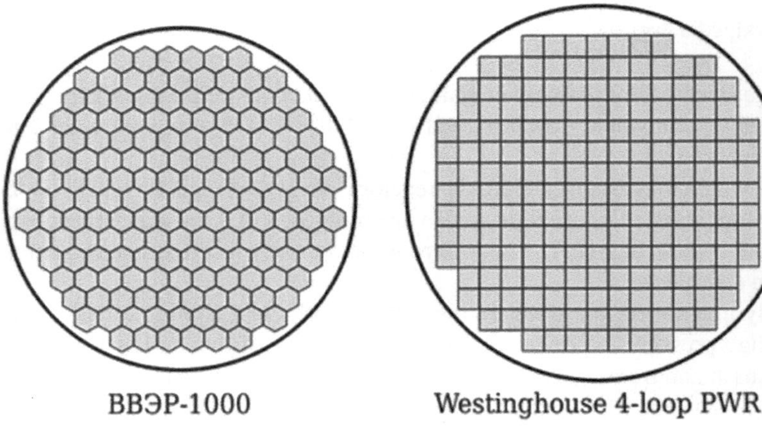

FIGURE 5.8 Fuel assembly infrastructure between Westinghouse PWR and VVW.

(Source: www.wikipedia.org)

One notable comment as illustrated in Figure 5.8 is the arrangement of hexahedral fuel assemblies compared to a Westinghouse PWR design. Note that there are 163 assemblies on this hexahedral arrangement and 193 on the Westinghouse arrangement.

5.5.2 Heterogeneous Core

Using Figures 5.9 and 5.10, the following factors have a major impact on the flux distribution in commercial reactor cores:

- **Heterogeneity of Fuel–Moderator Assembly**: Particularly in heterogeneous reactor cores, the shape of the core has a significant impact on the energy and spatial self-shielding. In summary, the unit cell's heterogeneous shape causes the neutron flow to fluctuate. Because of the large absorption cross-sections of the fuel nuclei, the flux in the fuel cell will be different (lower) than in the moderator cell. When compared to homogeneous cores, this phenomenon results in a notable increase in the *resonance escape probability* ("p" from the four-factor calculation).
- **Reactivity Feedback at Power Operation**: The flux distribution flattens as a result of the reactivity feedback since they act more strongly on locations with greater flux (i.e., above 1% of rated power). In commercial power reactors, the distribution of neutron flux is dependent on several other parameters, including the fuel loading pattern, the position of the control rods, and the possibility of short-term oscillations (for example, because of the spatial distribution of xenon nuclei). When the commercial power reactor is operating, it has neither cosine nor J_0.
- **Fuel Loading Pattern**: The fact that the core of PWR fuel cycles contains several fuel assemblies is their defining characteristic. Due to the possibility of varying enrichment and burnup, these assemblies exhibit distinct multiplication characteristics. A typical fuel assembly usually has enough energy to run at full capacity for around 4 years. Depending on how the working cycle is designed, the fuel remains in the core for 4 years after it is loaded. The reactor core has to be refueled over these 4 years. A portion of the fuel, often one-third or one-quarter of the core, is transferred to the spent fuel pool during refueling, which occurs every 12 to 18 months. Meanwhile, the rest is moved to a part of the core that is more appropriate for its remaining enrichment level. It is necessary to replace the withdrawn fuel (one-third or one-quarter of the core, or 40 assemblies) with new fuel assemblies.

Numerous fuel loading scenarios (patterns) have been examined, and it has generally been determined that the flatter the power distribution in the core, the more energy is taken from the fuel. These loading techniques can, in theory, be separated into two groups.

- **Out–In Loading Patterns**: The intermediate and high burnup batches are positioned in the middle of the core, while the fresh fuel batch is positioned at the core's perimeter in the out–in loading pattern. A new batch is loaded at the perimeter, the other batches are moved inside, and the batch with the highest burnup is released upon refueling. Because the power distribution is peaked at the perimeter and depressed in the core, it has been discovered that the out–in loading pattern goes too far. Producing a large number of fast neutrons at the periphery that leak from the core and harm the pressure vessel is another challenge.
- **In–Out Loading Patterns**: Core designers provide minimal leakage loading patterns to improve neutron and fuel efficiency. The outlying locations of the reactor core do not house fresh fuel assemblies. The gasoline with the greatest fuel burnup is placed in the peripheral places. These "high" burnup assemblies have a lower relative power by nature than ordinary assemblies (because of the lower and k_{eff} because they sense the presence of

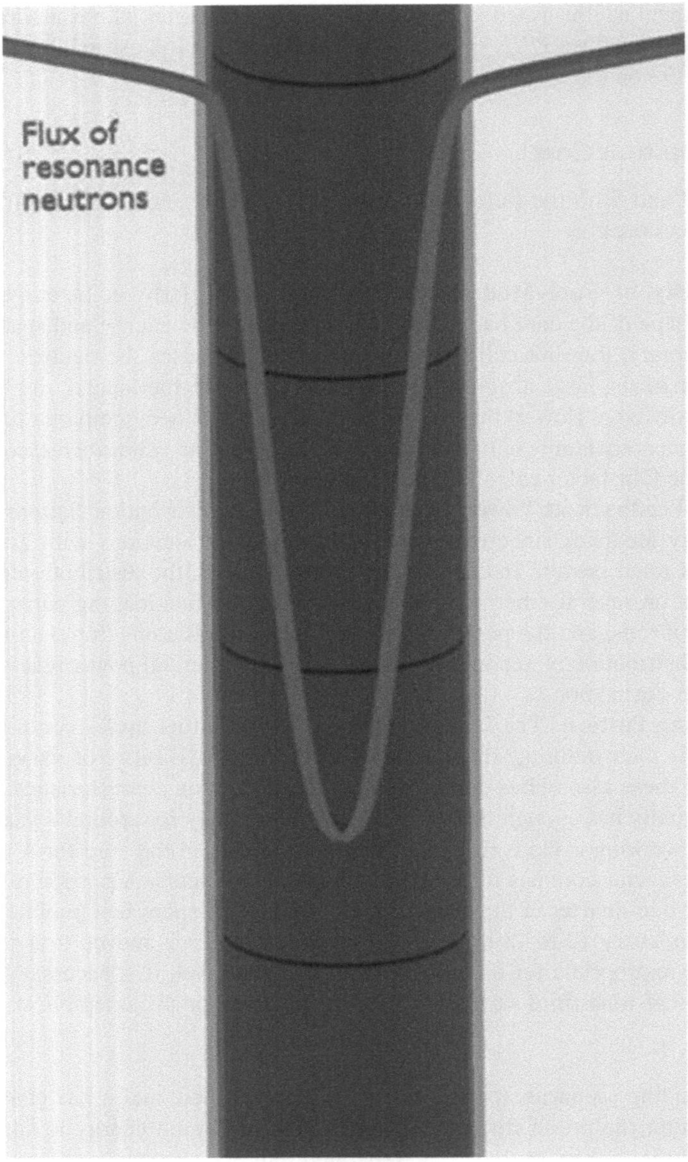

FIGURE 5.9 Flux of resonance neutrons.

(Source: www.wikipedia.org)

a non-multiplying environment – reflector). The flux distribution at the core's perimeter rises when fuel is depleted, particularly in patterns with low leakage loading. Reducing the variations in v between peripheral high-burnup assemblies and new fuel assemblies is what causes this process. These loading patterns achieve somewhat greater peaking factors than out–in loading patterns because of the low relative power of the peripheral assemblies. However, when refueling, less fresh or enriched fuel may be loaded due to improved neutron and fuel efficiency. Another advantage is that the neutron flux on the pressure vessel is decreased by packing the "high" burnup components at the perimeter. This gives

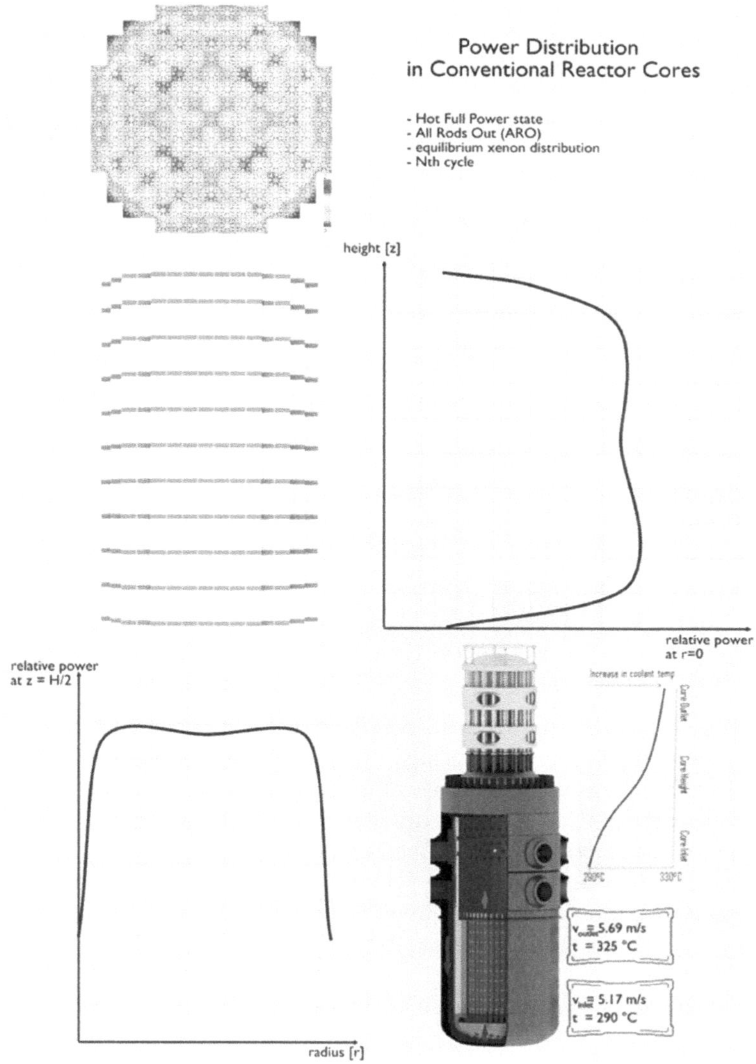

FIGURE 5.10 Power distribution in conventional reactor cores.

(Source: www.wikipedia.org)

the reactor vessel further defense against irradiation embrittlement, which is particularly brought on by rapid neutrons.

- **Burnable Absorbers (Burnable Poisons)**: Burnable absorbers (BAs) have a major impact on the power distribution pin-by-pin. BAs are substances that undergo radiative capture and change from having a high neutron absorption cross-section to having a relatively low absorption cross-section. Over core life, the BA's negative reactivity reduces as a result of the absorption material burning up. As the extra positive reactivity in the fuel is being reduced, these absorbers should ideally reduce their negative reactivity. BAs are employed in PWRs to lower the relative power of new fuel assemblies and lower the initial concentration of boric acid, which also lowers BOC MTC. Typically, fixed BAs are made of boron or gadolinium compounds that are added to fuel or formed into distinct lattice pins or plates.

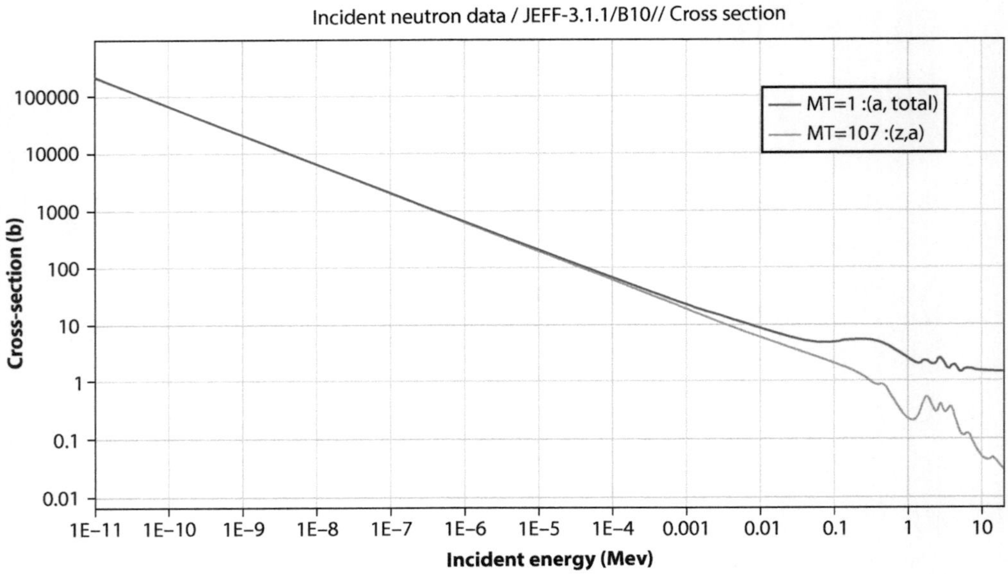

FIGURE 5.11 Incident neutron data/JEFF 1 1 ^{10}B cross-section.

(Source: U.S. Department of Energy); JANIS (Java-based Nuclear Data Information Software); The JEFF-3.1.1 Nuclear Data Library

These poisons are less disruptive to the core power distribution than control rods because they can often be applied more uniformly.

- **Boron as Burnable Absorber**: Because of the large neutron cross-section of isotope ^{10}B, boron is frequently utilized as a neutron absorber in the nuclear industry. For thermal neutrons, its (n, α) reaction cross-section is around 3,840 barns (for 0.025 eV neutrons). The absorption cross-section of isotope ^{11}B for thermal neutrons is around 0.005 barns (for 0.025 eV neutrons). Most thermal neutrons' (n, α) reactions are ^{10}B (n, α) ^{7}Li reactions, followed by 0.48-MeV gamma emission. Refer to Figure 5.11.

Boron-10. Comparison of total cross-section and cross-section for (n, α) reactions.

$$^{10}B + n_{th}(0.025eV) \longrightarrow {}^{4}He^{2+} + {}^{7}Li^{3+} + 2.79MeV(6\%)$$

$$^{10}B + n_{th}(0.025eV) \longrightarrow {}^{4}He^{2+} + {}^{7}Li^{3+} + 2.79MeV(6\%)$$

Moreover, ^{10}B isotope exhibits a high (n, α) reaction cross-section. At high energies, the cross-sections of the majority of other elements, including cadmium, get extremely tiny. As energy increases, the cross-section of ^{10}B drops monotonically. Its cross-section is around the size of a barn for fast neutrons. Another advantage of boron is that it absorbs neutrons. Lithium and helium, the reaction products (after a neutron absorption), are stable isotopes.

As a result, the reactor core's usage of BAs or decay heating control rods presents a few issues including the potential for reactivity swings during operation, challenges in long-term fuel management, and added complexity in thermal-hydraulic and safety analyses. However, when utilized as the absorbent material in control rods, helium generation may result in a notable rise in pressure (under rod cladding). Furthermore, as interactions with neutrons can sporadically result in the

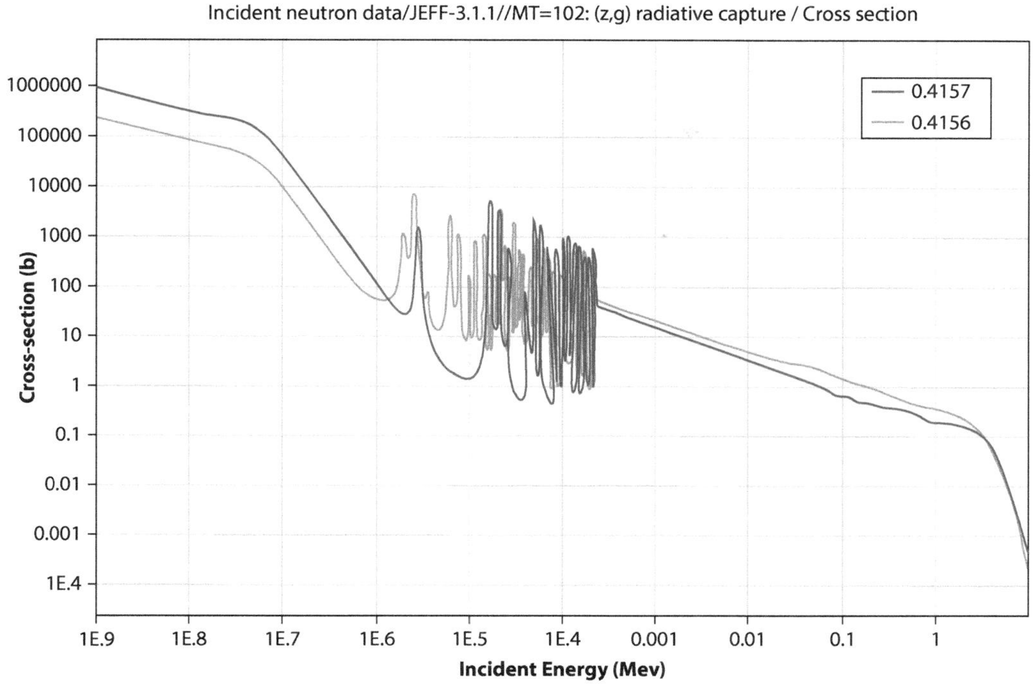

FIGURE 5.12 Gadolinium-155 and 157. Comparison of radiative capture cross-sections.

(Source: JANIS (Java-based Nuclear Data Information Software); The JEFF-3.1.1 Nuclear Data Library)

generation of radioactive tritium, ^{10}B is the main source of this element in the primary circuit of all PWRs (which utilize boric acid as a chemical shim).

- ^{10}B (n, 2 x α) ^{3}H Threshold reaction (~1.2 MeV)
- ^{10}B (n, α), ^{7}Li (n, α) ^{3}H Threshold reaction (~3 MeV)
- An absorbent that may be burned is gadolinium. Because of the extremely large neutron absorption cross-section of the two isotopes ^{155}Gd and ^{157}Gd, gadolinium is frequently utilized as a neutron absorber in the nuclear industry. Of all the stable isotopes, they have the largest absorption cross-sections. For thermal neutrons (for 0.025 eV neutrons), ^{155}Gd has 61,000 barns, whereas ^{157}Gd has even more, 254,000 barns. Because of this, gadolinium is frequently utilized as a BA in new fuel to offset the reactor core's excessive reactivity. When compared to other BAs, gadolinium exhibits the behavior of a fully black substance. As a result, gadolinium works wonders to offset excessive reactivity. However, an uneven neutron-flux density in the reactor core might result from an incorrect distribution of Gd-burnable absorbers. Refer to Figures 5.12 and 5.13.

Figure 5.13 illustrates this on boron letdown curves. The critical concentration of boric acid in the reactor core with BAs (red curve) and the reactor core without BAs (blue curve) at the start of a given fuel cycle differs substantially. The amount of BAs utilized determines the difference.

- A neutron reflector is present. Many neutrons that might otherwise escape are scattered back (or reflected) into the core by the neutron reflector, which lowers neutron leakage. The reflector raises k_{eff} and lowers the quantity of fuel required to keep the reactor critical for an extended amount of time by decreasing neutron leakage. By increasing the ratio of the

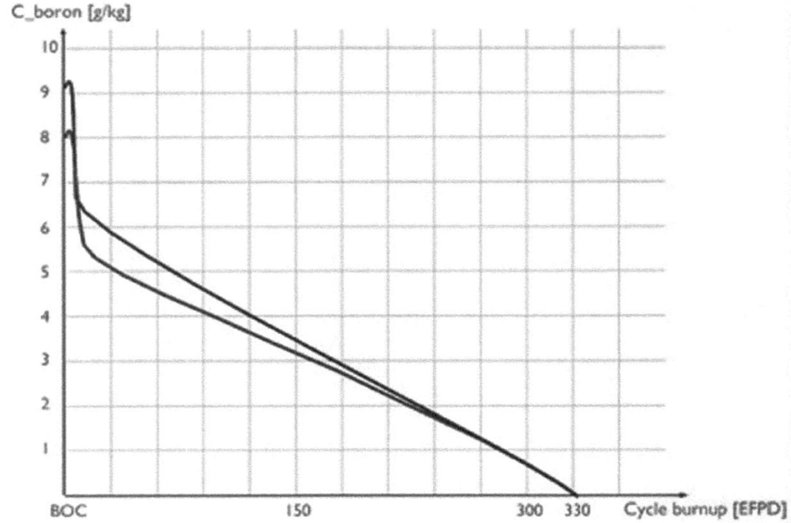

FIGURE 5.13 The effect of gadolinium burnable absorbers (BAs).

(Source: www.wikipedia.org)

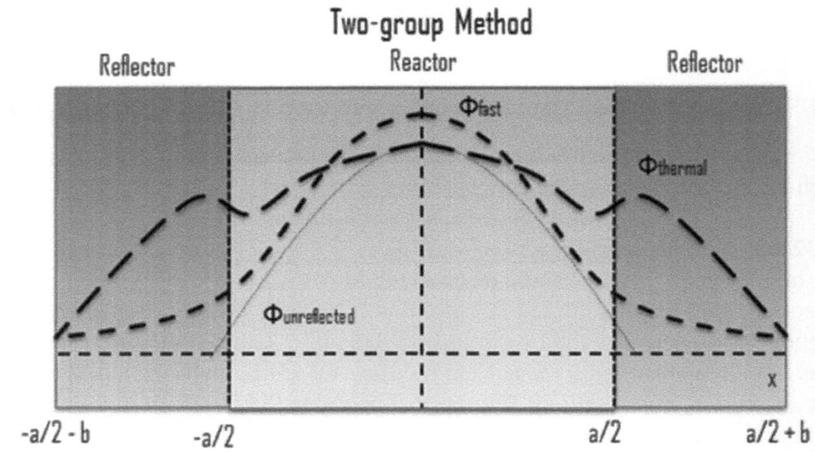

FIGURE 5.14 Presence of neutron reflector illustration.

(Source: www.wikipedia.org)

average flux to the highest flux, the neutron flux distribution is "flattened". Reflectors hence lessen the electricity distribution's irregularity. Refer to Figure 5.14.

The overall impact of reflection in the thermal reactor system is depicted in this graphic. Keep in mind that without altering the peak power, a reflector can boost the core–periphery power density and, consequently, the core average power level.

5.5.3 Operational Factors Affecting the Power Distribution

The power distribution as an operational factor is impacted by the following actions and processes.

5.5.3.1 Effect of Fuel Depletion

The amount of fissile material in the fuel assembly continuously drops while the reactor is operating. As a result, the assembly k_{inf} continuously drops. The assembly k_{inf} will drop the most where the power is highest during fuel depletion. High-burnup assemblies and new fuel assemblies have smaller k_{inf} discrepancies. Consequently, this mechanism will lead the power to migrate away from places with the greatest k_{inf} during cycle depletion. The initial assumption regarding the continuously declining assembly k_{inf} is likewise disrupted by this procedure, which relies on the usage of burnable absorbers.

5.5.3.2 Effect of Control Rods

An essential component of nuclear reactor safety and control systems is control rods. They are crucial because they act and respond to the reactor quickly. In a nuclear reactor, control rods are employed to keep the chain reaction in the appropriate state (subcritical, critical, or supercritical). They are an essential part of a SCRAM (emergency shutdown system).

To alter the reactor's reactivity (an increase or a decrease in the neutron flux) during startup and power operating mode, control rods are taken out of or put into the reactor core. Variations in neutron power are carried out by variations in reactivity. The thermal power of the reactor, the volume of steam produced, and ultimately the quantity of electricity produced are all impacted by this. Both axially and radially, this movement affects the neutron flux distribution. Although the flux depression is inherently greater close to the control rod that has been inserted, the motions of the control rod also have an impact everywhere (e.g., affect axial flux differential).

5.5.3.3. Effect of Flow Rate

The consequences listed below apply to reactors that use pressurized water. Variations in the intensity of boiling in channels are linked to variations in the flow rate in boiling reactors, making these impacts a more complicated problem. Refer to Figure 5.15.

The axial power distribution in PWRs is significantly impacted by changes in the flow rate via the primary loop. Nevertheless, it is uncommon for PWRs to alter the core's flow rate while power is operating.

In practice, the difference between inlet and outlet temperatures must rise when the reactor power is constant (e.g., at decreased power) and the flow rate abruptly changes (e.g., because the reactor coolant pump is disconnected). From the reactor coolant's fundamental energy equation, which is shown below.

$$P = \dot{m} \cdot c \cdot \Delta t \qquad \text{Eq. 5.2}$$

The steam generators' intake temperature is determined by the pressure. As a result, during the transient, the inlet temperature barely changes. As the flow rate varies, it follows that the output temperature must also shift dramatically. It makes sense that the average coolant (moderator) temperature would fluctuate considerably when the entrance temperature stayed relatively constant while the exit temperature changed dramatically. It occurs when the upper half of the core's temperature rises (as the flow rate decreases) more than the bottom half's temperature. The power from the top half will go to the lower half because the moderator temperature feedback has to be negative.

Be aware that the difference in normalized flux signals (axial flux difference; AFD) between the top and bottom parts of a two section excore neutron detector is known as the AFD, and it will decrease. $\Delta I = \uparrow \text{top} - \downarrow \text{bottom}$, where $\uparrow$ top and $\downarrow$ bottom are given as a proportion of rated

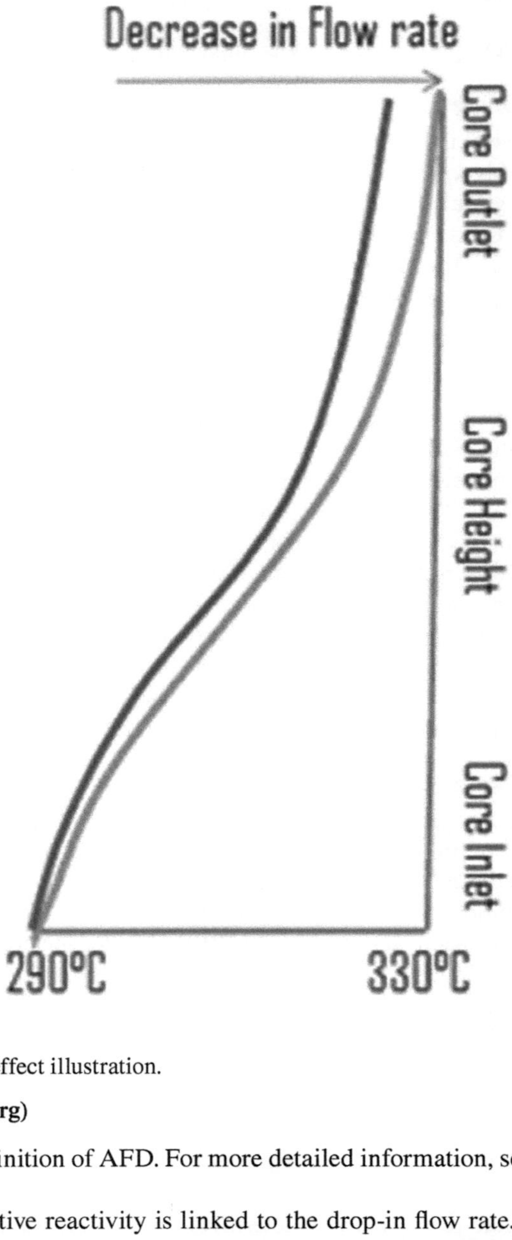

FIGURE 5.15 Flow rate effect illustration.

(Source: www.wikipedia.org)

thermal power, is the definition of AFD. For more detailed information, see Figure 5.16 and Section 5.6 below.

The insertion of negative reactivity is linked to the drop-in flow rate. If the flow rate suddenly increases (positive reactivity insertion), further care must be used. With the exception of the controlled reactor coolant pump connection, which can only be made under certain circumstances, such an increase in flow rate is not possible during regular operation.

5.5.3.4 Effect of Xenon Oscillations

See Figure 5.16, which illustrates the xenon-135 neutron absorption and scattering comparison cross-section situation.

A variety of effects on nuclear reactor performance are produced by the combination of high neutron capture cross-section and delayed generation. The following four nuclear reactor operation phases outline the methodology.

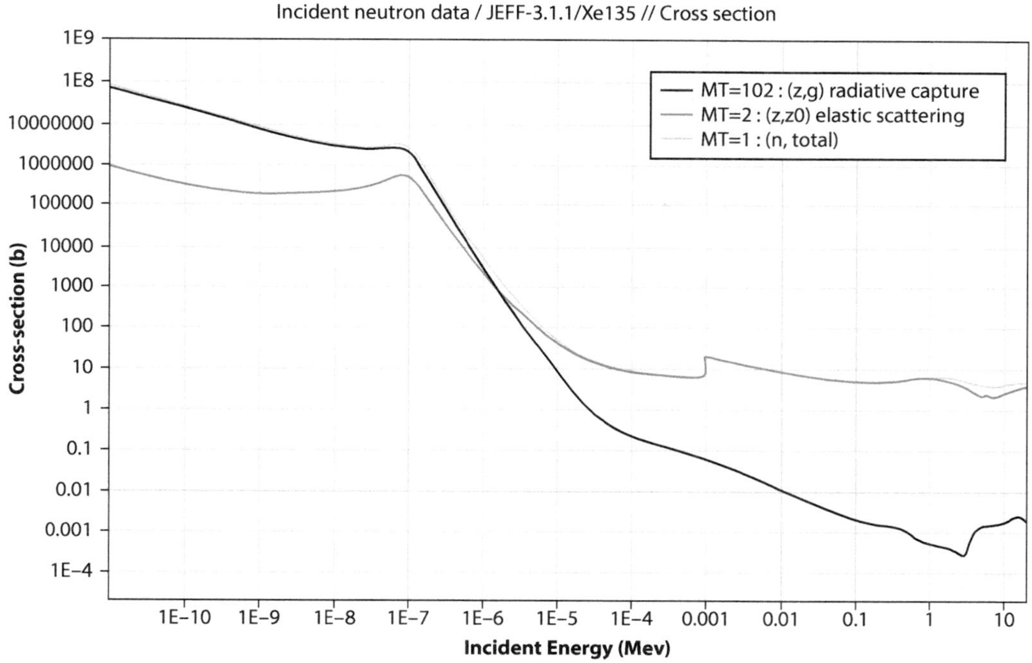

FIGURE 5.16 Xenon-135. Neutron absorption and scattering. Comparison of cross-sections.

(Source: JANIS (Java-based Nuclear Data Information Software); The JEFF-3.1.1 Nuclear Data Library)

1. An imbalance in fission rates within the reactor core, and consequently in the iodine-135 buildup and the xenon-135 absorption, results from an initial lack of symmetry (that is, the axial symmetry in the case of axial oscillations) in the core power distribution (for instance, due to significant control rod movement).
2. In xenon-135, burnout permits the flux to rise more in the high-flux zone, but an increase in xenon-135 results in a greater decrease in flux in the low-flux region. Where the flow is high, the iodine concentration rises, and where the flux is low, it falls. This change in the xenon distribution will improve the flux tilt by increasing (reducing) the multiplication characteristics of the area where the flux has increased (decreased).
3. Decomposition to xenon restores the original state as soon as the iodine-135 levels are high enough. Here, flux falls and power rises in the formerly low-flux zone.
4. When these patterns are repeated, xenon oscillations with durations of around 24 hours may start to move throughout the core.

These oscillations have the ability to drastically alter local power levels while causing minimal changes to the overall power level. The oscillations of xenon-135 are easily damped in a reactor system with high negative temperature coefficients. Reactors are designed with negative MTCs in part because of this. Local power distribution is impacted by this effect as it affects the core's global power distribution. However, the initial fluctuation in flux levels, which pushes the flux higher, is the issue. Large xenon spatial oscillations must be kept to a minimum in order to avoid exceeding F_Q restrictions since the local linear heat rate ($F_Q(z)$ limit) is quite limiting at higher heights.

To regulate xenon spatial oscillations, either the axial offset or the AFD is applied. The AFD restrictions are the most significant of them. AFD calculates the power or flux imbalance between the core's upper and lower portions. The outputs of the upper and lower excore neutron detectors, which are part of the so-called excore Nuclear Instrumentation System (NIS), are used to calculate the AFD.

$$\text{AFD or } \Delta I = I_{top} - I_{bottom} \qquad \text{Eq. 5.3}$$

where I_{top} and I_{bottom} are expressed as fractions of rated thermal power.

5.5.3.5 Effect of Thermal Power

Variations in the reactor's thermal output can cause a major shift in the power distribution. Temperature reactivity effects are crucial when power fluctuates at power operating mode (i.e., from roughly 1% to 100% of rated power).

The temperature of the fuel and moderator rises with an increase in the neutron population, causing the reactor's reactivity to decrease (almost all reactors are built with negative temperature coefficients).

This positive reactivity is counteracted by negative reactivity from temperature feedback, and the negative reactivity coefficient operates against the original positive reactivity insertion. Both local and global scales naturally experience this influence.

The temperature feedback works best when the power is highest during a thermal power rise. Because the feedback acts more strongly on locations with higher flux, this process flattens the flux distribution.

It should be mentioned that the axial power distribution is significantly impacted by changes in thermal power.

In practice, the difference between the inlet and exit temperatures must rise as the thermal power changes while the coolant flow rate stays constant. It is derived from the reactor coolant's fundamental energy equation, which is expressed in the form of Equation 5.2.

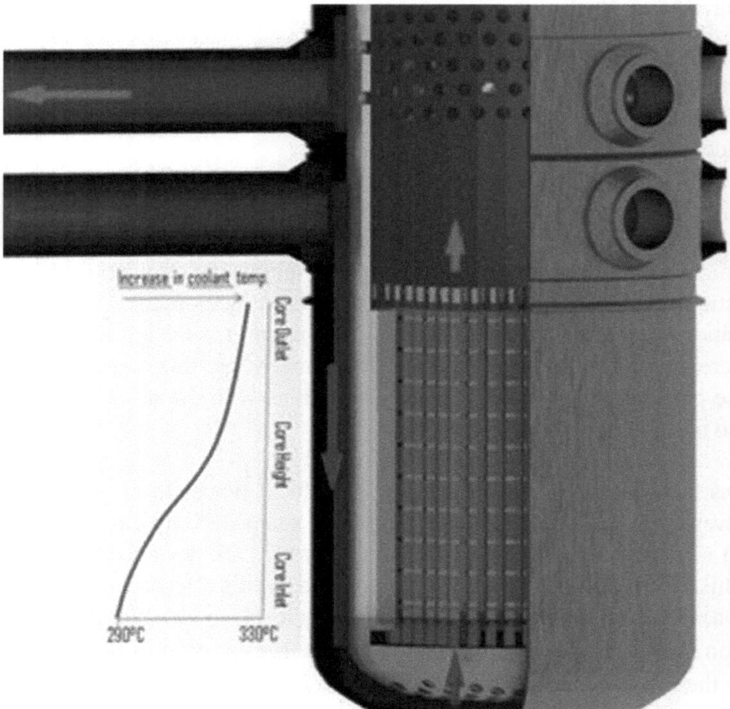

FIGURE 5.17 Half core infrastructure.

(Source: U.S. Department of Energy)

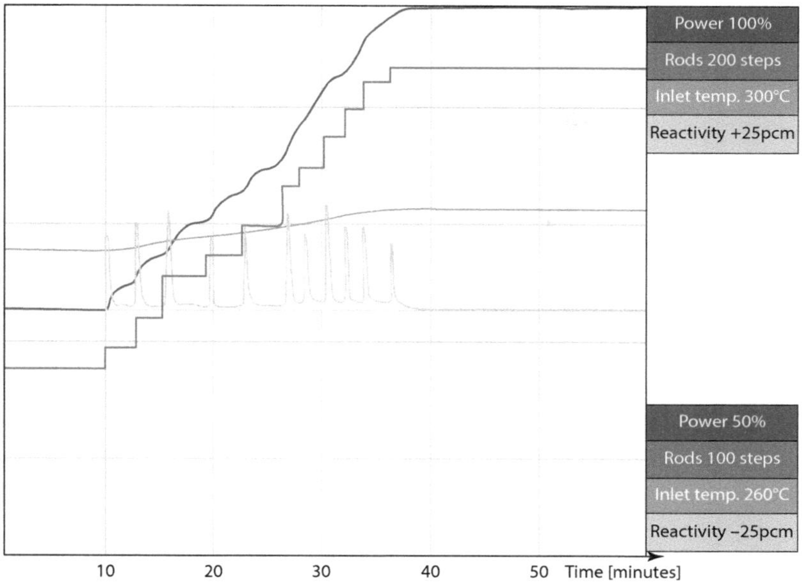

FIGURE 5.18 Power increase curve.

(Source: U.S. Department of Energy)

As can be seen from Figure 5.17, the temperature of the upper half of the core rises higher than that of the lower half. The power from the top half will go to the lower half because the moderator temperature feedback has to be negative.

The steam generator's pressure determines the intake temperature. As a result, when the thermal power varies, the input temperature barely changes. As the thermal power varies, it follows that the exit temperature must also fluctuate dramatically. It makes sense that the average coolant (moderator) temperature would fluctuate considerably when the entrance temperature stayed relatively constant while the exit temperature changed dramatically. As a result, the temperature of the upper half of the core rises higher than that of the lower half.

See Figure 5.18, where we assume that the reactor is critical at 75% of rated power and that the operator wants to increase power to 100% of rated power.

The power from the top half will go to the lower half because the moderator temperature feedback has to be negative. To put it briefly, the top half of the core is poorly moderated since it is chilled (moderated) by hotter coolant. The difference in normalized flux signals (AFD) between the top and bottom portions of a two-section excore neutron detector is known as the AFD, and it will therefore decrease. Equation 5.3 is the definition of AFD.

Once more, we might suppose that the plant operator wishes to raise power to 100% of rated power and that the reactor is critical at 75% of rated power.

5.6 AXIAL FLUX DIFFERENCE – AXIAL OFFSET CONTROL

Because the reactivity feedback operates more strongly on spots where the flux is larger, the flux distribution flattens at power operation (i.e., over 1% of rated power). As a result, variations must exist in an axial direction as well.

Assume, for instance, that the input temperature (T_{in}) is established by the steam generators' pressure. As a result, when the thermal power varies, the input temperature fluctuates very little. It follows that when the thermal power rises, the output temperature must also rise noticeably. It makes sense that the average coolant (moderator) temperature would fluctuate considerably when

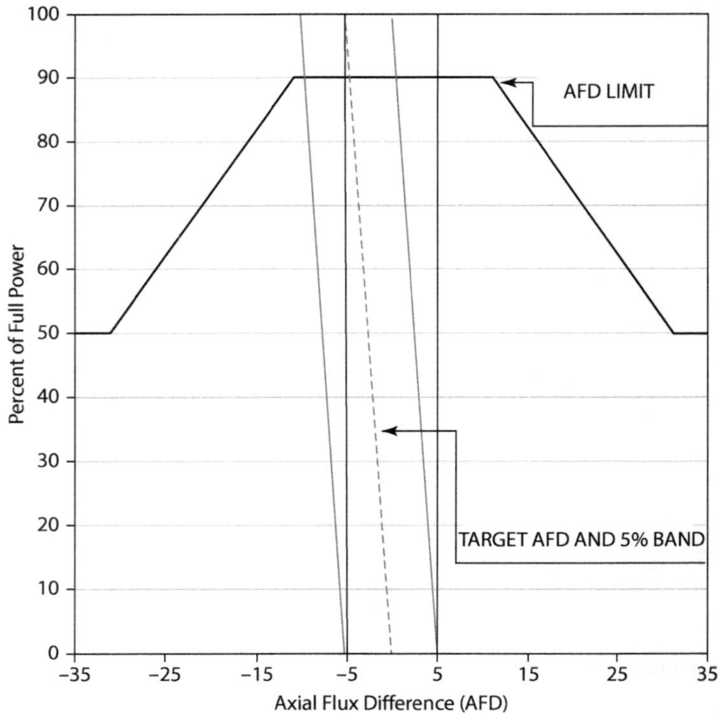

FIGURE 5.19 Axial flux difference (AFD) depiction.

(Source: U.S. Department of Energy)

the entrance temperature stayed relatively constant while the exit temperature changed dramatically. In the event of a power increase, it follows that the temperature of the upper half of the core rises more than the temperature of the bottom half. The power from the top half will go to the lower half because the moderator temperature feedback has to be negative. As a result, as shown in Figure 5.19, the axial flux differential – which is the difference in normalized flux signals (AFD) between the top and bottom portions of a two-section excore neutron detector – will decrease.

AFD is defined by Equation 5.3, where again I_{top} and I_{bottom} are expressed as fractions of rated thermal power, as an example, assuming the core is operating at 100% power. When the $I_{top} = 49\%$ and $I_{bottom} = 51\%$, the resulting $\Delta I = -2\%$. When the delta flux target is −5%, the core would then operate with a delta flux from −10% to 0%. The control rods can shift somewhat due to the +5% and −5% flux differential surrounding the target.

The AFD is kept within the designated bounds when the electricity is on. A very top- or bottom-skewed axial power distribution is avoided by adhering to the AFD restrictions.

Core peaking variables align with the safety studies' underlying assumptions by reducing the degree of power distribution skewing. One important aspect of axial power distribution management is xenon distribution skewing, which may be reduced by gradually limiting power distribution skewing.

In order to minimize the change of the axial peaking factor and axial xenon distribution during unit maneuvers, the operational scheme utilized to manage the axial power distribution keeps the AFD within a tolerance band around a burnup-dependent goal, sometimes referred to as the target flux difference.

The AFD limit includes a target band for plants that use "constant axial offset control, or CAOC". For instance, this target band lies between +5% and −5% of the goal value. The chance of starting a xenon transient increases with the length of time the core works beyond its intended range since the xenon concentration and distribution are flux- and time-dependent.

The target flux differential (ΔI_{target}) and its burnup-dependent values are calculated under equilibrium xenon circumstances. Control bank locations, core power level, axial burnup, axial xenon distribution, and, to a lesser degree, reactor coolant temperature and boron concentrations are among the numerous core-related characteristics that have an impact on the AFD.

A technique for managing axial power distribution during regular operations is called CAOC. In order to reduce the likelihood of starting xenon oscillations during power maneuvers, CAOC mandates that the AFD be managed within a limited tolerance band surrounding a burnup-dependent objective. Additionally, it reduces axial peaking factor fluctuation.

Axial offset is defined as

$$\frac{A}{O} = \frac{I_{top} - I_{bottom}}{I_{top} + I_{bottom}} \qquad \text{Eq. 5.4}$$

where I_{top} and I_{bottom} are expressed as fractions of rated thermal power.

5.6.1 Axial Flux Difference and Safety Analysis

The global power distribution of the core may be directly and continuously measured by the AFD and the quadrant power tilt ratio (QPTR). Global azimuthal power distribution is measured by QPTR, whereas global axial power distribution is measured by AFD. Acceptable peaking factors ($FQ(z)$ and $F\Delta\ H$) should be continually maintained by operating the control rods properly and staying within their bounds. By preventing an undetected change in the gross axial and radial power distribution, the AFD and QPTR restrictions guarantee that peaking factors ($FQ(z)$ and $F\Delta\ H$) stay below their limiting values.

Keep in mind that QPTR is defined as "the ratio of the maximum calibrated output from an upper excore detector to the average of the calibrated outputs from an upper excore detector, or the ratio of the maximum calibrated output from a lower excore detector to the average of the lower excore detector calibrated outputs, whichever is greater".

Four or six tandem (upper and lower) excore power range neutron detectors, which are part of the so-called excore NIS, are often used to provide QPTR. This parameter has to be regularly checked during regular operation. For QPRT and SPTR, the limit (limiting condition for operation, or LCO) is typically QPTR <1.02. Operation exceeding these bounds may render the fundamental power distribution hypotheses that were employed in the accident investigation incorrect.

To prevent core power distributions that surpass the safety analysis limits, the LCO limits on the AFD, the QPTR, the rod insertion limits, and the power distribution (i.e., the heat flux hot channel factor ($F_Q(z)$) and the nuclear enthalpy rise hot channel factor ($F_{\Delta H}^{N}$) are set together.

In conclusion, the worldwide power distribution of the core is directly and continuously measured by the AFD and the QPTR. Global azimuthal power distribution is measured by QPTR, whereas global axial power distribution is measured by AFD.

5.6.2 Axial Offset Anomaly

As stated, the QPTR and the axial offset-A/O (or AFD) are continuous, direct indicators of the worldwide power distribution of the core. Global azimuthal power distribution is measured by QPTR, whereas global axial power distribution is measured by A/O. Acceptable peaking factors ($FQ(z)$ and $F\Delta\ H$) should be continually maintained by operating the control rods properly and staying within their bounds. By preventing an undetected change in the gross axial and radial power distribution, the AFD and QPTR restrictions guarantee that peaking factors ($FQ(z)$ and $F\Delta\ H$) stay below their limiting values.

Small radial and particularly axial offset anomalies (AOAs) have already been observed in certain PWR-type nuclear power reactors (with LiOH water chemistry). This indicates that certain PWRs have seen a power shift from the top half to the bottom half, or an unanticipated and progressive movement in the target flux difference (ΔI_{target}) towards more negative values.

After much investigation, it was discovered that a buildup of crud on high-power fuel assemblies might result in enhanced subcooled nucleate boiling, which in turn can produce AOA. Crud accumulation results in a buildup of boron on the upper section of the assembly's surface since this subcooled nucleate boiling takes place there. The crud layer absorbs and concentrates lithium borate, which lowers the fission rate in the core's top region. The AOA has not occurred in every reactor core. It is now understood that AOA is primarily influenced by the chemistry of the reactor water and is brought on by a combination of high boiling intensity in the core and high amounts of crud in the coolant.

Consequently, the neutron flux moves from the top half to the bottom half during the fuel cycle, increasing local peaking factors. A decrease in the amount of SDM that is accessible (3–4 pcm/day) might also result from crud buildup. The burnup distribution is partially restored at the conclusion of the cycle when the power distribution shifts back toward the top half of the core due to excess burnup in the bottom of the core and decreased boron and lithium concentrations in the reactor coolant system.

For further information, the reader should refer to Information Notice No.97-85: Impact of Boron Deposition and Crud Buildup on Power Distribution and Shutdown Margin, US NRC, December 1997 [15].

REFERENCES

1. B. Zohuri, Nuclear Energy, Perspectives, Challenges and Future Directions, In Denver Morris (Ed.), *Nuclear Materials and Disaster Research*, NOVA Science Publishers, 2020.
2. B. Zohuri, P. McDaniel, *Advanced Smaller Modular Reactors, An Innovative Approach to Nuclear Power*, Springer Publisher, 2019.
3. B. Zohuri, *Nuclear Micro Reactors*, 1st edition, Springer Publishing Company, 2020.
4. B. Zohuri, P. McDaniel, *Combined Cycle Driven Efficiency for Next Generation Nuclear Power Plants: An Innovative Design Approach*, 2nd edition, Springer Publishing Company, 2017.
5. B. Zohuri, *Small Modular Reactors Renewable Energy Sources*, 1st edition, Springer Publishing Company, 2018.
6. P. Ottensmeyer, *Neutrons at the Core: Energy for Millennia A Gentle Introduction to Nuclear Energy from Neutrons to Fast-Spectrum Reactors*, Independently Published, 2021.
7. M. C. Brady, T. R. England, Delayed Neutron Data and Group Parameters for 43 Fissioning Systems, *Nucl. Sci. Eng.* 103, no. 2, 129–149 (1989).
8. A. Dos Santos, R. Diniz, L. C. C. B. Fanaro, R. Jerez, G. S. de Andrade e Silva, M. Yamaguchi, A Proposal of a Benchmark for βeff, β_{eff}/Λ, and Λ of Thermal Reactors Fueled with Slightly Enriched Uranium, *Ann. Nucl. Energy* 33, no. 9, 848–855 (2006).
9. M. Iqbal, T. Mahmood, Sh. Pervez, Flow of Kinetic Parameters in a Typical Swimming Pool Type Research Reactor, *Ann. Nucl Energy* 35, 518–524 (2008).
10. G. R. Keepin, *Physics of Nuclear Kinetics*, Addison Wesley Publishing Company, 1965.
11. J. F. Briesmeister, MCNP—A General Monte Carlo N-Particle Transport Code, Version 4C, LA-13709-M. Los Alamos National Laboratory, 2000.
12. R. K. Meulekamp, S. C. Van der Marck, Calculating the Effective Delayed Neutron Fraction with Monte Carlo, *Nucl. Sci. Eng.* 152, no. 2, 142–148 (2006).
13. Y. Nauchi, T. Kameyama, Proposal of Direct Calculation of Kinetic Parameters βeff and a Based on Continuous Energy Monte Carlo Method, *J. Nucl. Sci. Technol.* 42, no. 6, 503–514 (2005).
14. G. D. Spriggs, R. D. Busch, J. M. Campbell, Calculation of the Delayed Neutron Effectiveness Factor Using Ratios of k-Eigenvalues, *Ann. Nucl. Energy* 28, no. 5, 477–487 (2001).
15. Information Notice No.97-85: Impact of Boron Deposition and Crud Buildup on Power Distribution and Shutdown Margin, US NRC, December 1997

6 Two-Group Diffusion Theory for Bare and Reflected Reactors

6.1 INTRODUCTION

The fact that all neutrons are grouped into a single one-group, as shown in Figure 6.1 as an infinite slab reactor of thickness encircled on all sides by a reflector of extrapolated thickness, clearly limits the one-group approach b.

This type of reactor, as shown in Figure 6.1, has a uniform fuel and moderator combination in the center, while the reflector is only the moderator. All of the calculations pertaining to this reactor are provided in Chapter 2. The subscriber in this sort of reactor represented the center region, which correlates to the core of an actual reactor, and the subscript showed the reflector. All of them were shown by solution as all equations in Section 2.16 (of Chapter 2).

A more precise method, especially for thermal reactors, is to divide the neutrons into two groups. In this instance, the thermal neutrons are grouped into the slow group, which is also known as the thermal group, while the fast group contains all other neutrons. Therefore, the energy that separates the two groups is in the order of 5 kT [1].

As a simple but important application of the two-group form of effective multiplication factor and its diffusion equations is presented in Equation 6.1(a, b), which is limited to the analysis of a bare reactor as illustrated in Figure 6.2, and taking into consideration the effective multiplication factor $k_{eff} = k$ for a bare reactor as [2]

$$-\nabla \cdot D_1 \nabla \phi_1(\vec{r}) + (\Sigma_{a1} + \Sigma_r)\phi_1(\vec{r}) = \frac{1}{k}[\nu\Sigma_{f1}\phi_1(\vec{r}) + \nu\Sigma_{f2}\phi_2(\vec{r})] \qquad \text{Eq. 6.1}$$

$$-\nabla \cdot D_2 \nabla \phi_2(\vec{r}) + \Sigma_{a2}\phi_2(\vec{r}) = \Sigma_r \phi_1(\vec{r}) \qquad \text{Eq. 6.2}$$

The cross-section Σ_r, introduced in Equation 6.1, representing the net slowing down of fast neutrons into the thermal group, is also often referred to as a removal cross-section in the Pressurized Water Reactor (PWR) literature and should not be confused with Σ_{R1}. Equation 6.2 may also be obtained without the formal machinery of Equations 6.2–6.4 of the steady state multi-group diffusion (MGD) equation as shown below.

$$-\nabla \cdot D_n \nabla \phi_n(\vec{r}) + \Sigma_{Rn}\,\phi_n(\vec{r}) = \frac{\chi_n}{\lambda}\sum_{j=1}^{N} \nu\,\Sigma_{fj}\,\phi_j(\vec{r}) + \sum_{j \neq N}^{N} \Sigma_{j\to n}\,\phi_j(\vec{r}) \qquad \text{Eq. 6.3}$$

Equation 6.3 of steady state MGD equation in a matrix form for a $N-group$ formulation is given as

DOI: 10.1201/9781003590194-6

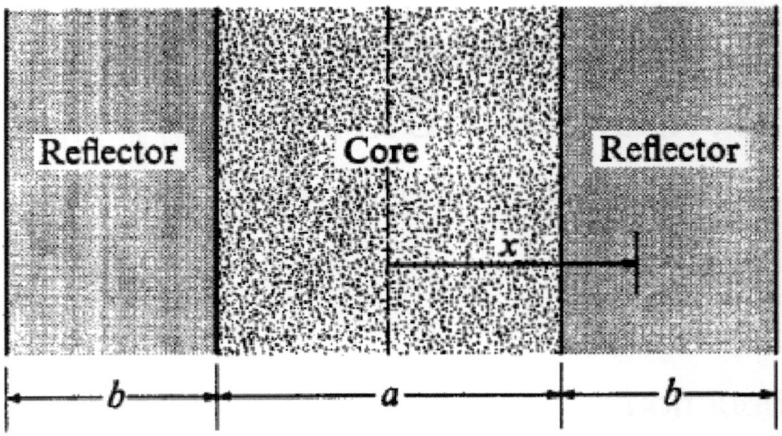

FIGURE 6.1 The reflected slab reactor.

(Source: U.S. Department of Energy)

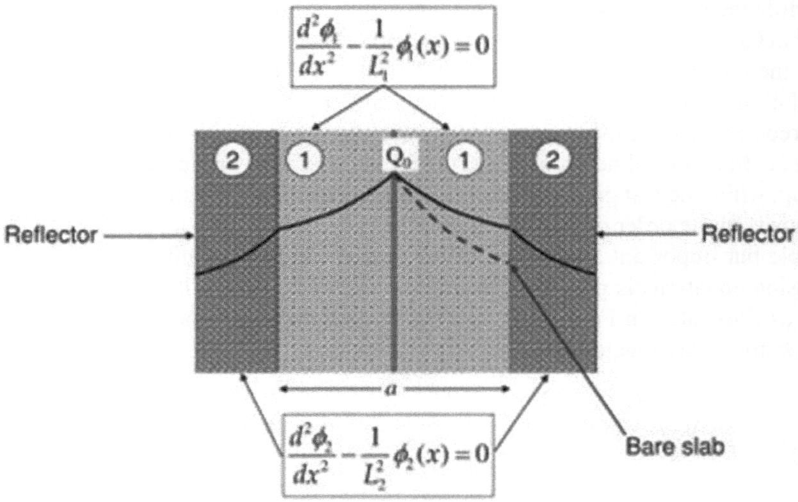

FIGURE 6.2 Plane infinite source with reflector and diffusion of neutron.

(Source: U.S. Department of Energy)

$$L\vec{\Phi} = \frac{1}{\lambda}\chi\vec{F}^T\vec{\Phi} = \frac{1}{\lambda}M\vec{\Phi} \qquad \text{Eq. 6.4}$$

where

$$L = \begin{bmatrix} -\nabla\cdot D_1\nabla + \Sigma_{R1} & -\Sigma_{2\to 1} & -\Sigma_{3\to 1} & \cdot & -\Sigma_{N\to 1} \\ -\Sigma_{1\to 2} & -\nabla\cdot D_2\nabla + \Sigma_{R2} & -\Sigma_{3\to 2} & \cdot & -\Sigma_{N\to 2} \\ -\Sigma_{1\to 3} & -\Sigma_{2\to 3} & -\nabla\cdot D_3\nabla + \Sigma_{R3} & \cdot & -\Sigma_{N\to 3} \\ \cdot & \cdot & \cdot & \cdot & \cdot \\ -\Sigma_{1\to N} & -\Sigma_{2\to N} & -\Sigma_{3\to N} & \cdot & -\nabla\cdot D_N\nabla + \Sigma_{RN} \end{bmatrix}$$

$$\chi = \begin{bmatrix} \chi_1 \\ \chi_2 \\ \cdot \\ \cdot \\ \chi_3 \end{bmatrix} \quad \vec{F} = \begin{bmatrix} \nu\Sigma_{f1} \\ \nu\Sigma_{f2} \\ \cdot \\ \cdot \\ \nu\Sigma_{fN} \end{bmatrix} \quad \vec{\Phi} = \begin{bmatrix} \phi_1 \\ \phi_2 \\ \cdot \\ \cdot \\ \phi_3 \end{bmatrix}$$

and

$$M = \begin{bmatrix} \chi_1\nu\Sigma_{f1} & \chi_1\nu\Sigma_{f2} & \cdot & \cdot & \chi_1\nu\Sigma_{fN} \\ \chi_2\nu\Sigma_{f1} & \chi_N\nu\Sigma_{f1} & \cdot & \cdot & \chi_2\nu\Sigma_{fN} \\ \cdot & \cdot & \cdot & \cdot & \cdot \\ \cdot & \cdot & \cdot & \cdot & \cdot \\ \chi_N\nu\Sigma_{f1} & \chi_N\nu\Sigma_{f2} & \cdot & \cdot & \chi_N\nu\Sigma_{fN} \end{bmatrix}$$

Note that the steady state MGD Equation 6.4 is widely used to determine global flux and power distributions, as well as the eigenvalue or the effective multiplication factor $\lambda = k_{eff} = k$, in steady state reactor configurations. Equation 6.4 also forms the basis for fuel depletion analysis in multidimensional core geometries [2].

We should also note that Equation 6.3 for each group is in the form of Equation 6.5, as the steady state one-group diffusion equation is given as

$$\nabla \cdot \vec{J}(\vec{r}) + \Sigma(\vec{r})\phi(\vec{r}) = S(\vec{r}) \qquad \text{Eq. 6.5}$$

with the recognition that in the one-group model, $\Sigma = \Sigma_a$ and the source term $S(\vec{r}) = \nu\Sigma_f\,\phi(\vec{r})/\lambda$, where λ is the eigenvalue of the system.

6.2 TWO-GROUP CORE DIFFUSION EQUATIONS

We take into account a thermal energy group in the two-group theory approach and merge all higher-energy neutrons into a fast energy group.

The following balancing equations for the fast and thermal groups result from taking into account the fast and thermal energy fluxes.

Fast Group

–[Leakage] – [Removal by scattering] – [Absorptions] + [Fast fissions] + [Thermal fissions] = 0

$$D_1\nabla^2\phi_1 - \Sigma_{1s}\,\phi_1 - \Sigma_{1a}\,\phi_1 + \nu\,\Sigma_{1f}\,\phi_1 + \nu\,\Sigma_{2f}\,\phi_2 = 0 \qquad \text{Eq. 6.6}$$

Thermal Group

–[Leakage] – [Absorptions] + [Down scattering neutrons from fast group] = 0

$$D_2\nabla^2\phi_2 - \Sigma_{2a}\,\phi_2 + \Sigma_{1s}\,\phi_1 = 0 \qquad \text{Eq. 6.7}$$

The thermal fission term connects the equations

$$\Sigma_{2f}\,\phi_2 = 0 \qquad \text{Eq. 6.8}$$

and the fast removal term

$$\Sigma_{1s}\,\phi_1 = 0 \qquad \text{Eq. 6.9}$$

6.3 CRITICALITY EQUATION FOR TWO-GROUP THEORY AND BARE REACTORS

We assume that the fluxes in the core have geometrical buckling B^2 that satisfies

$$\begin{cases} \nabla^2\phi_1 + B^2\phi_1 = 0 \\ \nabla^2\phi_2 + B^2\phi_2 = 0 \end{cases} \qquad \text{Eq. 6.10}$$

The thermal and fast fluxes are proportional everywhere for the bare reactor as B^2 is the same for both. You may rewrite Equations 6.6 and 6.7 as follows:

$$-D_1B^2\phi_1 - \Sigma_{1s}\,\phi_1 - \Sigma_{1a}\,\phi_1 + \nu\,\Sigma_{1f}\,\phi_1 + \nu\,\Sigma_{2f}\,\phi_2 = 0 \qquad \text{Eq. 6.11}$$

and

$$-D_2B^2\phi_2 - \Sigma_{2a}\,\phi_2 + \Sigma_{1s}\,\phi_1 = 0 \qquad \text{Eq. 6.12}$$

From Equation 6.12, the ratio of thermal to fast flux is given as

$$\frac{\phi_2}{\phi_1} = \frac{\Sigma_{1s}}{D_2B^2 + \Sigma_{2a}} = \left(\frac{\Sigma_{1s}}{\Sigma_{2a}}\right)\left(\frac{1}{1 + L_2^2B^2}\right) \qquad \text{Eq. 6.13}$$

where $L_2^2 = \dfrac{D_2}{\Sigma_{2a}}$.

Now considering the two terms in Equation 6.11, we can write

$$\nu\,\Sigma_{1f}\,\phi_1 + \nu\,\Sigma_{2f} = \frac{\nu\,\Sigma_{1f}\,\phi_1 + \nu\,\Sigma_{2f}\,\phi_2}{\nu\,\Sigma_{2f}\,\phi_2}\nu\,\Sigma_{2f}\,\phi_2 = \varepsilon\nu\,\Sigma_{2f}\,\phi_2 \qquad \text{Eq. 6.14}$$

where ε is the fast fission factor and is written as

$$\varepsilon = 1 + \frac{\Sigma_{1f}\,\phi_1}{\Sigma_{2f}\,\phi_2} \qquad \text{Eq. 6.15}$$

Since the ratio of thermal flux to fast flux is not constant in a reflected reactor core, this is not the case.

Substituting Equation 6.14 into Equation 6.11, we can write

$$+D_1B^2\phi_1 + \left(\Sigma_{1s}\,\phi_1 + \Sigma_{1a}\right)\phi_1 = \varepsilon\nu\,\Sigma_{2f}\,\phi_2 \qquad \text{Eq. 6.16}$$

Defining the fast diffusion as

$$L_1^2 = \frac{D_1}{\Sigma_{1s} + \Sigma_a} \qquad \text{Eq. 6.17}$$

then, we can write

$$\begin{aligned}\left(1 + L_1^2 B^2\right) &= \frac{\varepsilon \nu \Sigma_{2f}}{\Sigma_{1s} + \Sigma_{1a}} \frac{\phi_2}{\phi_1} \\ \left(1 + L_1^2 B^2\right)\left(1 + L_2^2 B^2\right) &= \varepsilon \nu \frac{\Sigma_{1s}}{\Sigma_{1s} + \Sigma_{1a}} \frac{\Sigma_{2f}}{\Sigma_{2a}}\end{aligned} \qquad \text{Eq. 6.18}$$

Since the ratio

$$\frac{\Sigma_{1a}}{\Sigma_{1s} + \Sigma_{1a}} \qquad \text{Eq. 6.19}$$

is the percentage of fast neutrons eliminated by fast absorptions, the resonance escape probability p may be written as follows:

$$p = 1 - \frac{\Sigma_{1a}}{\Sigma_{1s} + \Sigma_{1a}} = \frac{\Sigma_{1s}}{\Sigma_{1s} + \Sigma_{1a}} \qquad \text{Eq. 6.20}$$

Since

$$\begin{aligned}\eta f &= \nu \frac{\Sigma_{2f}}{\Sigma_{2a}} \\ k_\infty &= \eta \varepsilon p f\end{aligned} \qquad \text{Eq. 6.21}$$

Equation 6.18's second term may be rewritten as follows:

$$k_{eff} = k_\infty \left[\frac{1}{\left(1 + L_1^2 B^2\right)} \frac{1}{\left(1 + L_2^2 B^2\right)} \right] = 1 \qquad \text{Eq. 6.22}$$

In order for Fermi Theory to be accepted, it would be required that

$$e^{-B^2 t} = \frac{1}{\left(1 + L_2^2 B^2\right)} \qquad \text{Eq. 6.23}$$

which can be true only in weakly absorbing media.

6.4 CRITICALITY EQUATION FOR THE CORE OF A REFLECTED REACTOR

The fast flux to thermal flux ratio in a reflected system is not constant, hence the approach used for a naked, unreflected reactor must be changed. It is possible to rewrite Equation 6.11 as follows:

$$\begin{aligned}D_1 B^2 \phi_1 + (\Sigma_{1s} + \Sigma_{1a} - \nu \Sigma_{1f})\phi_1 &= \nu \Sigma_{2f} \phi_2 \\ D_1 B^2 \phi_1 + \Sigma_{1n} &= \nu \Sigma_{2f} \phi_2\end{aligned} \qquad \text{Eq. 6.24}$$

where

$$\Sigma_{1n} = (\Sigma_{1s} + \Sigma_{1a} - \nu\Sigma_{1f}) \qquad \text{Eq. 6.25}$$

is the cross-section of quick net removal from the fast group to the thermal group. One way to rewrite Equation 6.24 is as follows:

$$\left(1 + L_n^2 B^2\right)\frac{\phi_1}{\phi_2} = \nu\frac{\Sigma_{2f}}{\Sigma_{1n}} \qquad \text{Eq. 6.26}$$

where $L_n^2 = \frac{D_1}{\Sigma_{1n}}$.

Similarly, Equation 6.13 can be written as

$$\left(1 + L_2^2 B^2\right)\frac{\phi_1}{\phi_2} = \frac{\Sigma_{1s}}{\Sigma_{2a}} \qquad \text{Eq. 6.27}$$

Multiplying the sides of Equations 6.26 and 6.27, we obtain the following result:

$$\left(1 + L_2^2 B^2\right)\left(1 + L_n^2 B^2\right) = \frac{\Sigma_{1s}}{\Sigma_{2n}}\frac{\nu\Sigma_{2f}}{\Sigma_{1n}} = k_n \qquad \text{Eq. 6.28}$$

where the changed multiplication factor is denoted by k_n. Only in the absence of rapid fissions would it equal the infinite medium multiplication factor with

$\Sigma_{1f} = 0$, and $\varepsilon = 1 \Rightarrow k_n = k_\infty$

Equation 6.28 is a quadratic equation in B^2 given as

$$(B^2)^2 L_2^2 L_n^2 + (L_2^2 + L_n^2)B^2 - (k_n - 1) = 0$$

$$(B^2)^2 + \left(\frac{1}{L_n^2} + \frac{1}{L_2^2}\right)B^2 - \frac{(k_n - 1)}{L_2^2 L_n^2} = 0 \qquad \text{Eq. 6.29}$$

If we denote

$$a = 1, b = \left(\frac{1}{L_n^2} + \frac{1}{L_2^2}\right), \quad c = \frac{(k_n - 1)}{L_2^2 L_n^2} = 0$$

Then, the roots of Equation 6.29 are

$$+\mu^2 = \frac{-b + \sqrt{b^2 + 4c}}{2} \qquad \text{Eq. 6.30}$$

and

$$-\nu^2 = \frac{-b - \sqrt{b^2 + 4c}}{2} \qquad \text{Eq. 6.31}$$

By adding Equations 6.30 and 6.31, we get

$$\nu^2 = \mu^2 + b \qquad \text{Eq. 6.32}$$

The major buckling is the root μ^2, while the alternative buckling is the root ν^2. Both roots have positive values. The sole gas utilized in a bare reactor is μ^2. μ^2 will represent the difference between two almost equal numbers and will not be numerically precise if 4 is tiny in relation to b^2. The following binomial expansion can be used in this situation instead.

$$(1+x)^n \approx 1+nx+\frac{n(n-1)}{2!}x^2+\frac{n(n-1)(n-3)}{3!}x^3+\cdots$$

Hence, we have

$$\begin{aligned}\mu^2 &= \frac{1}{2}\left[-b+b\sqrt{1+\frac{4c}{b^2}}\right] \\ &= \frac{1}{2}\left[-+b\left(1+\frac{4c}{b^2}\right)^{\frac{1}{2}}\right] \\ &= \frac{1}{2}\left[-b+b+2\frac{c}{b}-2\frac{c^2}{b^2}+4\frac{c^3}{b^5}+\cdots\right] \\ &= \frac{c}{b}\left[1-\frac{c}{b^2}+2\left(\frac{c}{b^2}\right)+\cdots\right]\end{aligned} \qquad \text{Eq. 6.33}$$

6.5 THE CORE FLUX DISTRIBUTIONS

Having determined that the buckling B^2 has two values μ^2 and ν^2 as given by Equations 6.30 and 6.31, the flux distributions now have two solutions. For primary buckling, we have

$$\begin{aligned}\nabla^2\phi_{1c}+\mu^2\phi_{1c} &= 0 \\ \nabla^2\phi_{2c}+\nu^2\phi_{2c} &= 0\end{aligned} \qquad \text{Eq. 6.34}$$

The answers to these two equations, taking into account spherical geometry, are

$$\phi_{1c} = AX, \quad X = \frac{\sin(\mu r)}{r} \qquad \text{Eq. 6.35}$$

$$\phi_{2c} = A'X \qquad \text{Eq. 6.36}$$

The relationship between A and A' is found by substituting Equations 6.35 and 6.36 in Equation 6.12, yielding

$$-D_2\mu^2 A'X - \Sigma_{2a}A'X + \Sigma_{1s}AX = 0 \qquad \text{Eq. 6.37}$$

Thus, we can write for the principle coupling coefficient as

$$S_1 = \frac{A'}{A} = \frac{\Sigma_{1s}}{\mu^2 D_2 + \Sigma_{2a}} \qquad \text{Eq. 6.38}$$

Equations 6.35 and 6.36 can be written as

$$\phi_{1c} = AX$$
$$\phi_{2c} = AS_1X \quad \text{Eq. 6.39}$$

For the solution corresponding to the alternate buckling, similarly

$$\begin{cases} \nabla^2\phi_{1c} - v^2\phi_{1c} = 0 \\ \nabla^2\phi_{2c} - v^2\phi_{2c} = 0 \end{cases} \quad \text{Eq. 6.40}$$

$$\begin{cases} \phi_{1c} = CY, \qquad Y = \dfrac{\sinh(vr)}{r} \\ \phi_{2c} = C'Y \end{cases} \quad \text{Eq. 6.41}$$

Equation 6.12 may be substituted to yield the alternative coupling coefficient as follows:

$$\begin{cases} +D_2v^2C'Y - \Sigma_{2a}C'Y + \Sigma_{1s}CY = 0 \\ S_2 = \dfrac{C'}{C} = \dfrac{\Sigma_{1s}}{\Sigma_{2a} - v^2D_2} \end{cases} \quad \text{Eq. 6.42}$$

In the event that the terms in the denominator disagree slightly, we allow

$$\begin{aligned} v^2 &= \mu^2 + b \\ &= \mu^2 + \frac{1}{L_n^2} + \frac{1}{L_2^2} \\ &= \mu^2 + \frac{1}{L_n^2} + \frac{\Sigma_{2a}}{D_2} \end{aligned} \quad \text{Eq. 6.43}$$

and use

$$\begin{aligned} S_2 &= \frac{\Sigma_{1s}}{\Sigma_{2a} + v^2D_2} \\ &= \frac{\Sigma_{1s}}{\Sigma_{2a} + \mu^2D_2 - \dfrac{D_2}{L_n} - \Sigma_{2a}} \\ &= \frac{\Sigma_{1s}}{D_2\left(\mu^2 - \dfrac{1}{L_n^2}\right)} \end{aligned} \quad \text{Eq. 6.44}$$

The alternate solutions then become

$$\begin{cases} \phi_{1c} = CY \\ \phi_{2c} = CS_2Y \end{cases} \quad \text{Eq. 6.45}$$

From Equations 6.39 and 6.41, the total solutions become

$$\begin{cases} \phi_{1c} = AX + CY \\ \phi_{2c} = AS_1X + CS_2Y \end{cases} \quad \text{Eq. 6.46}$$

A spherical reactor for the two-group fluxes in the core is shown as

$$\begin{cases} \phi_{1c} = A\dfrac{\sin(\mu r)}{r} + C\dfrac{\sinh(\nu r)}{r} \\ \phi_{2c} = AS_1\dfrac{\sin(\mu r)}{r} + CS_2\dfrac{\sinh(\nu r)}{r} \end{cases} \quad \text{Eq. 6.47}$$

For a bare reactor, $C = 0$ since the flux disappears at the extrapolated boundary. C does not equal 0 for a reflected reactor, and the flux does not disappear at the extrapolated border.

The following relationship may be used to determine a bare reactor's critical size.

$$\mu^2 = \left(\frac{\pi}{R_{ex}}\right)^2 \quad \text{Eq. 6.48}$$

But for the reflected reactor, the situation is different.

Using both one-group and two-group models of $R_{out} = R + T + d$, Figure 6.3 shows the flux distributions in the core and reflector of a two-region reactor.

6.6 SOLUTION IN THE REFLECTOR IN TWO-GROUP THEORY

There is no neutron source in the reflector, and the fast and thermal groups' equations are as follows:

$$\begin{cases} D_{1r}\nabla^2\phi_{1r} - \Sigma_{1sr}\,\phi_{1r} - \Sigma_{1ar}\,\phi_{1r} = 0 \\ D_{2r}\nabla^2\phi_{2r} - \Sigma_{2ar}\,\phi_{2r} - \Sigma_{1sr}\,\phi_{1r} = 0 \end{cases} \quad \text{Eq. 6.49}$$

Here, the removal phrase from the fast group serves as the source term in the thermal group. Determining the diffusion areas by dividing by the diffusion coefficients

$$\begin{cases} L_{2r}^2 = \dfrac{D_{2r}}{\Sigma_{2ar}} \\ L_{1r}^2 = \dfrac{D_{1r}}{\Sigma_{1sr} + \Sigma_{1ar}} \end{cases} \quad \text{Eq. 6.50}$$

The equations can be written as

$$\nabla^2\phi_{1r} - \frac{1}{L_{1r}^2}\phi_{1r} = 0 \quad \text{Eq. 6.51}$$

$$\nabla^2\phi_{2x} - \frac{1}{L_{2r}^2}\phi_{2r} + \frac{\Sigma_{1sr}}{D_{2r}}\phi_{1r} = 0 \quad \text{Eq. 6.52}$$

Equation 6.51 has a solution in spherical geometry

$$\phi_{1r} = FZ_1$$
$$Z_1 = \frac{e^{-\frac{r}{L_{1r}}}}{r} \quad \text{Eq. 6.53}$$

for an infinite reflector, while the following solution is better for a finite reflector with thickness T and an extrapolation distance d.

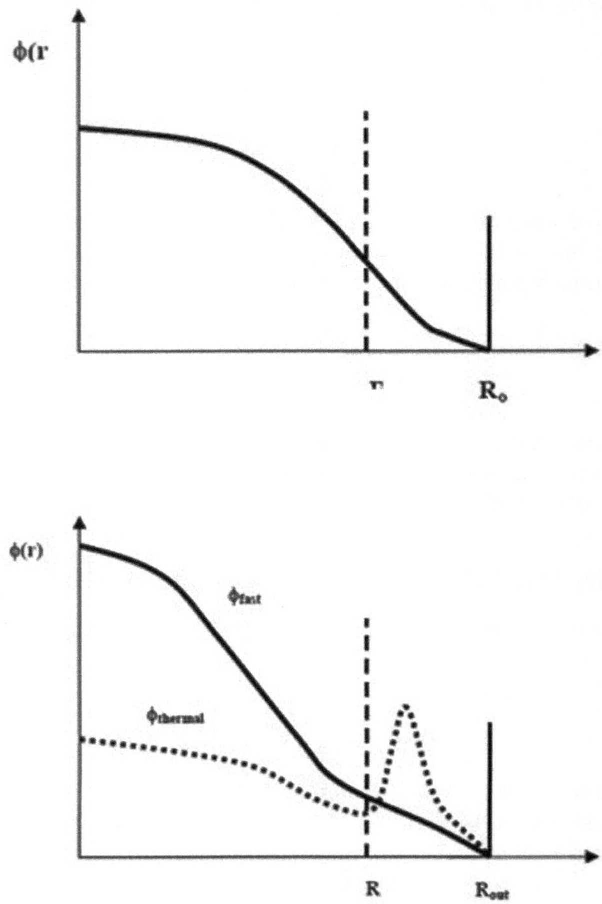

FIGURE 6.3 Flux distributions in the core and reflector of a two region reactor using the One-Group and the Two-Group models, $\boldsymbol{R_{out} = R + T + d}$

$$Z_1 = \frac{\sinh(R + T + d - r)}{r} \tag{Eq. 6.54}$$

For the thermal flux solution in the reflector, we can write

$$\phi_{2r} = GZ_2 + S_3\phi_{1r} = GZ_2 + S_3FZ_1 \tag{Eq. 6.55}$$

where

$$Z_2 = \frac{e^{-\frac{r}{L_{2r}}}}{r}$$

or

$$Z_2 = \frac{\sinh(R + T + d - r)}{r} \tag{Eq. 6.56}$$

Here, S_3 is a coupling coefficient to be determined. It can be noticed that in any geometry

$$\begin{cases} \nabla^2 Z_1 = \dfrac{1}{L_{1r}^2} Z_1 \\ \nabla^2 Z_2 = \dfrac{1}{L_{2r}^2} Z_2 \end{cases} \qquad \text{Eq. 6.57}$$

Substituting in Equation 6.52, we get

$$G\frac{Z_2}{L_{2r}^2} + S_3\frac{FZ_1}{L_{1r}^2} - G\frac{Z_2}{L_{2r}^2} - S_3\frac{FZ_1}{L_{1r}^2} + \frac{\Sigma_{1sr}}{D_{2r}} FZ_1 = 0 \qquad \text{Eq. 6.58}$$

From which

$$S_3 = \frac{\dfrac{\Sigma_{1sr}}{D_{2r}}}{\dfrac{1}{L_{2r}^2} - \dfrac{1}{L_{1r}^2}} \qquad \text{Eq. 6.59}$$

6.7 CRITICALITY CONDITION

It is possible to express the continuity of the fluxes and currents at the reactor core–reflector contact as follows:

$$\begin{aligned} \phi_{1c} &= \phi_{1r} \\ \phi_{2c} &= \phi_{2r} \\ -D_{1c}\nabla\phi_{1c} &= -D_{1r}\nabla\phi_{1r} \\ -D_{2c}\nabla\phi_{2c} &= -D_{2r}\nabla\phi_{2r} \end{aligned} \qquad \text{Eq. 6.60}$$

Equations 6.46, 6.53, and 6.55 may be substituted in Equation 6.60 to produce the following outcome:

$$\begin{aligned} & AX + CY = FZ_1 \\ & S_1AX + S_2CY = S_3FZ_1 + GZ_2 \\ & D_{1c}AX + D_{1c}CY = D_{1r}FZ_1 \\ & D_{2c}S_1AX + D_{2c}S_2CY = D_{2r}S_3FZ_1 + D_{2r}GZ_2 \end{aligned} \qquad \text{Eq. 6.61}$$

where the functions are evaluated at $r = R$ and the primes indicate a gradient operation.

These equations can be written as follows in matrix notation.

$$\begin{bmatrix} X & Y & -Z_1 & 0 \\ S_1X & S_2Y & -S_3Z_1 & -Z_2 \\ D_{1c}X & D_{1c}Y & -D_{1r}Z_1 & 0 \\ D_{2c}S_1X & D_{2c}S_2Y & -D_{2r}S_3Z_1 & -D_{2r}Z_2 \end{bmatrix} \begin{bmatrix} A \\ C \\ F \\ G \end{bmatrix} = 0 \qquad \text{Eq. 6.62}$$

The four unknowns in these four simultaneous linear algebraic equations are A, C, F, and G.

Explicit formulas for all four constants cannot be obtained, and these equations are of the homogeneous kind, meaning that the right-hand side (RHS) is equal to zero. There will always be one unknown constant that represents the reactor's flux or power level, just like in other steady state reactor issues. Therefore, the fourth may be used to determine three of these unknowns.

This linear system of equations can only be solved non-trivially if the determinant of the coefficients disappears.

$$\Delta \equiv \begin{bmatrix} X & Y & -Z_1 & 0 \\ S_1X & S_2Y & -S_3Z_1 & -Z_2 \\ D_{1c}X & D_{1c}Y & -D_{1r}Z_1 & 0 \\ D_{2c}S_1X & D_{2c}S_2Y & -D_{2r}S_3Z_1 & -D_{2r}Z_2 \end{bmatrix} \equiv 0 \qquad \text{Eq. 6.63}$$

Given that it includes both the group constants and the reactor's dimensions, this equation might be considered the criticality condition for the reflected reactor in the two-group formulation. The above relation can be used to determine the other based on the parameters of the fuel concentration or the core size.

6.8 EVALUATION OF CRITICALITY DETERMINANT

Although simple, evaluating the criticality connection is a time-consuming process. Making use of the fact that many terms in the determinant are insensitive to changes in the fuel concentration or the size of the core helps simplify the computation to some extent.

The value of the determinant remains unchanged when the first column is divided by X, the second by Y, the third by Z_1, the fourth by Z_2, the third row by D_{1c}, and the fourth row by D_{2c}. The results are as follows:

$$\Delta \equiv \begin{vmatrix} 1 & 1 & -1 & 0 \\ S_1 & S_2 & -S_3 & -1 \\ \alpha & \beta & -\rho_1\gamma & 0 \\ S_1\alpha & S_2\beta & -\rho_2S_3\gamma & -\rho_2\delta \end{vmatrix} = 0 \qquad \text{Eq. 6.64}$$

where, $\alpha = \frac{X'}{X}$, $\beta = \frac{Y'}{Y}$, $\gamma = \frac{Z_1'}{Z}$, $\delta = \frac{Z_2'}{Z_2}$, $\rho_1 = \frac{D_{1r}}{D_{1r}}$, and $\rho_2 = \frac{D_{2r}}{D_{2r}}$.

Moving on with the solution using the fourth column's minors, we get

$$\Delta = (-1)\begin{pmatrix} 1 & 1 & -1 \\ \alpha & \beta & -\rho_1\gamma \\ S_1\alpha & S_2\beta & -\rho_2S_3\gamma \end{pmatrix} - (\rho_2\delta)\begin{pmatrix} 1 & 1 & -1 \\ S_1 & S_1 & -S_3 \\ \alpha & \beta & -\rho_1\gamma \end{pmatrix} = 0 \qquad \text{Eq. 6.65}$$

When we extend the determinants using the scissors rule, we obtain

$$\alpha = \frac{\rho_2\delta C_1 + \rho_1\gamma C_2 + \beta C_3}{C_1 + C_2 + C_3} \qquad \text{Eq. 6.66}$$

where

$$C_1 = S_1(\rho_1\gamma - \beta)$$
$$C_2 = S_2(\beta - \rho_2\delta)$$
$$C_3 = S_3\rho_2(\delta - \gamma)$$

Only the critical dimension can satisfy this two-group critical equation. In order to use Equation 6.66 for a particular reactor size, the computational process entails assessing each side of Equation 6.13 and applying different fuel concentration values.

Plotting the left-hand side (LHS) and RHS as a function of fuel concentration allows one to determine the critical concentration by locating the junction of the two curves.

Finding the critical reactor size for a specific reactor composition would be another kind of computation. Table 6.1, which presents solution functions for reflected cores in spherical geometry, lists the functions *X*, *Y*, Z_1, and Z_2 along with their gradients in spherical geometry. *R* stands for core radius, *T* for reflector thickness, and *d* for extrapolation length.

6.9 DETERMINATION OF THE FLUX DISTRIBUTIONS

If the reactor critical composition or size are determined, the constants C , F , and g can now be evaluated in terms of A . We first rewrite Equation 6.61 in the form

$$AX + CY - FZ_1 = 0 \qquad \text{Eq. 6.67}$$

$$S_1AX + S_2CY - S_3FZ_1 - GZ_2 = 0 \qquad \text{Eq. 6.68}$$

$$AX' + CY' - \rho_1 FZ_1' = 0 \qquad \text{Eq. 6.69}$$

$$S_1AX' + S_2CY' - \rho_2 S_3 FZ_1' - \rho_2 GZ_2' = 0 \qquad \text{Eq. 6.70}$$

TABLE 6.1
Solution Functions for Reflected Cores in Spherical Geometry

$X(\vec{r})$	$\dfrac{\sin(\mu r)}{r}$
$X'(\vec{r})$	$\dfrac{\mu\cos(\mu r)}{r} - \dfrac{\sin(\mu r)}{r^2}$
$Y(\vec{r})$	$\dfrac{\sinh(\nu r)}{r}$
$Y'(\vec{r})$	$\dfrac{\nu\cosh(\nu r)}{r} - \dfrac{\sinh(\nu r)}{r^2}$
$Z_1(\vec{r})$	$\dfrac{\sinh\left(\dfrac{R+T+d-r}{L_{1r}}\right)}{r}$
$Z_1'(\vec{r})$	$\dfrac{-\dfrac{1}{L_{1r}}\cosh\left(\dfrac{R+T+d-r}{L_{1r}}\right)}{r} - \dfrac{\sinh\left(\dfrac{R+T+d-r}{L_{1r}}\right)}{r^2}$
$Z_2(\vec{r})$	$\dfrac{\sinh\left(\dfrac{R+T+d-r}{L_{2r}}\right)}{r}$
$Z_2'(\vec{r})$	$\dfrac{-\dfrac{1}{L_{2r}}\cosh\left(\dfrac{R+T+d-r}{L_{2r}}\right)}{r} - \dfrac{\sinh\left(\dfrac{R+T+d-r}{L_{2r}}\right)}{r^2}$

Equation 6.69 divided by Equation 6.67 yields the following outcome:

$$\frac{AX'+CY'}{AX+CY}=\frac{\rho_1 FZ'}{FZ_1}=\rho_1\gamma \qquad \text{Eq. 6.71}$$

from which, we can estimate C as

$$C=\left(\frac{\rho_1\gamma-\alpha}{\beta-\rho_1\gamma}\right)\left(\frac{X}{Y}\right)A \qquad \text{Eq. 6.72}$$

Additionally, we are able to derive F as follows by substituting Equations 6.67 and 6.68 into Equation 6.69.

$$F=\left(\frac{\beta-\alpha}{\beta-\rho_1\gamma}\right)\left(\frac{X}{Z_1}\right)A \qquad \text{Eq. 6.73}$$

Additionally, we are able to derive G as follows by substituting Equations 6.67 and 6.68 into Equation 6.70.

$$G=\left[\frac{S_1(\beta-\rho_1\gamma)+S_2(\rho_1\gamma-\alpha)-S_3(\beta-\alpha)}{\beta-\rho_1\gamma}\right]\left(\frac{X}{Z_2}\right)A \qquad \text{Eq. 6.74}$$

We now know C, F, and G in terms of A, and Equation 6.70 does not provide any new information. The following method can be used to calculate the constant A as a function of reactor power level. Reactor power can be expressed as follows:

$$\begin{aligned}P[\text{Watts}]&=\left(\Sigma_{1f}\,\bar{\phi}_{1c}+\Sigma_{2f}\,\bar{\phi}_{2c}\right)\frac{\text{fissions}}{\text{cm}^3\ \text{s}}\cdot V\ \text{cm}^3\cdot 200\frac{\text{MeV}}{\text{fission}}1.6\times10^{-13}\frac{\text{J}}{\text{MeV}}\\&=3.2\times10^{-11}\left(\Sigma_{1f}\,\bar{\phi}_{1c}+\Sigma_{2f}\,\bar{\phi}_{2c}\right)V\\&=3.2\times10^{-11}\varepsilon\,\Sigma_{2f}\,\bar{\phi}_{2c}V\end{aligned} \qquad \text{Eq. 6.75}$$

where ε is the fast fission factor and

$$\begin{aligned}\bar{\phi}_{1c}&=\frac{\int_V \phi_{1c}dV}{\int_V dV}=\frac{A\int_V \phi_{1c}dV+C\int_V YdV}{\int_V dV}\\ \bar{\phi}_{1c}&=\frac{\int_V \phi_{2c}dV}{\int_V dV}=\frac{A\int_V XdV+CS_2\int_V YdV}{\int_V dV}\end{aligned} \qquad \text{Eq. 6.76}$$

In spherical geometry the element of volume dV is written as $dV=4\pi r^2dr$, from which we can derive the following equation:

$$\begin{aligned}\bar{\phi}_{1c}&=\frac{4\pi A\int_0^R r\sin(\mu r)dr+4\pi\int_0^R r\sin(\nu r)dr}{\frac{4}{3}\pi R^3}\\&=\frac{3A}{\mu^2R^3}\left(\sin(\mu R)-\mu R\cos(\mu R)\right)+\frac{3C}{\nu^2R^3}\left(-\sinh(\nu R)+\nu R\cosh(\nu R)\right)\end{aligned} \qquad \text{Eq. 6.77}$$

Equation 6.77 can be rewritten as

$$\bar{\phi}_{1c} = \left(\frac{3}{R}\right)\left(\frac{A}{\mu^2}\right)\left(\frac{\sin(\mu R}{R}\right)\left[\begin{array}{l}\left(\frac{1}{R} - \mu\coth(\mu R)\right) \\ \quad + \left(\frac{3}{R}\right)\left(\frac{C}{\nu^2}\right)\left(\frac{\sinh(\nu R)}{R}\right)\left(-\frac{1}{R} + \mu\cosh(\mu R)\right)\end{array}\right] \quad \text{Eq. 6.78}$$

Since

$$\begin{aligned}\alpha &= \frac{X'}{X} = \mu\coth(\mu R) - \frac{1}{R} \\ \beta &= \frac{Y'}{Y} = \nu\coth(\nu R) - \frac{1}{R}\end{aligned} \quad \text{Eq. 6.79}$$

Thus, using Equation 6.72 for expression C, we get

$$\begin{aligned}\bar{\phi}_{2c} &= -\left(\frac{3}{R}\right)\left(\frac{A}{\mu^2}\right)\left(\frac{\sin(\mu R)}{R}\right)\alpha + \left(\frac{3}{R}\right)\left(\frac{1}{\nu^2}\right)\left(\frac{\rho_1\gamma - \alpha}{\beta - \rho_1\gamma}\right)\left(\frac{X}{Y}\right)(A)\left(\frac{\sinh(\nu R)}{R}\right)\beta \\ &= -\left(\frac{3}{R}\right)\left(\frac{A}{\mu^2}\right)X\alpha + \left(\frac{3}{R}\right)\left(\frac{1}{\nu^2}\right)\left(\frac{\rho_1\gamma - \alpha}{\beta - \rho_1\gamma}\right)\left(\frac{X}{Y}\right)AY\beta \\ &= \left[-\frac{\alpha}{\mu^2}(\beta - \rho_1\gamma) + \frac{\beta}{\nu^2}(\rho_1\gamma - \alpha)\right]\frac{3X}{R(\beta - \rho_1\gamma)}A\end{aligned} \quad \text{Eq. 6.80}$$

Similarly, the average thermal flux value in the core is determined to be

$$\bar{\phi}_{2c} = \left[-\frac{S_1\alpha}{\mu^2}(\beta - \rho_1\gamma) + \frac{S_2\beta}{\nu^2}(\rho_1\gamma - \alpha)\right]\frac{3X}{R(\beta - \rho_1\gamma)}A \quad \text{Eq. 6.81}$$

From Equation 6.75, we now get

$$P[\text{Watts}] = 3.2\times10^{-11}\varepsilon\Sigma_{2f}\left[-\frac{S_1\alpha}{\mu^2}(\beta - \rho_1\gamma) + \frac{S_2\beta}{\nu^2}(\rho_1\gamma - \alpha)\right]\frac{3X}{R(\beta - \rho_1\gamma)}(A)\frac{4\pi R^3}{3}$$

From which we have

$$\begin{aligned}A &= \left(\frac{P}{3.2\times10^{-11}\varepsilon\Sigma_{2f}\cdot 4\pi R^2}\right)\left(\frac{(\beta - \rho_1\gamma)}{X}\right)\left(\frac{1}{-\frac{S_1\alpha}{\mu^2}(\beta - \rho_1\gamma) + \frac{S_2\beta}{\nu^2}(\rho_1\gamma - \alpha)}\right) \\ &= \left(\frac{P}{3.2\times10^{-11}\varepsilon\Sigma_{2f}\cdot 4\pi R^2}\right)\left(\frac{(\beta - \rho_1\gamma)}{X}\right)\left(\frac{1}{(S_1 m_1 + S_2 m_2)}\right)\end{aligned} \quad \text{Eq. 6.82}$$

It is now possible to visualize the flux distributions in the core and the reflector by knowing A in terms of reactor power and the constants C, F, and G in terms of A.

Since the neutron flux distribution dictates the heat production distribution, they would be required for the heat transfer computations. They would also be required for materials concerns, radiation damage, and fuel burnup and management.

REFERENCES

1. J. R. Lamarsh, *Introduction to Nuclear Engineering*, Addison-Wesley Publishing Company, 1983.
2. C. J. Lee, *Nuclear Reactor, Physics and Engineering*, 1st edition, John Wiley Publishing Company, 2020.

7 Gamma Thermometer Instrument

7.1 INTRODUCTION

The Gamma Thermometers (GTs) rely on gamma rays (~90%) and high-energy neutrons (~10%) heating a metal block. The specific power of the adjacent fuel rods in the reactor of interest determines how much heat is produced. As a suggested method of measuring the heat created, only a controlled heat channel with strictly constrained dimensions can escape from the metal block to a sink. The temperature dips along the heat path are precisely proportional to the heater's heat rate (Watts/g), which means that power – rather than neutron flux – in the nearby fuel rods determines the temperature drops. A signal proportionate to this power is subsequently produced by a thermocouple, or thermocouples, positioned to monitor the temperature drop along the regulated heat channel. The typical building material is stainless steel. At maximum power, the stainless steel's heat rate varies between 0.5 and 7.5 Watts/g. At full power, temperature reductions between 15°C and 250°C have been chosen. The instruments are still repeatable beyond 500°C, but as the radiation component of heat loss increases, they tend to become non-linear with power. For more than 15 years, GTs have been used to measure power distribution in big heavy water-moderated reactor cores. The following benefits have been established by this GT deployment experience.

1- Signal proportional to local power
2- Non-changing calibration
3- Ruggedness
4- Long service life

In a way, GTs for power measurement rely on fission and fission product gamma rays heating the instrument rod. The linear power of adjacent fuel rods determines how much heat is produced. Only a well-regulated conduction channel and temperature decrease, as detected by a difference thermocouple, allow heat to escape to the coolant sink. The signal, not neutron flux per se, is proportional to fuel power! Figure 7.1 illustrates all of these elements. The signal is robust and consistent with fuel burnup.

7.2 GAMMA THERMOMETER OVERVIEW

The Automatic Fixed In-core Probe (AFIP) system includes the GT. A GT is a simple solid state instrument used to quantify the thermal effects of strong gamma ray fluxes. By offering a cost-effective substitute for the Traversing In-core Probe (TIP) system, GTs hold significant potential for streamlining the Boiling Water Reactor (BWR) in core monitoring for future facilities.

The TIP system provides measurements for calibrating the Local Power Range Monitors (LPRMs) and for adjusting (enhancing) the power distribution from the Core Monitoring System (CMS), as shown in Figure 7.2. It can be based on either gamma- or neutron-sensitive detectors. TIP systems have substantial operating and maintenance (O&M) expenses even though they perform their intended function relatively well.

DOI: 10.1201/9781003590194-7

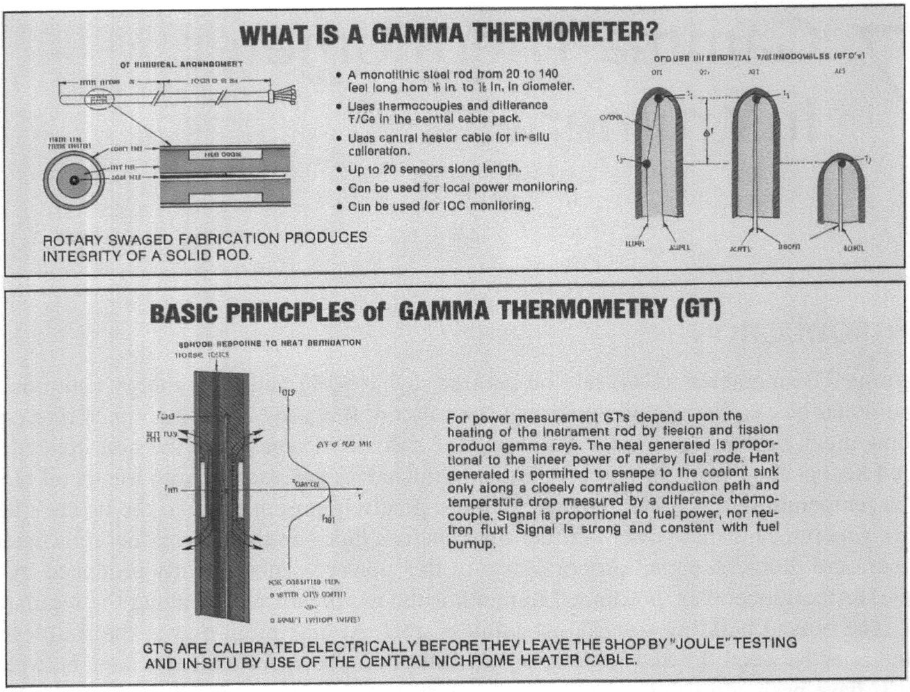

FIGURE 7.1 Illustration of gamma thermometer functionality and capability.

(Source: www.wikipedia.org)

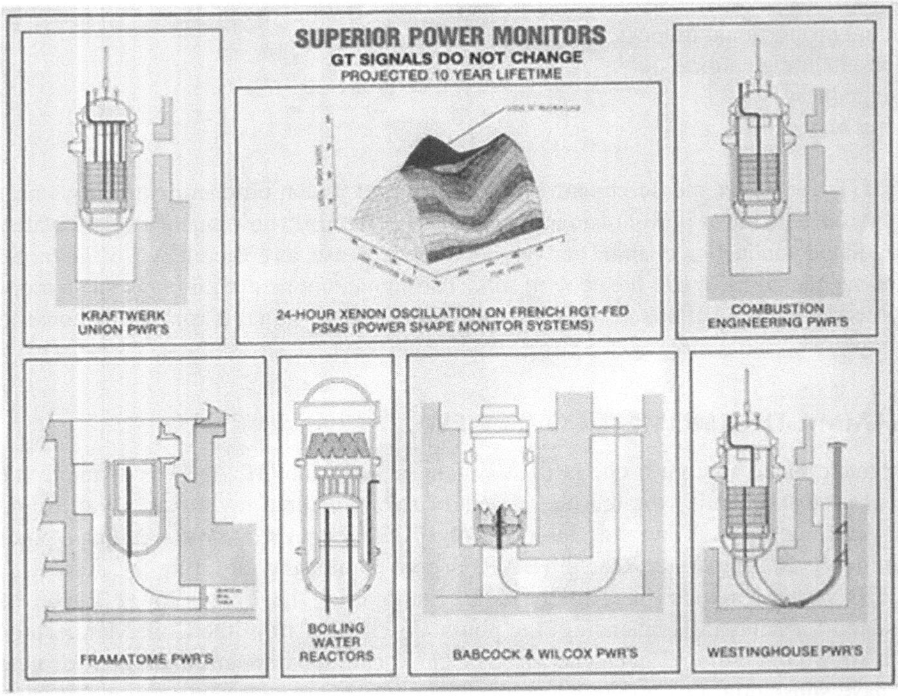

FIGURE 7.2 Superior power monitors driven by gamma thermometer.

(Source: www.wikipedia.org)

This is caused by the system's numerous moving parts, the intricate undervessel tubing that frequently needs to be removed for maintenance, the radiation exposure of maintenance staff near the TIP probes that are being stored, and other issues. The TIP system's duties are carried out by a GT system, which is predicted to be very dependable while lacking undervessel tubing, moving components, and radiation issues.

For the Economic Simplified Boiling Water Reactor (ESBWR), a GT-based instrument system has been chosen over a TIP system, and it is probably going to be used in all next BWR designs.

As illustrated in Figure 7.2, we observe that integration and instrument such as GT within a core of a reactor as a CMS, in various reactors as shown in the Figure 7.2, provide the best Superior Power Monitors (SPMs), where GT signals do not change and have a projected lifecycle of 10 years. The various reactor cores in this figure include KRAFTWERK Union's Pressurized Water Reactor (PWR), Combustion Engineering PWRs, Framatome PWRs, BWRs, Babcock and Wilcox PWRs, and Westinghouse PWRs.

Recently, in the past two decades, General Electric (GE), Boiling Water Reactors (BWR-5) Unit-1 and Unit-2 also integrated with these GTs, located at Laguna Verde Nuclear Power Plant (LVNPP) at Alto Lucero, Veracruz, Mexico. See Figure 7.3.

The LVNPP is situated near Alto Lucero, Veracruz, Mexico, on the Gulf of Mexico coast. It generates around 4.5% of Mexico's electrical energy and is the nation's only nuclear power plant [1]. It uses low enriched uranium (3%) as fuel and is made up of two GE Boiling Water Reactors (BWR-5), each of which has an installed capacity of 682 MW. Operation of Unit-1 (U-1) began on July 29, 1990. Operation of Unit-2 (U-2) began on April 10, 1995. Burns and Roe Inc. were the plant's original architects in 1975. Ebasco Services oversaw the project's design and construction. Mitsubishi Electric was the manufacturer of the steam turbine and other parts. The Mexican government's national electric corporation, Comisión Federal de Electricidad (CFE), owns and runs the facility.

Because of its ability to generate large amounts of electricity, regulate frequency and voltage, and have the lowest operating costs, Laguna Verde has been regarded as a strategic asset for the National Electricity System (SEN). The National Energy Control Center (CENACE), its only customer, receives all of the electricity produced. Planning, directing, and overseeing the transmission

FIGURE 7.3 Laguna Verde nuclear plant.

(Source: www.wikipedia.org)

and distribution of electricity to end customers is the responsibility assigned to CENACE. Since the start of its operations, LVNPP has been categorized by CENACE as a base load power plant.

During the past 5 years, the LVNPP has generated an average of 10.5 TWh of electricity annually, which is enough to cover the needs of almost 4 million residents. Furthermore, the Ministry of Energy approved an extension of U-1's operational license in 2020 for a further 30 years, bringing its total lifespan to 60 years until 2050.

7.3 DESCRIPTION AND PRINCIPAL OPERATION OF A GAMMA THERMOMETER

One tool for determining the gamma flux in a nuclear reactor is a GT. Figure 7.4 shows a cross-section of a normal GT. Short segments of the stainless steel rod's length are thermally isolated from the reactor coolant. In reaction to gamma radiation deposition, the insulation – typically, an argon gas chamber – allows the temperature in the insulated area to increase. The temperature differential

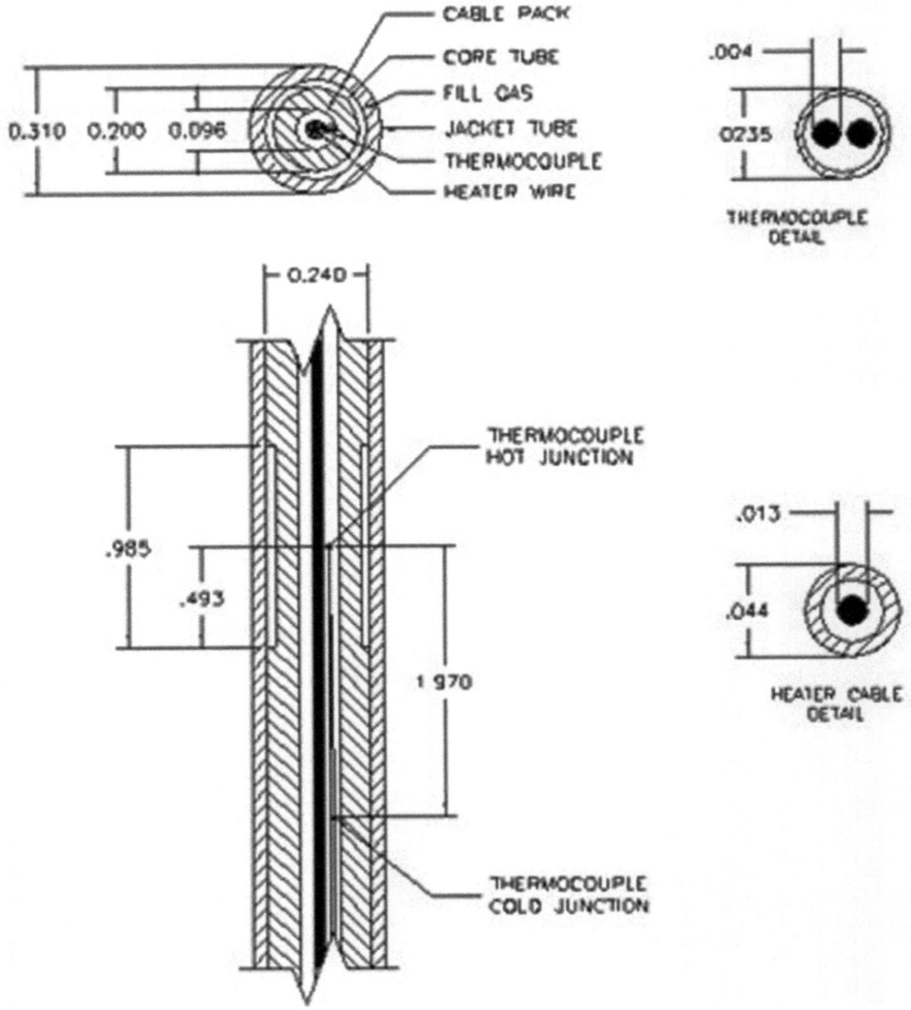

FIGURE 7.4 Cross-section of a gamma thermometer.

(Source: U.S. Nuclear Regulatory Commission)

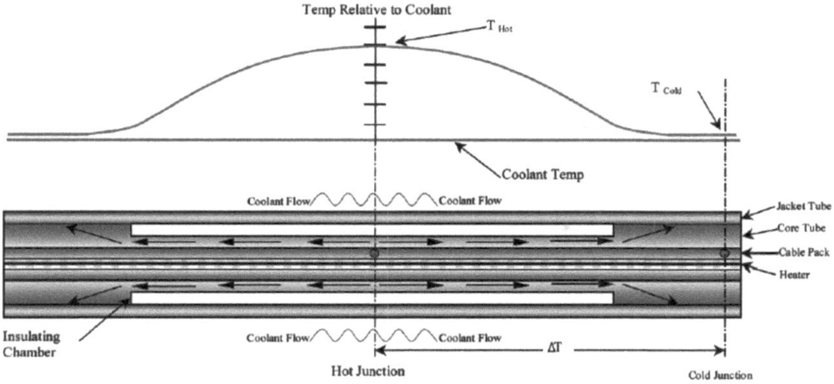

FIGURE 7.5 Gamma thermometer heat flow path.

(Source: U.S. Nuclear Regulatory Commission)

between the rod's insulated and non-insulated parts is measured by a two-junction thermocouple. Thus, there is a clear relationship between the thermocouple reading and the gamma flux. The gamma flux and, thus, the thermocouple reading may be used to indirectly infer the fission density in the surrounding fuel after being appropriately corrected for the quantity and spectrum of gamma rays generated by fission and neutron capture.

Figure 7.4 shows the nominal dimensioning of the jacket tube, core tube, cable pack housing tube, heater cable, thermocouple, and finally gas gap length.

A non-depleting method of determining a nuclear reactor's power level is to use GTs. As seen in Figure 7.5, their monolithic structure enables an anticipated lifespan of more than 10 years.

GTs use fission and fission product gamma rays to heat the instrument rod in order to measure power. The linear power of adjacent fuel rods determines how much heat is produced. Only a carefully regulated conduction channel allows the produced heat to escape to the coolant sink, and a differential thermocouple measures the temperature reduction. Fuel power, not neutron flow, determines the signal! With gasoline burnup, the signal is strong and consistent.

7.3.1 Gamma Thermometer Calibration

Accurate results are always possible thanks to in-situ calibration. Years of worry-free operation are made possible by a robust mechanical design. There is no need to use complex burnup procedures for the output result because there is no fissile material to be concerned about.

The architecture of GTs makes them ideal tools for applications involving inadequate core cooling monitoring (ICCM). An illustration of this type of inadequate core cooling (ICC) monitoring for GTs is shown in Figure 7.6.

Furthermore, use of GT as a collapsed level monitor for ICCM, we may state that the following points also apply:

- As a level indicator
- As shutdown power monitors
- As heat transfer monitors
- To measure reactor temperature

Information on the primary coolant level in a nuclear reactor vessel may be obtained quickly and accurately with the use of GTs. When water is present on the GT's border, a heater within the GT produces a distinctive temperature differential between the differential thermocouples. An

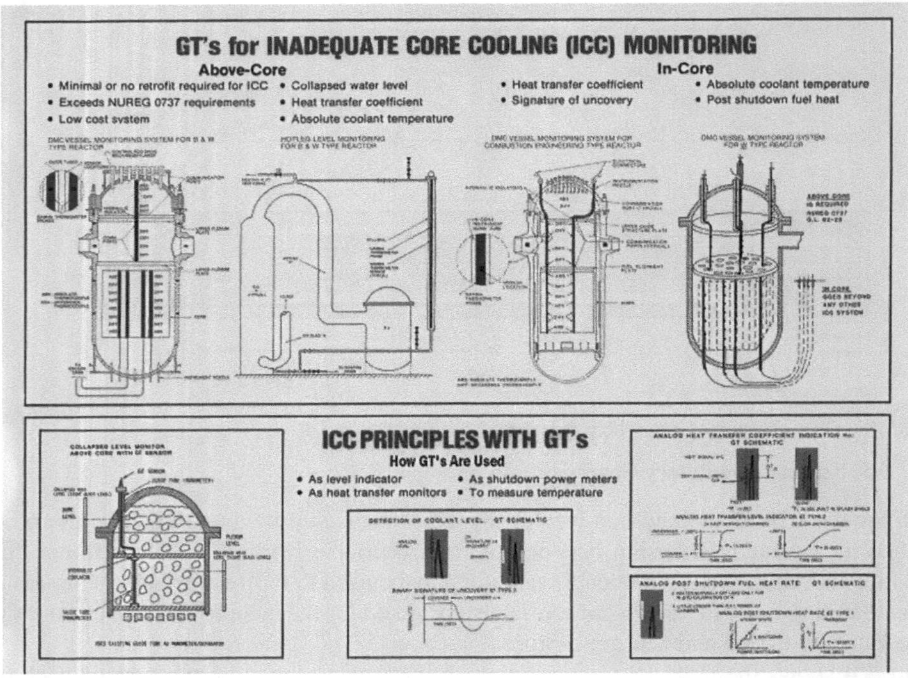

FIGURE 7.6 Gamma thermometers for inadequate core cooling monitoring applications.

(Source: www.wikipedia.org)

emergency reactor shutdown can be triggered by the temperature differential that quickly rises when the rod becomes exposed due to a decrease in the primary coolant level.

Bear in mind, in the case of a reactor such as General Electric Hitachi (GEH) BWRX-300, there would be NO Integrated Plant Design (IPD) unless we have a functionally adequate GT solution for the AFIP for LPRM power adaption and diverse reactor power measurement as an input to the analog only Diverse Protection System (DPS) in place.

7.3.1.1 Gamma Thermometer Assembly

A typical LPRM/TIP assembly is intended to be replaced by the GT assembly. Figure 7.7 shows a schematic of this assembly. A GT rod, many GT sensors (seven in the prototype), and the standard complement of four LPRMs make up the GT assembly.

In order to calibrate the thermometers, which are anticipated to gradually lose sensitivity over the first few months of operation, an Ohmic heating wire is positioned in the middle of the GT rod. A known current is sent through the heater wire as part of the calibration process, and the rise in the sensor response is noted. When calibrated correctly in this way, the GT may carry out the two typical tasks of a TIP: giving the plant monitoring system power shape information and calibrating the LPRMs.

As mentioned in this subsection, Figure 7.7 displays a schematic of an LPRM/GT assembly intended to take the place of a conventional LPRM/TIP assembly. The standard complement of four LPRMs and a GT rod with seven GT sensors make up the LPRM/GT assembly.

Suggested GT material details are listed here.

1) Core Tube and Jacket Tube Materials
   - 316L SS: The materials are seamless and are procured in accordance with ASTM A-213, A-269, or equivalent standards.
2) Cable Pack Material
   - 316L SS

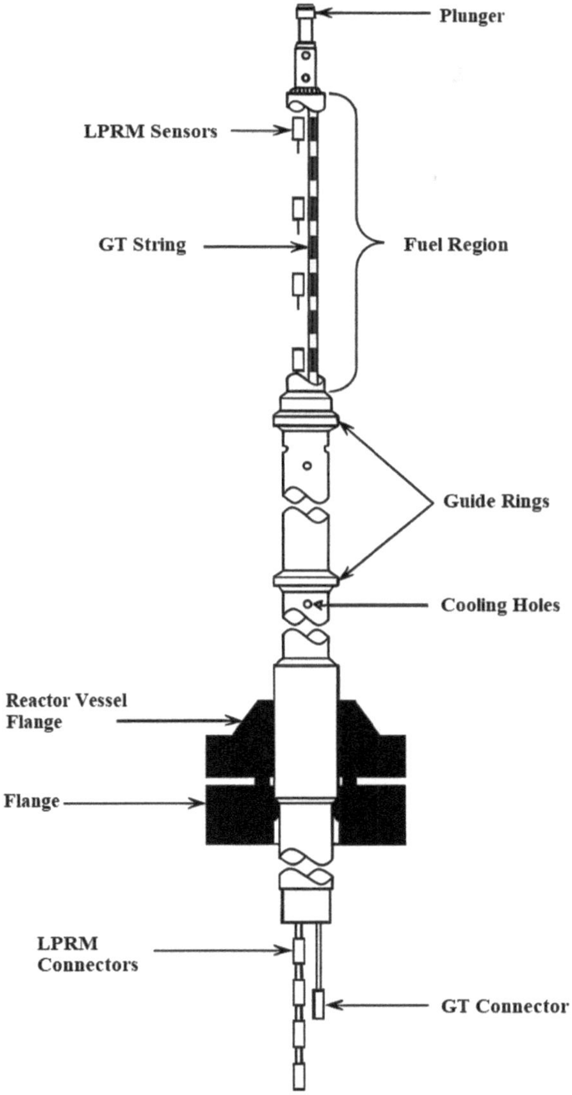

FIGURE 7.7 Typical LPRM/gamma thermometer assembly.

(Source: U.S. Nuclear Regulatory Commission)

3) Heater Cable Materials
- Sheath is 316L SS.
- Heater wire is Nichrome V alloy (80% Ni, 20% Cr).
- Heater wire is copper over approximately 83% of the length and alloy Ni-200 over 17% of the length in the area of the GT sensors.
- Insulation material is Al_2O_3 of 99.8% minimum purity.

4) Thermocouple Materials
- Sheath is 316L SS.
- Thermoelement materials are Chromel and Alumel (Type K calibration per ISA MC 96.1).
- Insulation material is Al_2O_3 with the same specifications as those for the Heater Cable.

- The fill gas pressure is 10 bar or 145 psi at 20°C and thus will be 283.6 psi at a core coolant temperature of 300°C.

7.3.1.2 Gamma Thermometer Prototype

GTs have been used on an experimental basis a few times at BWR plants. The last installation was in Laguna Verde Unit 2 in 2007. For this implementation of GTs, the Gamma Thermometer Data Acquisition System (GT-DAS) operational test equipment was developed. The GT-DAS was created to verify that the design, operability, accuracy, and lifetime of the GT sensors are maintained over the expected life of the LPRM assembly, which houses the GTs. The test equipment is designed to collect operational GT data and compare that data to the flux measurement of the adjacent LPRMs to determine the accuracy and lifetime (minimum 10 years) of the GT sensors.

In the ESBWR, the AFIP system will be used to generate calibration data for the LPRMs. The GT-DAS monitors six assemblies of seven GTs (42 differential thermocouples), each of which provides sufficient data to characterize the performance and accuracy of the GT sensor. A three-domain monitoring (3DM)-GT data-logger communicates with the GT-DAS PLC to collect data. The 3DM-GT data-logger receives these data along with LPRM data from the neutron monitoring system (NMS) through a 3DM process computer.

Generally speaking, the axial elevations of the GTs are tightly controlled during manufacturing and later confirmed by post-manufacturing examination so that the location of every GT sensor is obtained within a narrow tolerance. Bear in mind that after the testing phase of the GTs in the GT-DAS system is completed, the design of the prototype AFIP system will commence. The AFIP system will be utilized on the ESBWR and all other future BWR designs.

For more than a couple of decades, engineers and scientists looked at the application of GT-based reactors and liquid level detectors [1]. The development of this method has a broad connection to liquid level monitoring, specifically with regard to liquid level detectors used to track the coolant level in the core of a liquid-cooled nuclear reactor. It is the outcome of an agreement with the U.S. Department of Energy.

The total amount of coolant in the reactor core vessel is a major consideration while operating a PWR. The core may sustain serious damage if the coolant level drops and exposes it. There is currently no instrument in use for direct monitoring of the reactor core coolant, mainly because of the hostile environment inside the reactor vessel level.

It should be noted that it has long been standard practice to determine the core coolant level in the core coolant loop by using external temperature and pressure sensors. Because measurements collected outside of the reactor core may not always represent core conditions, this approach can occasionally lead to inaccurate conclusions about the coolant level in the core. It is obvious that a direct method of determining the coolant level in a nuclear reactor core that is liquid-cooled is required.

Given the aforementioned requirement, one goal of this strategy is to offer a technology for the direct measurement of a nuclear reactor's core coolant level.

Providing a liquid-coolant level monitor for nuclear reactor cores that does not necessitate the use of extra space inside the reactor core is another goal of this innovation.

Additional goals, benefits, and innovative aspects of the invention will be partially outlined in the description that follows, and partially revealed to experts in the field upon reviewing the following or maybe discovered by the invention's use. The instrumentalities and combinations specifically mentioned in the appended claims may be used to realize and achieve the invention's goals and benefits.

The nuclear reactor core liquid level detector of this invention may include at least one GT placed inside the reactor core in order to accomplish the aforementioned goals, as well as others, and inline with the current invention's objective, as embodied and comprehensively disclosed herein. The GT's sensing zone has a dual differential thermocouple that is electrically designed and positioned geometrically to produce two voltage signals at distinct outputs. The ratio of these signals indicates the amount of surface heat transfer at the sensing region. The presence or absence of core coolant at the sensing region level is indicated by an output signal proportionate to the ratio of the two voltage signals, which is produced by a signal processing circuit.

The initial purpose of the GT prototype was to detect local power production in a water-cooled reactor. These thermometers respond to gamma ray heating and are dispersed throughout the reactor core. They are still used exclusively for that purpose today.

Consider the technology that is offered that uses a modified GT to measure the amount of liquid coolant in a nuclear reactor core. In order to acquire a clear indication of the presence or absence of coolant liquid at the GT position, thermocouple connections modify the GT, which is often used to monitor the local core heat generating rate (reactor power). By comparing the signals from the thermocouples at the thermometer position, a signal processor produces a signal based on the thermometer surface heat transfer coefficient. When coolant liquid goes below the thermometer location, a change in surface heat transfer causes a signal to be created that directly indicates coolant loss. Reactor power at the thermometer position has no bearing on the coolant indicator loss. Furthermore, the standard power monitoring function may still be performed using the same thermometer.

In conclusion, the development of a reactor core liquid level detector based on a GT has to do with liquid level monitoring in general and liquid level detectors for checking the coolant level in a liquid-cooled nuclear reactor core in particular. It is the outcome of an agreement with the U.S. Department of Energy.

7.3.1.3 Advantages of the Gamma Thermometer System

The GT system has many advantages over the TIP system.

- No moving parts to wear out
- Reduced occupation exposure
- Reduced radioactive waste
- No open tube penetrations to the containment
- Reduced space requirements in the reactor building
- More frequent LPRM calibrations
- More frequent adaptive monitoring calculations

7.4 GAMMA THERMOMETER SYSTEM REQUIREMENTS

Following are the general approaches to the GT system requirements.

7.4.1 Design Requirements

There are two general functional requirements for the GT system.

1. Give precise information so that the LPRMs may be calibrated. Such calibration must take place in accordance with the technical criteria of each given plant, but it may happen considerably more frequently, even several times in a single day, when the GT system is used. Only under steady state core circumstances is this function carried out.
2. Provide the core monitoring system axial shape information. Only under steady state core circumstances is this information used.

7.4.1.1 Codes and Standards

All relevant U.S. Codes and Standards must be followed in the design, production, and testing of LPRM/GT assemblies.

LPRM/GT assemblies must also adhere to the laws and regulations of the country in which they are produced if they are intended for use outside of the United States.

7.5 GAMMA THERMOMETER SYSTEM DEFINITION

The following subsections define and describe the GT system definition both from hardware and software points of view.

7.5.1 Typical Hardware Components

Figure 7.8 displays the hardware setup for a standard GT core monitoring system. New parts such as the LPRM/GT assemblies, a control cabinet, the Data Acquisition System (DAS), and the Heater Power Supplies (HPSs) are included in the hardware. Table 7.1 provides a list of the new components. The quantity of each kind of component needed for an average system is also mentioned.

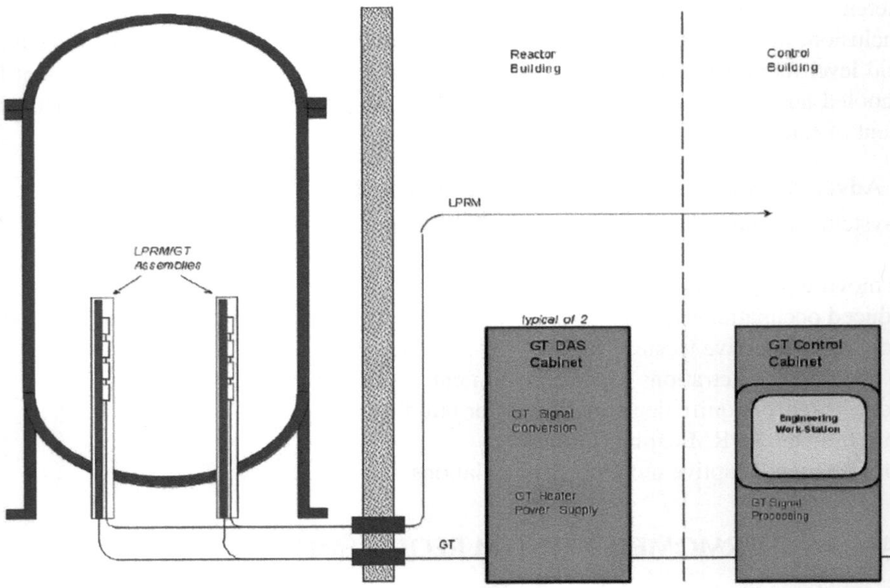

FIGURE 7.8 Typical gamma thermometer core monitoring system configuration.
(Source: U.S. Nuclear Regulatory Commission)

TABLE 7.1
Gamma Thermometer Core Monitoring Component List

Component Name	Number (Typical)
LPRM/GT Assemblies	65
GT Sensor per Assembly	7
Data Acquisition System Cabinet	2
Heater Power Supplies	16
GT Control Cabinet (including Work Station)	1

7.5.1.1 Local Power Range Monitor/Gamma Thermometer Assembly

In Figure 7.7, a typical LPRM/GT assembly is displayed. The LPRM/GT assemblies are made to satisfy all of the standards and have four LPRMs, just like regular LPRM assemblies. Each assembly in the ESBWR arrangement includes seven GT sensors on a single GT string. Each pair of LPRMs has one GT sensor next to it and another in the middle. Nine GT sensors are employed in alternative designs, and they are positioned as follows: one next to each LPRM, one halfway between each pair of LPRMs, one halfway between the lowest LPRM and the bottom of the core, and one halfway between the top of the core and the highest LPRM. Each of the four GT sensor configurations is placed next to an LPRM.

The operating temperature, neutron flux, gamma dose rate, and seismic loadings are all rated according to the same environmental design criteria as the conventional LPRM assemblies. At least as long as the LPRMs, the GT sensors are made to last.

A calibration tube is incorporated inside the assembly for demonstration systems only in order to maintain the TIP system's functionality. This makes it possible to compare the TIP and GT systems directly.

7.5.1.2 Data Acquisition System

Converting the analog signal from the GT measurements to a digital value is the primary duty of the DAS. The environmental design ratings are comparable to those of the reactor building's other electronics. To eliminate noise, the DAS may also use digital filtration.

7.5.1.3 Heater Power Supply

During calibration, the HPS's function is to provide the internal GT heaters with a DC electrical current. For each GT sensor to be calibrated accurately, the current must be large enough. For cost-effective reasons, the HPS could be multiplexed such that a single power source can support many GT assemblies in succession. The HPS is part of the DAS cabinets in Figure 7.8.

7.5.1.4 Gamma Thermometer Control Cabinet

The main interface to the GT system is an engineering work station (EWS) located in the GT control cabinet. Via the plant's communications system, the EWS interacts with the DAS, ATLM, and 3D MONICORETM cabinets, among other devices.

7.5.2 Typical Software Components

The typical software for the GT CMS is shown in Figure 7.9. It consists of four main modules: a GT monitor and signal conditioning module, a GT calibration module, a 3D simulator, and a user interface module. Each of these modules is described in the sections that follow.

7.5.2.1 Gamma Thermometer Monitor Module

The GT monitor module (GTMM) is responsible for conditioning the GT signals. This conditioning will occur continuously while the reactor is in operation. The time step for digital filtration will be 1 second or less. The specific functions are as follows:

1. Digitally filter the GT signals.
2. Digitally filter the GT heater current and voltage readings during GT calibration.
3. Block transmittal to the 3D simulator or otherwise mark as unusable GT signals associated with GT strings that are being calibrated.

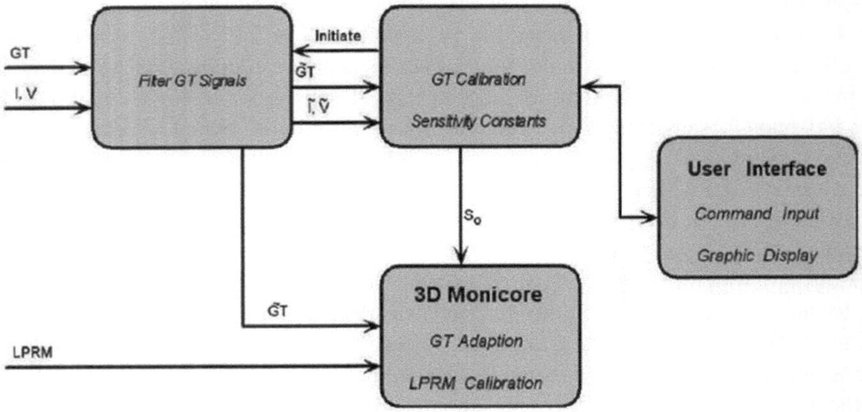

FIGURE 7.9 Typical gamma thermometer core monitoring software diagram.

(Source: U.S. Nuclear Regulatory Commission)

7.5.2.2 Gamma Thermometer Calibration Module

The GT calibration module (GTCM) is responsible for maintaining the accuracy of the system by initiating calibration of the GTs at appropriate times. This initiation is automatically performed, or it may be requested by the user. The specific functions are as follows:

1. Calibrate the GTs.
2. Initiate the GT calibration by turning on the heater power supplies.
3. Closely monitor and record the GT sensor readings during calibration.

Calculate the sensitivity constant S_0 .

7.5.2.3 3D Simulator

The plant 3D simulator is responsible for calculating the core power distribution and the associated thermal limits, such as the critical power ratio and the linear power density. Some of the specific functions are

1. Compute the core power, void, and exposure distributions.
2. Calculate calibration constants for the LPRMs.
3. Adapt the power distribution to the current GT readings.
4. Adapt to the LPRMs (if necessary, due to plant transient state).
5. Compute GT and LPRM exposures.

7.5.2.4 User Interface

The user interface is responsible for accepting user command input and displaying calculated results in the most useful possible way. Some of the major functions of the user interface are

1. Accept user commands.
2. Provide color graphic displays.
3. Initiate GT calibration.
4. Display calibration results.
5. Display simulation results.

7.5.3 Gamma Thermometer Response Model

A different aspect of the GT response model has been described and defined in the following subsections.

7.5.3.1 Gamma Thermometer Response to Gamma Energy Deposition

The response of a GT is the relationship between the voltage across the thermocouple pair to the energy deposition in the sensor. The energy deposition in the sensor may be due to either gamma rays or internal resistance heat during calibration or both.

The GT response, denoted by the symbol U, is the voltage across the thermocouple pair as measured in millivolts. The response is proportional to the gamma energy deposition, w, in units of Watts per gram (W/g):

$$U = S \cdot w \quad \text{Eq. 7.1}$$

where S is the sensitivity (mV-g/W). Due to the variation in the thermal properties of the thermometer and the non-linearity of the thermocouple response to temperature, the sensitivity is somewhat dependent on temperature. This is normally accounted for with a small correction term, as follows:

$$U = S_0 \left(1 + \alpha U\right) \cdot w \quad \text{Eq. 7.2}$$

where S_0 is the base sensitivity and α is a correction factor, which has a negative sign and a small magnitude (less than 0.05). The value of α is determined for each GT sensor during factory calibration. Solving this equation for U results in

$$U = \frac{S_0 \cdot w}{1 - \alpha (S_0 \cdot w)} \quad \text{Eq. 7.3}$$

In normal operation, the thermocouple output is known, and it is the energy deposition that needs to be determined. In this case, the previous equation is solved for w.

$$w = \frac{U}{S_0 \left(1 + \alpha U\right)} \quad \text{Eq. 7.4}$$

For the equations above, the GT software utilizes the specific correction factor, α, for each GT sensor.

7.5.3.2 Gamma Thermometer Factory Calibration

The main purpose of the factory calibration is to determine the ratio of the heater wire resistance (Ohms/cm) to the mass of the sensor (g/cm). The calibration of GT sensors is an essential step in maintaining the accuracy. Each GT is individually calibrated in the factory by both the internal heater wire method and the Joule method.

The heater calibration method consists of passing a known electrical current through the GT heater wire and recording the thermocouple response as demonstrated in Figure 7.10 (shape a). This same method is used during normal in-plant operation. The specific heat rate is limited to approximately 1 W/g. Multiple measurements are made for each sensor, ranging from zero to approximately 1 W/g.

The Joule calibration method consists of passing a known electrical current through the entire GT rod, including the core tube and the jacket tube, as seen in Figure 7.10 (shape b). This results in uniform internal heating. Very high specific heat rates are possible with this method. Multiple measurements are made for each sensor corresponding to heat rates from zero to greater than 4 W/g.

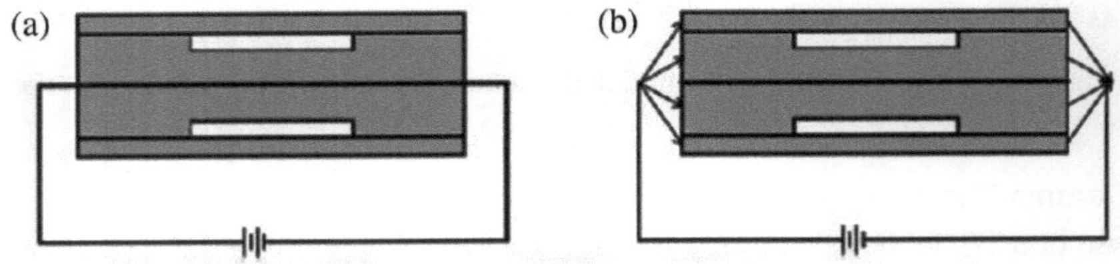

FIGURE 7.10 Gamma thermometer factory calibration methods.

(Source: U.S. Nuclear Regulatory Commission)

The factory heater calibration results may be used alone, without reference to the Joule results. However, in the current study, the heater ratings were adjusted by the ratio of the Joule calibration sensitivity to the heater calibration sensitivity. This has the effect of making both calibration methods consistent.

The sensitivities determined by either factory method are higher than the sensitivities determined in an operating BWR that sensitivity changes are accounted for by in-plant calibrations as discussed in the next subsection. This is due to the difference in temperature between the factory at about 20°C and the reactor at about 286°C.

The higher temperatures cause changes in basic material properties that affect GT sensitivity. These include the thermal conductivity of the core and jacket tubes and the response of the thermocouples.

7.5.3.3 Gamma Thermometer In-Plant Calibration

GTs must be periodically recalibrated due to small changes in their physical properties as a result of exposure to high temperatures and intense radiation. This calibration is a simple procedure involving the application of a known electric current to the built-in heater wire and recording the thermocouple response. The calibration procedure is non-disruptive of plant operation and may be even performed when the plant is not in operation as long as sufficient cooling flow is available. The only requirement is that the plant must be in a steady state, constant power mode.

Based on the experience of the in-plant tests, the preferred method is the "Fixed Alpha Procedure". In this procedure, α , the non-linear correction constant, is predetermined from the factory tests for each GT sensor prior to installation in the plant. The base sensitivity, S_0 , is then determined from the following procedure:

Let I represent the heater current in amps that is supplied to the thermometer by the heater power supply. The energy deposition, P_H , to the GT in units of Watts per gram is equal to the square of the current times the heater resistance rating, R_H :

$$P_H = I^2.R_H \quad \text{Eq. 7.5}$$

The heater rating is determined during manufacturing by dividing the resistance of the heater wire in Ohms per cm by the linear density of the thermometer in grams per centimeter. After the initial application of the current, a wait period is required for the thermocouple response to reach thermal equilibrium (at least five-time constants) prior to collecting data. A typical calibration procedure is as follows:

To address the thermal lag (per GT system requirements the time constant is less than 20 seconds, see Table 7.2) during GT calibration, the following process is utilized:

1. A known electrical current is applied to the GT heater cable.
2. The electrical current is held constant for a minimum of five-time constants.

TABLE 7.2
System Requirements Specification (Specific for the Tokai Plant Test)

	Description	Required Value	Actual Value (as-built)	Meets Requirement
1	Expected Gamma Heating Range	2.4 w/g	Test to 4.3w/g	√
2	Thermocouple Type	K	K	√
3	GT Sensitivity (beginning-of-life) at 286°C	1.5 mV-g/W ± 20%	1.36–1.50 mVg/W	√
4	GT Sensitivity (beginning-of-life) at 20°C	2.0 mV-g/W ± 20%	1.99 mV-g/W ± 5%	√
5	GT Heater Heat Rate (286°C, 3A)	1.0 W/g ± 20%	0.97–0.99 W/g @ 25°C	√
6	Minimum Heater Current Capability	3.0 A	3.0 A	√
7	Minimum GT Response to Rated Heater Current	1.0 mV	1.7–2.0 mV	√
8	Detector Thermal Response Time	<20 s	18–19 s	√
9	Accuracy of Heater Wire Linear Resistance (as measured for an individual GT sensor)	± 1.0% Ω/cm (1σ)	±0.48%	√
10	Accuracy of Sensor Linear Mass (as measured for an individual GT sensor)	±1.0% g/cm (1σ)	±0.28%	√
11	Minimum Cooling	No boiling when LPRMs are in use		NT
12	Thermocouple (TC) resistance	<1.5 kΩ	730–990 Ω	√
13	Minimum TC Insulation Resistance lead-to-sheath (20°C)	1 x 10^9 Ω	2–5 x 10^9 Ω	√
14	Minimum Heater Wire Insulation Resistance (20°C)	1 x 10^9 Ω	2 x 10^{11} Ω	√

3. After five-time constants and while continuing to hold the current constant, a statistical average of each GT sensor output is determined.
4. Different levels of electrical current may be applied during the calibration process (e.g., 1, 2, and 3 Amps) that utilize this same process for each level of current.

The output of the thermocouple after it stabilizes, U_c, depends on the sum of the gamma energy deposition and the heater energy deposition, shown as follows:

$$w + P_H = \frac{U_c}{S_0\left(1+\alpha U_c\right)} \qquad \text{Eq. 7.6}$$

Substituting the expression for w from the previous equation and solving for the sensitivity results, we get

$$S_0 = \frac{\left[\dfrac{U_c}{1+\alpha U_c} - \dfrac{U}{1+\alpha U}\right]}{P_H} \qquad \text{Eq. 7.7}$$

The sensitivities determined in this way, along with the α 's determined by the factory tests for each GT sensor, are used in Equation 7.4 to calculate the gamma energy deposition.

Based on the analysis of in-plant test data and to ensure the root mean square (RMS) error is maintained at less than 1%, the optimum GT calibration interval is as follows:

- **Procedure**

1. During the first 500 hours of in-plant operation of a new GT, perform GT calibration every 2 days.
2. Between 500 and 1,000 hours of in-plant operation of a new GT, perform GT calibration every week.
3. Between 1,000 and 2,000 hours of in-plant operation of a new GT, perform GT calibration every 2 weeks.
4. After 2,000 hours, perform GT calibration every month.
5. Before every LPRM gain adjustment, perform GT calibration.
6. At the beginning of a new cycle, when new GTs are installed; repeat steps 1 through 5 for each new GT string.

7.5.3.4 Void Fraction Response and Bypass Subcooling

The next subsections address the topics of bypass subcooling and void fraction response in relation to subcooling.

7.5.3.4.1 Void Fraction Response

At other facilities, such as GEH, voiding is not anticipated to happen within the LPRM/GT assembly under typical operating circumstances. To address the sensitivity of the GT reaction to voiding, a study was conducted. The total gamma cross-section typically rises exponentially by atomic numbers, and the GT can detect gamma rays from a variety of energy spectra. The bypass vacuum has less of an impact on the gamma attenuation effect. The energy deposition into the GT sensors is compared to two distinct bypass void circumstances in the Monte Carlo nuclear code's gamma transport calculation in order to validate this effect.

Using a [[70%{3}]] void history, a standard GE14E lattice [2] is chosen, with the level-D LPRM situated assuming the maximum [[5%{3}]] bypass void with [90%] in-channel instantaneous void. This state aligns with ESBWR's feedwater temperature/power operational domain. The in-channel void situation mentioned above is present in the other comparable instance, but there is no bypass void. The impact of various gamma source distributions brought on by various bypass vacuum situations is also included in the comparison. In GT sensors, the relative energy deposition difference is [0.56%].

The literature has another study on GT sensitivity to cavities in the instrument tube [3]. Toshiba used the Multiple-Use Safety Experimental (MUSE) facility to conduct thermohydraulic testing on an experimental GT assembly, as detailed in this reference.

Saturation temperature, saturation pressure, and vacancy fraction are BWR's core conditions that may be replicated at the MUSE facility. Verifying the GT's sensitivity to variations in fluid conditions was the aim of the test. The test item was subjected to voiding caused by an intake heater. With the exception of length, the test piece was produced in accordance with the identical requirements as the in-plant test specimen. Figure 7.11, taken from Reference [3], illustrates the connection between the GT output signal and heating power.

Across all measured heating powers, the change in the GT output signal is insignificant, even at the maximum void percentage of 55%.

7.5.3.4.2 Bypass Subcooling

The ESBWR will use feedwater temperature control in addition to control rods to regulate reactor output. However, a GT calibration must be carried out as outlined in Section 7.5.3.3 and specified core operating conditions must be set as outlined in Section 4.5 before LPRM calibrations may take

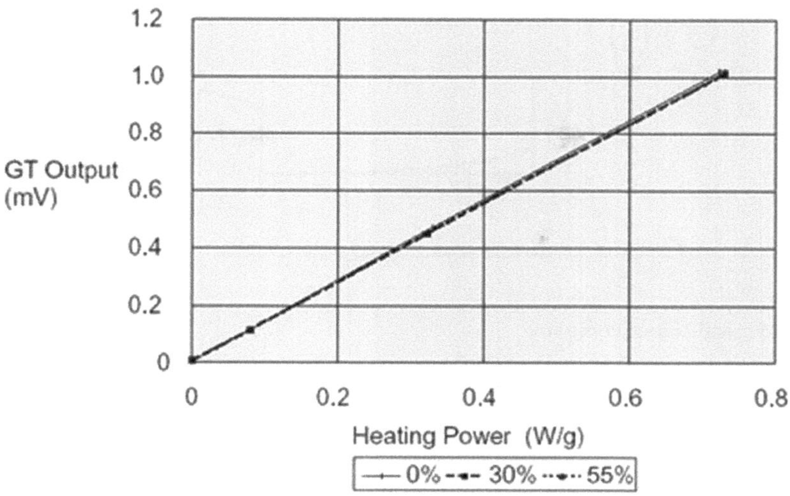

FIGURE 7.11 Gamma thermometer output signal versus heating power under varying void fraction [3].

place. Consequently, any sensitivity variations brought on by power and coolant temperature are taken into consideration throughout the GT calibration process for the LPRM calibration procedure.

7.6 DELAYED GAMMA COMPENSATION

For the purpose of LPRM calibration, the non-steady state response of the GT system is not significant since this operation is performed only in the quasi-steady state. However, where core monitoring is concerned, transient response is very important, since poor response might limit or restrict the use of GTs in this application.

One factor that limits the transient response of GTs is the effect of delayed gammas. Although most of the gammas that affect the GTs are in fact prompt and deposit their energy immediately, the delayed gammas are sufficiently numerous to delay the GT transient response. Fortunately, it is possible to compensate for this effect and improve the response.

A model for delayed gamma compensation (DGC) was developed in order to compensate for the time lag in the GT response. In this model, allowance is made for 10 groups of delayed gammas, each with its own time constant τ_n ranging from a low of 8 seconds to a high of slightly over 1 day. The response of each GT is divided into 11 modes (including one prompt mode) as follows:

$$\overline{GT}_i(t) = a_0 GT_i(t) + \sum_{n=1}^{10} \frac{a_n}{\tau_n} \int_{-\infty}^{t} dt' GT_i(t') e^{-(t-t')/\tau_n} \qquad \text{Eq. 7.8}$$

where $\overline{GT}_i(t)$ is the measured GT reading $GT_i(t)$ is the corrected reading and $\sum_{n=0}^{10} a_n = 1$. Each mode represents one or more fission product nuclides that have similar half-lives. The time constant τ_n is an average for the nuclides in the group n and the constant a_n is the sum of the product of the cumulative yield times the gamma energy. The constants τ_n and a_n may be obtained from any of several sources. See Figure 7.12.

Each mode is preconditional with a digital filter with a filter time constant appropriate to the mode and subjected to digital compensation using the model time constant. The 11 modes are then summed to provide the corrected GT response. The following table shows the values of the constants used for the purpose of this chapter as Table 7.3 and this subsection as well.

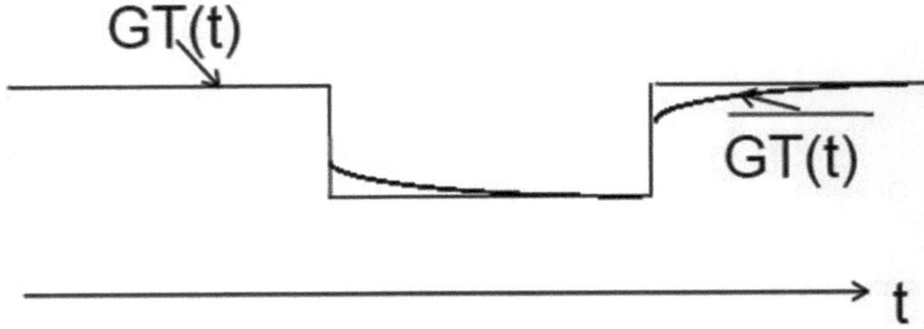

FIGURE 7.12 Several source concept.

(Source: U.S. Nuclear Regulatory Commission)

TABLE 7.3
Constants for Delayed Gamma Compensation

Mode	a_n	τ_n (seconds)
0	0.76519	1.83800E+01
1	0.04129	8.48320E+00
2	0.03725	2.90994E+01
3	0.04250	8.50196E+01
4	0.02054	2.77277E+02
5	0.01488	7.16178E+02
6	0.02867	1.59724E+03
7	0.02294	5.28430E+03
8	0.01214	1.81574E+04
9	0.00802	4.76849E+04
10	0.00559	1.00060E+5

In Table 7.3, the average time constant (Joule method) of the detectors in the Joint Study.

There are four transient response situations that were examined very carefully during the in-plant testing by companies like GEH, such as

- Startup
- Flow change
- Control blade movement
- Power down at end-of-cycle (EOC)

The constants in Table 7.3 are based on ANSI/ANS-5.1-1994, decay heat power in Light Water Reactors (LWRs), Table 7.2, Parameters for ^{235}U Thermal Fission (α and λ in 23 groups). Time constants are simply $1/\lambda$. Groups 4 through 13 were used, covering time constants between roughly 8 seconds to a little more than 1 day.

Although a DGC model was discussed in this section and demonstrated utilizing the table above in Section 7.6.1, it will not be considered a part of the design basis of the GTs. Additionally, the GTs will not be utilized during non-steady state conditions as this function is performed by the LPRMs. To ensure the effects of delayed gamma and the thermal lag are accounted for, the following procedure is to be utilized:

- **Core Requirements for Gamma Thermometer Use**

1. Establish and maintain core, as close to quasi-steady state conditions and xenon equilibrium as practical, prior to and during GT data acquisition.
2. No core power changes greater than 1% for a minimum of 1 hour prior to and during GT data acquisition.
3. No changes in control rod positions 1 hour prior to and during GT data acquisition.
4. No APRM gain adjustments during GT acquisition time.
5. Ensure the minimum number of GT sensors and strings are operable per Section 7.6.2.

7.6.1 Response during Non-Steady State Conditions

Although the GTs are not considered for LPRM calibration during non-steady state conditions, the following sections provide an analysis of the GT response using the DGC model discussed in Section 4.5. Again, this model is not considered for the design basis of the GT.

The observation of the GT response during non-steady state conditions was an important part of the Tokai 2 in-plant test. As mentioned previously, non-steady state response is not important for the calibration of LPRMs as this is only performed during steady state conditions as discussed in Section 4.5 (core monitoring is only performed during steady state conditions also). However, for providing continuous core monitoring it is very important.

One of the factors that limit the non-steady state response of GTs is the effect of delayed gammas. The model for DGC which was used in this study was described previously in Section 4.5.

There are four non-steady state conditions examined in the in-plant test: startup, flow change, control blade movement, and power down at EOC. Each of these situations can be considered in a separate subsection of Section 7.2.6 of reference [4].

7.6.2 Gamma Adaptive Core Monitoring Accuracy

The next sections examine the power uncertainties related to the GT-based monitoring system. The list of uncertainties outlined in [5] is consistent with the uncertainty components. Establishing the uncertainties relevant to the linear heat generation rate (LHGR) and minimum critical power ratio (MCPR) computations for the ESBWR is the aim of the investigations, which are based on simulated GT measurements. The list of LHGR uncertainties is shown in the next section "Conclusion". Keep in mind that the GT-based monitoring system is not the only source of uncertainty.

7.6.2.1 Minimum Monitoring Configuration

As with the TIP system, a minimum monitoring configuration is required to accurately provide LPRM calibration and core monitoring data. Option 1 for the GT adaptation defines the minimal GT-based monitoring configuration, as described in chapter 8 of [4], with the following description:

1. At least one-third of the GT strings have to be fully operable, that is, all seven sensors' signals have to be available in 33% of the GT strings for the monitoring system.
2. It is allowed to have up to one-third of the total number of GT strings with two failed GT sensors. The remaining operable GT sensors in those strings contribute their signals to calculate core average adapting parameters.
3. Up to one-third of the GT strings can be declared inoperable. The monitoring system will supply the adapting information from core average values with no impact on the total bundle power uncertainty.

TABLE 7.4
LHGR Calculation Uncertainties ESBWR GE14E Core Loading

		ESBWR Revised		
LHGR Uncertainty	**NEDE-3260P-A NEDC-3269-A**	**NEDC-33239 Rev 2**	**Core Loading GE14E Lattice**	
Infinite Lattice Pin Power	1.44%	1.48%	1.48%	1.87% [b] [c]
Manufacturing	1.25%		1.25	
Gradient Effect	2.70%		2.30% [a]	
Channel Bow	1.80%		2.00% [a]	
PAL Uncertainty	1.56%			
PAL Uncertainty GT			1.56%	2.36% [a] [b]
TIP Random	1.20%			
GT Random Uncertainty			1.21%	
P4B Uncertainty TIP	2.70%			
P4B Uncertainty GT			2.70%	
Update/Failure Uncertainty	2.00%			
Update Uncertainty			1.34%	
Failure Uncertainty			1.16%	
GT Interpolation			2.03%	
Total Uncertainty	5.24%		**5.61%**	**5.99%**

[a] Value from NEDC-33173P revision.
[b] Value obtained with 95% upper tolerance limit.
[c] Response to RAI 4.3-2 S01 from Reference [6].

If those conditions are met, the total update/failure uncertainty for the MCPR and LHGR calculations will bound the thermal margin calculations with a calibration interval of 750 MWd/T. See Table 7.4.

7.7 CONCLUSION

A thorough in-plant test program has successfully assessed the GT technology as a substitute for the TIP technology. The test program's main goals have been accomplished.

1. GT sensor accuracy relative to gamma TIP, neutron TIP, and LPRM measurements has been evaluated.
2. GT sensor reliability under BWR operating conditions has been established.

Two BWRs have had their GT sensitivity trends monitored for a total of three operating cycles. In the most recent test, the sensitivity trend showed a steady fall for the remainder of the cycle after a somewhat quick initial climb during the first 500 hours of operation. However, any variations in sensitivity may be adjusted for by in-plant calibration.

The steady state GT response has been compared to the LPRM response, gamma TIP, and neutron TIP responses. The LPRM reaction and the GT response under various scenarios (starting, flow change, control blade change, and power down) have been compared. The accuracy of the GT sensors was assessed using a mix of core monitoring, in-plant, and manufacturing tests:

1. For the Tokia 2 test, the factory tests proved that the GT sensors met all of the requirements of the system requirements specification.
2. The in-plant tests proved the accuracy, linearity, and range of the GT sensors with respect to the TIPs and LPRMs.

The GT in-plant test program's final conclusion is that the GT system can be utilized for both power shape monitoring and LPRM calibration in place of a TIP system.

Reference [7] provides further details, and nuclear instrumentation is used in the core of reactors like the PWR and BWR, which are members of the LWR family.

ACRONYMS AND ABBREVIATIONS

Term:	**Definition**
3DM:	Three-Domanial Monitoring
AFIP:	Automatic Fixed In-core Probe
ANFM V:	Advances in Nuclear Fuel Management V
BOC:	Beginning of cycle
BWR:	Boiling Water Reactor
CFE:	Comisión Federal de Electricidad
CFR:	Code of Federal Regulations
CMS:	Core Monitoring System
CPR:	Critical Power Ratio
DACS:	Data Acquisition and Calibration System
DAS:	Data Acquisition System
DLC:	Data Logging Computer
DPS:	Diverse Production Protection System
EOC:	End Of Cycle
EWS:	Engineering Work-Station
ESBWR:	Economic Simplified Boiling Water Reactor
FS:	Full Scale
GETAB:	General Electric Thermal Analysis Basis
GNF:	Global Nuclear Fuel Americas LLC
GT:	Gamma Thermometer
GTCC:	Gamma Thermometer Control Cabinet
GTCM:	Gamma Thermometer Calibration Module
GTMM:	Gamma Thermometer Control Module
GTCU:	Gamma Thermometer Control Unit
GTM:	Gamma Thermometer Monitoring Unit
HPS:	Heater Power Supply
HSU:	Heater Switching Unit
LHGR:	Linear Heat Generation Rate
LPRM:	Local Power Range Monitor
LTR:	Licensing Topical Report
LV2:	Laguna Verde 2
ICC:	Inadequate Core Cooling
ICCM:	Inadequate Core Cooling Monitoring
MCPR:	Minimum Critical Power Ratio
MLHGR:	Maximum Linear Heat Generation Rate
MOC:	Method of Characteristics
MUSE:	Multiple-Use Safety Experimental
NA:	Not Applicable

NMS: Neutron Monitoring System
NRC: Nuclear Regulatory Commission
O&M: Operation and Maintenance
RMS: Root Mean Square
PANAC11: PANACEA, GEH BWR Core Simulator
SRS: System Requirements Specifications
TC: Thermocouple
TIP: Traversing In-core Probe
V&V: verification and validation

REFERENCES

1. T. J. Burns, Gamma-Thermometer-Based Rector-Core Liquid-Level Detector, Patents-US—274121, DE*2 011763, February 27, 1979.
2. NEDC-33239P, Rev. 4, GE14 for ESBWR Nuclear Design Report, Global Nuclear Fuel, March 21, 2009, Agencywide Documents Access, and Management System (ADAMS Accession Nos. ML090970167, ML090970168, ML090970169).
3. NUREG-0800, Standard Review Plan, Section 15.0.2, Review of Transient and Accident Analysis Methods, U.S. Nuclear Regulatory Commission, December 2005 (ADAMS Accession No. ML053550265).
4. GEH document of NEDE-33197P-A, Revision 3 Class III October 2010, eDRF Section 0000-0041-9907 Licensing Topical Report, Gamma Thermometer System for LPRM Calibration and Power Shape Monitoring, Copyright 2008-2010, GE-Hitachi Nuclear Energy Americal, LLC.
5. Power Distribution Uncertainties for Safety Limit MCPR Evaluations, NEDC-32694PA, August 1999 of GEH.
6. MFN 08-293, Response to Portion of NRC Request for Additional Information Letter NO.1 06 – Related to ESBWR Design Certification Application – RAI Numbers 4.2-12 Supplement, 2 and 4.3-2 Supplement 2, April 2008.
7. A. Zamani Paydar, S. K. Mousavi Balgehshiri, B. Zohuri, *Advanced Reactor Concepts (ARC): A New Nuclear Power Plant Perspective for Energy*, 1st edition, Elsevier, Academic Press, 2023.

8 Nuclear Reactor Statics

8.1 INTRODUCTION

A nuclear reactor is a device used to start and manage nuclear fusion or fission nuclear chain reactions. It was formerly known as an atomic pile (see Figure 8.1). Both nuclear power plants (NPPs) and nuclear marine propulsion use nuclear reactors to generate electricity. A working fluid, such as gas or water, receives heat from nuclear fission and passes it through steam turbines. These power the propellers on a ship or rotate the shafts of electrical generators.

In theory, steam produced by NPPs can be utilized for district heating or industrial process heating. Certain reactors are used to create weapon-grade plutonium or isotopes for industrial and medical applications. According to the International Atomic Energy Agency (IAEA), there are 220 nuclear research reactors and 437 nuclear power reactors operating worldwide as of early 2021 [1 –3].

Nuclear reactors transform the energy released by controlled nuclear fission (see Figure 8.2) into thermal energy that can then be further transformed into mechanical or electrical forms, much like traditional thermal power plants do by using the thermal energy released from burning fossil fuels to create electricity.

A Uranium-235 atom's nucleus absorbs a neutron, which causes it to split into free neutrons and quickly move lighter elements known as fission products. Although nuclear chain reactions are the basis of both reactors and nuclear weapons, the rate of reactions in a reactor is far slower than in a bomb.

FIGURE 8.1 Core of crocus, a small nuclear reactor used for research at the EPFL in Switzerland.

(Source: www.wikipedia.org)

DOI: 10.1201/9781003590194-8

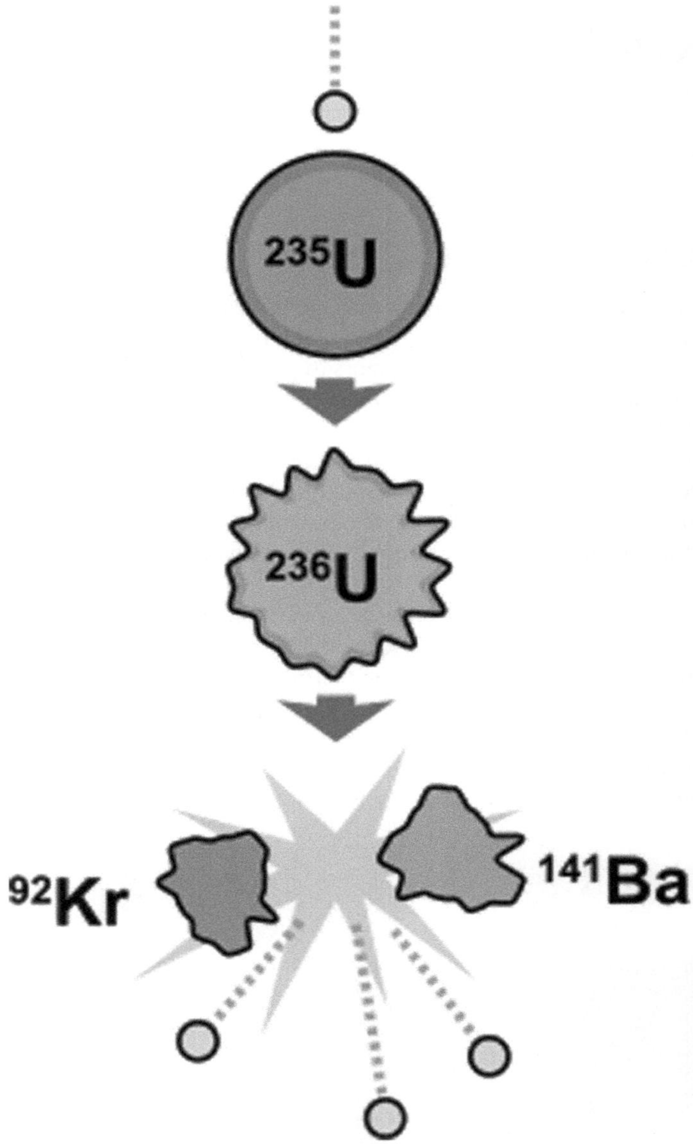

FIGURE 8.2 An example of an induced nuclear fission event.

(Source: www.wikipedia.org)

Nuclear fission can occur when a neutron is absorbed by a large fissile atomic nucleus, such as that of uranium-235 or plutonium-239. Free neutrons, gamma radiation, and kinetic energy are released when the heavy nucleus breaks into two or more lighter nuclei, or fission products. Other fissile atoms may absorb some of these neutrons, which could lead to more fission events that release more neutrons, and so on. We call this a nuclear chain reaction.

Control rods with neutron poisons and moderators can alter the percentage of neutrons that will continue to induce additional fission, thereby controlling such a nuclear chain reaction [4]. In the event that monitoring or instrumentation identify dangerous situations, nuclear reactors typically contain both automated and manual procedures to stop the fission process [5].

The reactor core generates heat in a number of ways.

When these nuclei collide with neighboring atoms, the kinetic energy of fission products is transformed into thermal energy. A portion of the gamma rays created during fission are absorbed by the reactor, which then transforms their energy into heat.

Fission products and materials activated by neutron absorption decay radioactively to produce heat. Even when the reactor is shut down, this decay heat source will continue to exist for a while.

About three million times as much energy is released by a kilogram of uranium-235 (U-235) transformed by nuclear processes as by a kilogram of coal burned conventionally (7.2×10^{13} joules per kilogram of uranium-235 against 2.4×10^{7} joules per kilogram of coal) [6, 7] [first study].

One kilogram of uranium-235 releases roughly 19 billion kilocalories when it fissions, meaning that 1 kilogram of uranium-235 releases the same amount of energy as burning 2.7 million kilograms of coal.

Additionally, to absorb the heat produced, a nuclear reactor coolant – typically water, but occasionally a gas, liquid metal (such as liquid sodium or lead), or molten salt – is pumped around the reactor core. After leaving the reactor, the heat is transferred and used to create steam. The majority of reactor systems, such as the Pressurized Water Reactor (PWR), use a cooling system that is physically isolated from the water that will be boiled to create pressured steam for the turbines. However, in some reactors, such as the Boiling Water Reactor (BWR), the reactor core directly boils the water for the steam turbines [8].

By regulating the number of neutrons capable of causing additional fission events, the rate of fission processes within a reactor core can be changed. Several neutron control techniques are commonly used in nuclear reactors to modify the reactor's power output. When some of these techniques are procedures built into the reactor design for a specific purpose, others are a natural consequence of the physics of radioactive decay and are simply taken into account when the reactor is operating.

Moving the control rods is the quickest way to change the amount of fission-inducing neutrons in a reactor. Neutrons are absorbed by control rods because they are composed of neutron poisons. A control rod absorbs more neutrons than the material it displaces, which is frequently the moderator, as it is put farther into the reactor. As a result, the reactor produces less power and there are fewer neutrons available to initiate fission. On the other hand, removing the control rod will cause the power and the rate of fission events to increase.

Neutron populations in a reactor are also impacted by the physics of radioactive decay. Several neutron-rich fission isotopes exhibit delayed neutron emission as one such process. The remaining neutrons (referred to as "prompt neutrons") are released instantly upon fission, while these delayed neutrons make up around 0.65% of the total neutrons produced in fission. Determining precisely when a reactor hits the critical point takes a long time since the fission products that create delayed neutrons have half-lives for their decay by neutron emission that vary from milliseconds to as long as several minutes. Mechanical devices or human operators can control a chain reaction in "real time" by keeping the reactor in the zone of chain reactivity where delayed neutrons are required to achieve a critical mass state. If this was not the case, the time between achieving criticality and nuclear meltdown due to an exponential power surge from the normal nuclear chain reaction would be too short to allow for intervention.

The quick critical point is the final level at which delayed neutrons are no longer needed to maintain criticality. Criticality can be expressed numerically using a scale where the immediate critical point is one dollar, bare criticality is zero dollars, and other process points are interpolated in cents.

The coolant in certain reactors also modifies neutrons. By making the fast neutrons emitted from fission lose energy and transform into thermal neutrons, a moderator boosts the reactor's power. Fission is more likely to be caused by thermal neutrons than by fast neutrons. Temperature variations may alter the coolant's density and, consequently, power output if the coolant functions as a moderator. Because it would be less thick, a coolant with a higher temperature would be a less effective moderator.

Similar to how the control rods absorb neutrons, the coolant in other reactors also functions as a poison. Heating the coolant in these reactors can boost power production by making the poison less thick. In the event of an emergency shutdown, nuclear reactors typically contain both human and automatic methods to scram the reactor. In the event that hazardous conditions are identified or predicted, these devices inject significant quantities of poison – typically boron in the form of boric acid – into the reactor to stop the fission reaction [9].

The majority of reactor types are susceptible to a process known as iodine pit or xenon poisoning. Xenon-135, a typical fission product, absorbs neutrons and functions as a neutron poison, which eventually causes the reactor to shut down. Maintaining power levels high enough to eliminate xenon-135 by neutron absorption as quickly as it is created will help control its accumulation. Iodine-115 is another byproduct of fission that decays to fresh xenon-115 with a half-life of 6.57 hours. Iodine-135 continues to decay to xenon-135 after the reactor is shut down, making it more difficult to restart for a day or two since xenon-135 decays into cesium-135, which has a half-life of 9.2 hours and is not nearly as toxic as xenon-135.

This transient condition is known as the "iodine pit". Restarting the reactor is possible if it has enough additional reactivity capacity. A "xenon burn-off (power) transient" occurs in the reactor within a few hours as the excess xenon-135 is converted to xenon-136, which is far less of a neutron poison. To replace the lost xenon-135's neutron absorption, control rods must be inserted further. One crucial element in the Chernobyl catastrophe was the improper application of such a method [10].

Nuclear marine propulsion reactors, particularly nuclear submarines, frequently require a very long core life without refueling and are frequently unable to operate continuously around the clock like land-based power reactors. Because of this, a lot of designs use highly enriched uranium yet include fuel rods that can produce burnable neutron poison [11]. The reactor can be built with more fissionable material because of this. However, early in the reactor's fuel burn cycle, the neutron-absorbing material makes it relatively safe. Later, normally produced long-lived neutron poisons (much longer-lived than xenon-135) take their place, and these poisons gradually build up over the operating life of the fuel load.

The power obtained from the nuclear fission process is the fundamental component of nuclear energy. A big uranium atom gets bombarded by a neutron during fission, and when it splits, more neutrons are released, which hit other atoms to cause a chain reaction. The energy released when uranium atoms fission may raise the temperature of water to exceptionally high levels, and the steam from this process is utilized to power turbines that are connected to electricity generators.

Around 389.5 GW of nuclear power could be generated globally in 2021 from 437 operating reactors spread across 32 nations. Nuclear power capacity is expected to increase over the next few decades, particularly in Asia, despite increased safety concerns following accidents like Chernobyl and Fukushima.

8.2 HOW DOES A NUCLEAR REACTOR WORK?

They confine and regulate nuclear chain reactions, or fission, which is a physical process that generates heat. Utilizing that heat, steam is produced, which turns a turbine to generate power. As seen in Figure 8.3, the core component of an NPP is a nuclear reactor [https://www.energy.gov/].

With over 440 operational reactors throughout the globe, including 92 in the United States, nuclear power remains one of the most abundant and consistent sources of carbon-free electricity.

8.2.1 Nuclear Fission Creates Heat

A reactor's primary function is to contain and regulate nuclear fission, which is the process by which atoms break and release energy. As an alternative, in the instance of nuclear fusion, two light hydrogen atoms are fused to one another.

Uranium is used as nuclear fuel in reactors. Small ceramic pellets made from uranium are piled in enclosed metal tubes known as fuel rods. To create a fuel assembly, more than 200 of these rods

FIGURE 8.3 Core of nuclear reactor power plant.

(Source: www.wikipedia.org)

are usually grouped together. Depending on the power level, a reactor core usually consists of several hundred assemblies.

The water that envelopes the fuel rods inside the reactor vessel works as a moderator and coolant. In order to maintain the chain reaction, the moderator aids in slowing down the neutrons created by fission. After that, control rods can be removed to speed up the reaction or put back into the reactor core to slow it down.

Fission heat converts water into steam, which drives a turbine to generate carbon-free power.

8.2.2 Fission and Fusion: What Is the Difference?

We generate all of our energy through fundamental physical and chemical processes.

Throughout history, this has primarily been achieved by burning carbon-based materials like coal, gas, and wood, or by using energy from the sun, wind, and water.

Two physical processes that generate enormous amounts of energy from atoms are nuclear fission and nuclear fusion.

Through nuclear reactions, they provide energy that is millions of times more than that from other sources.

Figure 8.4 shows how the two differ from one another.

1 **Fission**

When a neutron hits a larger atom, it causes the atom to excite and divide into two smaller atoms, which are referred to as fission products. Also, more neutrons are released that have a chance to start a chain reaction.

An enormous quantity of energy is produced when each atom divides.

Due to their ease of initiation and control, uranium and plutonium are most frequently utilized for fission reactions in nuclear power reactors.

In these reactors, water is heated to produce steam using the energy released by fission. To generate carbon-free electricity, a turbine is spun by the steam.

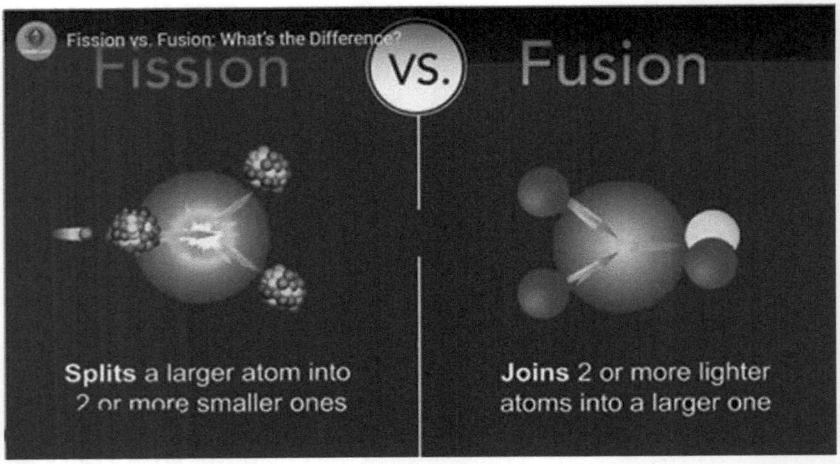

FIGURE 8.4 Fission versus fusion.

(Source: U.S. Department of Energy; https://youtu.be/2W-GEE6YU4M)

2 **Fusion**

As in the case of two hydrogen atoms fusing to generate one helium atom, fusion happens when two atoms collide to create a heavier atom.

The sun generates enormous quantities of energy through this same process, which is many times more powerful than fission. Furthermore, it does not generate fission products that are extremely radioactive.

Scientists are researching fusion reactions, but they are challenging to maintain for extended periods of time due to the extreme pressure and temperature required to fuse the nuclei together.

8.3 TYPES OF LIGHT WATER REACTORS IN THE UNITED STATES

All Light Water Reactors (LWRs) are the only kind of commercial nuclear reactors in the United States. This indicates that they employ regular water as a neutron moderator and cooling.

In the United States, LWRs are available in two varieties.

8.3.1 Pressurized Water Reactors

PWRs, make up over 65% of all commercial reactors in the United States. To keep the water from boiling, these reactors inject water at high pressure into the reactor core (Figure 8.5).

Nuclear fission heats the water in the core, which is subsequently fed into pipes within a heat exchanger. To produce steam, the pipes heat a different supply of water. After that, an electric generator is turned by the steam to create power.

The cycle is repeated with the core water cycling back to the reactor for reheating.

8.3.2 Boiling Water Reactors

BWRs constitute around one-third of all reactors in operation in the United States. Refer to the depiction in Figure 8.6.

BWRs generate steam and heat water inside the reactor vessel. Fission heats the water as it is forced up through the reactor core. The steam is then sent straight to a turbine via pipes to generate power.

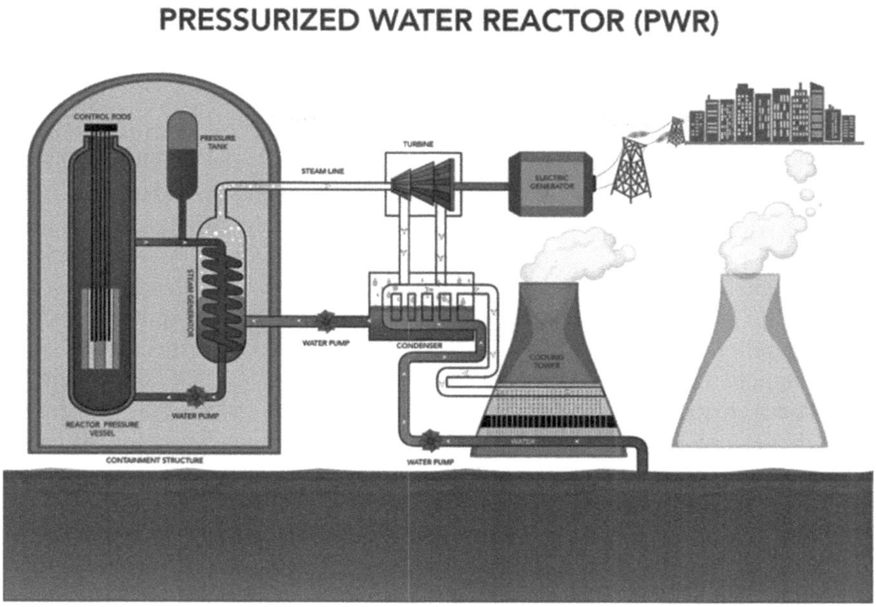

FIGURE 8.5 Pressurized water reactor (PWR) artistic schematic.

(Source: Graphic by Sarah Harman I U.S. Department of Energy)

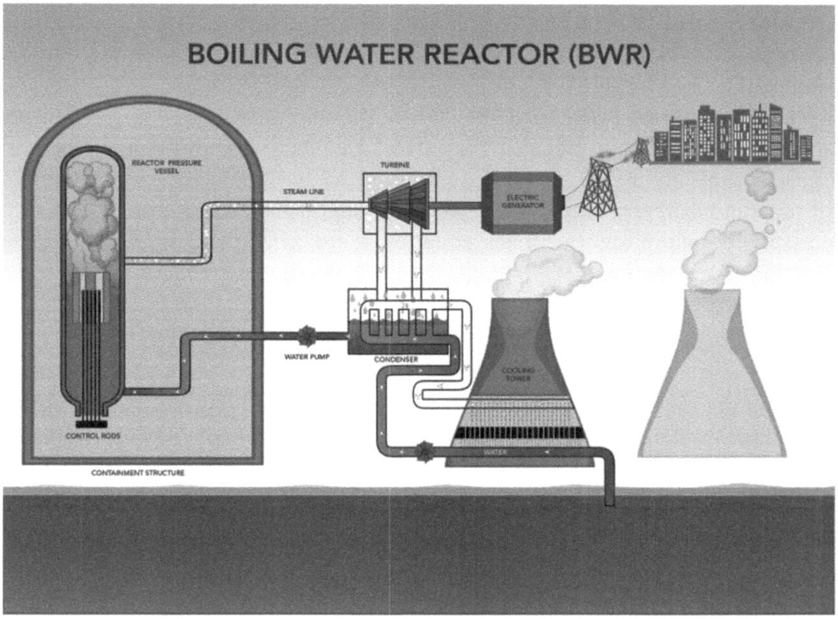

FIGURE 8.6 Boiling water reactor (BWR) artistic schematic.

(Source: Graphic by Sarah Harman I U.S. Department of Energy)

After that, the leftover steam condenses back into water and is utilized again for heating.

8.4 NUCLEAR REACTOR STABILITY

Particularly, the arithmetic sign of α_T determines how a reactor reacts to temperature changes (i.e., the overall reactor stability α_T). Reactors with negative α_T are naturally stable when their temperature and thermal power fluctuate, whereas those with positive α_T are naturally unstable.

In accordance with Criterion 11 of the Code of Federal Regulations (10 CFR), the main design requirements for a proposed facility must be included in a construction permit application in accordance with §50.34. The major design requirements for a planned facility must be included in an application for a design certification, combination license, design approval, or manufacturing license, as required by 10 CFR 52.47, 52.79, 52.137, and 52.157, respectively. For structures, systems, and components that are critical to safety—meaning those that help ensure the facility can operate without creating an unacceptable risk to public health and safety—the principal design criteria define the requirements for their design, fabrication, construction, testing, and performance.

For water-cooled NPPs that are similar to those for which the Commission has granted building licenses in terms of both site and design, these General Design Criteria set minimum standards for the major design criteria. The General Design Criteria are meant to serve as a guide for developing the primary design criteria for other kinds of nuclear power units, and they are also thought to be broadly relevant to them [12].

According to the Code of Federal Regulation (CFR), the following statement applies:

"In the power operating range, the net effect of the prompt inherent nuclear feedback characteristics tends to make up for a rapid increase in reactivity". This is the design for the reactor core and related coolant systems.

We shall now talk about reactor stability when operating at power. At power operation, there is always a sufficient number of neutrons to produce heat. Large heat generation is the primary function of power reactors, which results in changes to the system's temperature and material densities (due to thermal expansion).

Reactivity coefficients are used to define these alterations in reactivity, which are typically referred to as reactivity feedback. Because they affect the reactor's stability, the reactivity feedback and their time constants are a crucial component of reactor design. Reactivity and reactor power as a function of time and different temperature feedback are displayed in Figure 8.7.

We have assumed a simplified feedback equation in the article (Point Dynamics Equations) in Chapter 9, Section 9.1 (Equation 9.1).

$$\rho(t)=\rho_i(t)-|\alpha_f|\left[\overline{T}_f(t)-\overline{T}_f(0)\right]-|\alpha_{fc}|\left[\overline{T}_c(t)-\overline{T}_c(0)\right] \qquad \text{Eq. 8.1}$$

The dependency of the reactivity on different factors is expressed by this equation. However, in this instance, just the fuel and coolant temperatures are influenced. Since temperature instability is the primary source of most instabilities, temperature stability is crucial to overall stability for PWRs. We will use a further simplified model for demonstration, assuming that the fuel and moderator have a single temperature coefficient α_T.

$$\alpha_T \approx \frac{1}{k}\frac{dk}{kdT} \qquad \text{Eq. 8.2}$$

The algebraic sign of α_T is particularly important for determining how a reactor reacts to temperature changes (i.e., overall reactor stability).

While a reactor with a positive α_T is intrinsically unstable, one with a negative α_T is unaffected by to variations in temperature and heat output. We will use the next two instances to illustrate the issue.

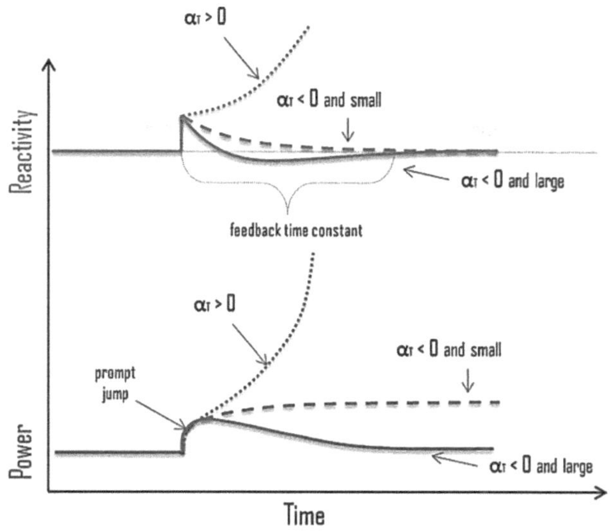

FIGURE 8.7 Reactor power and reactivity in response to time and different temperature feedback.

(Source: U.S. Nuclear Regulatory Commission)

8.5 NEGATIVE REACTIVITY FEEDBACK – $\alpha_T < 0$

The scenario described above is very different for $\alpha_T < 0$. In this instance, the core gains negative reactivity when the moderator's temperature is raised. Reactor power drops as a result of this negative reactivity, which prevents temperature or power increases from continuing. The reactor reaches the critical condition (steady state) when the power coefficient, which is similarly dependent on the sign of α_T, acts against the decline in thermal power. Reactor power self-stabilizes. Figure 8.8 illustrates this impact. Assume that variations in the core inlet temperature are what cause all of the changes. Refer to Figure 8.8.

It should be mentioned at this stage that a reactor core's temperature does not fluctuate consistently. For instance, since the majority of thermal power is produced in this area, an increase in fuel temperature is the first indication of a rise in thermal power. Until the heat has been transmitted from the fuel to the reactor coolant, neither the coolant temperature nor, in thermal reactors, the moderator temperature changes. Heat transmission time to the moderator is often expressed in seconds (~5 s).

There is typically a delay in heat transmission from the fuel to the coolant. This is due to several resistance barriers, such as the clad, gap, fuel, and stain layer on the clad. The radial temperature gradient is increased by gaps, which typically have the nethermost thermal conductivity coefficient.

Therefore, it is necessary to specify the component whose temperature changes.

8.5.1 Fuel Temperature Coefficient and Fuel Design Temperature Coefficient

The change in reactivity per degree change in fuel temperature is known as the fuel temperature coefficient (FTC) or design temperature coefficient (DTC) α_f. Refer to Equation 8.3.

$$\alpha_f = \frac{d\rho}{dT f} \qquad \text{Eq. 8.3}$$

The fuel composition, particularly the fuel enrichment, is the primary determinant of the fuel temperature coefficient's size and sign (positive or negative). The Doppler coefficient in power reactors

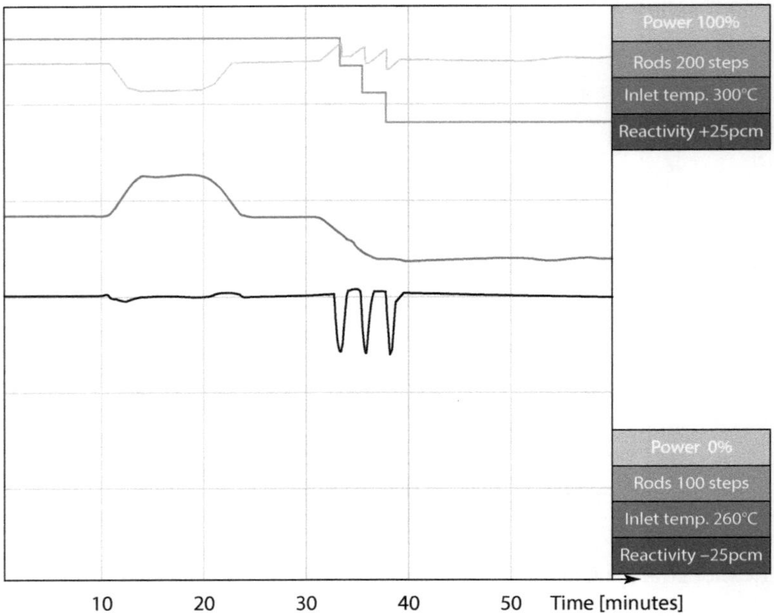

FIGURE 8.8 Critical condition (steady state).

(Source: U.S. Nuclear Regulatory Commission)

that employ low-enriched fuel (such as PWRs and BWRs, which need 3% to 5% of ^{235}U) is always negative. For instance, the Doppler coefficient in PWRs might vary between –5 and –2 pcm/°C. In terms of reactor stability, this coefficient is the most crucial. Most people agree that the FTC is even more significant than the Moderator Temperature Coefficient (MTC). The first and most crucial feedback that will make up for the added positive reactivity is the FTC, particularly in the case of all reactivity-initiated accidents (RIAs).

The FTC thus becomes nearly immediate. Because it produces an instantaneous reaction to variations in fuel temperature, this coefficient is also known as the quick temperature coefficient.

Generally speaking, the fuel enrichment has a major influence on the FTC. The Doppler effect is linked to two particular occurrences. In the resonance zone, it enhances the capture of neutrons by fuel nuclei, both fissile and fissionable. On the other side, it causes fissile nuclei to produce additional neutrons.

DTC could be advantageous as well. Higher enrichments of fissile nuclei are necessary for a net positive Doppler coefficient. According to certain estimates, somewhat positive DTC is linked to enrichments of more than 30%; nonetheless, this enrichment produces a harsher neutron spectrum. Since oxygen nuclei scatter elastically, LEU oxide fuels often have a significant negative DTC (softened spectrum).

Moderator nuclei added directly into the fuel matrix can also soften the neutron spectra and increase the amount of negative DTC. For instance, uranium zirconium hydride (UZrH) fuel, which has a significant negative FTC, is used in the TRIGA reactor (Figure 8.9).

Keep in mind that Cherenkov radiation is the source of the blue light. When a charged particle travels through a dielectric material faster than the phase velocity of light in that medium, electromagnetic radiation known as Cherenkov radiation is released. The distinctive blue light of an underwater nuclear reactor is a well-known illustration of Cherenkov radiation. Refer to Figure 8.10 for illustration. Its cause is comparable to that of a sonic boom, which is the harsh sound produced when movement happens more quickly than sound. Pavel Cherenkov, a Soviet scientist, is honored by the phenomenon's name.

FIGURE 8.9 Picture of a TRIGA reactor core.

(Source: www.wikipedia.org)

A thermal neutron in the fuel element is more likely to gain energy from an excited state of an oscillating hydrogen atom in the lattice when the temperature of the hydride rises. As the neutrons gain energy from the ZrH, the fuel element's thermal neutron spectrum shifts to a higher average energy (i.e., the spectrum is hardened). The negative temperature coefficient is created in a different way using this spectrum hardening.

The Nuclear Regulatory Commission in the United States will not grant a reactor a license unless the α_f prompt (FTC) is negative. This guarantees the inherent stability of all licensed reactors. Many other nations also adhere to this criterion.

8.5.2 Moderator Temperature Coefficient and Reactor Design

As previously stated, Figure 8.11 shows that the moderator-to-fuel ratio (N_{H2O}/N_{Fuel} ratio) is the primary determinant of the MTC. The ratio between the number of moderator nuclei and fuel nuclei in a reactor core's volume is known as the moderator-to-fuel ratio. Fuel volume and number density generally stay the same as the core temperature rises.

The moderator's numerical density falls with thermal expansion, while its volume stays constant. The thermal expansion of water (particularly at 300°C; see Density of Water) causes the ratio of the moderating atoms (water molecules) to decrease as the moderator temperature rises. Its density dramatically drops.

As a result, the neutron spectrum in the reactor core becomes harder, which raises resonance absorption (lower p). Neutrons remain at a higher energy for longer as the moderator's density decreases, increasing the likelihood that these neutrons will be captured without fission.

The MTC is determined by three mechanisms, including this one. The neutron leakage chance is linked to the second step, while the thermal utilization factor is linked to the third.

In particular, the moderator-to-fuel ratio has a significant impact on

FIGURE 8.10 Cherenkov radiation glowing in the core of the advanced test reactor.

(Source: www.wikipedia.org)

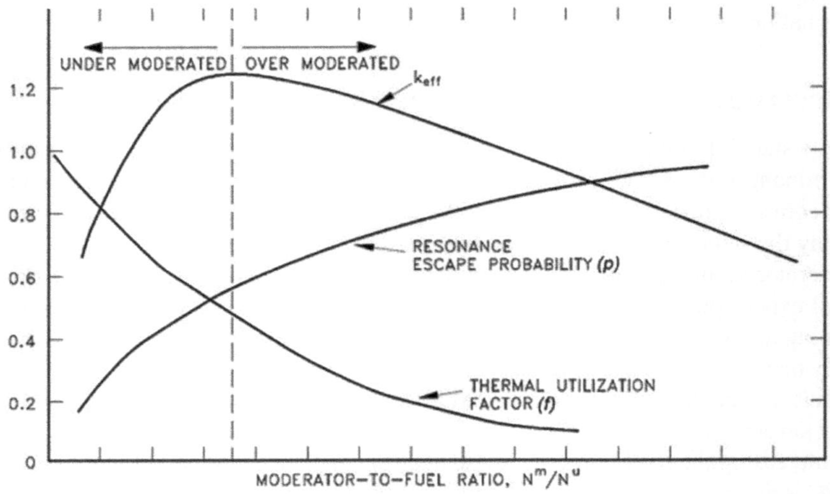

FIGURE 8.11 Moderator temperature coefficient versus moderator-to fuel ratio.

(Source: U.S. Nuclear Regulatory Commission)

- **Resonance Escape Probability**: The likelihood of a resonance escape increases as the moderator-to-fuel ratio rises. Neutrons may easily slow down to thermal energy without running into a resonant absorption at the resonance energies when more moderator molecules are introduced in comparison to fuel molecules. For further information, see Chapter 1 of this book.
- **Thermal Utilization Factor**: The thermal utilization factor decreases as the moderator-to-fuel ratio rises. The number of thermal neutrons absorbed in the fuel (all nuclides) divided by the number of thermal neutrons absorbed in the core's constituent material yields the thermal utilization factor's value. For further information, see Chapter 1 of this book.
- **Thermal and Fast Non-Leakage Probability**: The non-leakage probability rises when the moderator-to-fuel ratio rises because it shortens the migration length. For further information, see Chapter 1 of this book.

The image illustrates that at low moderator-to-fuel ratios, the resonance escape probability is modest, resulting in a tiny product of all six components (k_{eff}). The so-called "optimal moderation" occurs when k_{eff} reaches its highest value at the ideal moderator-to-fuel ratio. Because of the low thermal utilization factor, k_{eff} is tiny at large ratios as well.

8.5.3 Under-Moderated versus Over-Moderated Reactor

From the moderator-to-fuel ratio point of view, any multiplication system may be built as

- **Under-moderated**: Less than the ideal quantity of moderator between fuel plates or fuel rods is referred to as under-moderation. An increase in voids and moderator temperature lowers the system's k_{eff} and introduces negative reactivity. A core that is not regulated would produce the negative temperature and vacuum feedback necessary for a stable system.
- **Over-moderated**: An excess of moderator between fuel plates or fuel rods is referred to as over-moderation. The system's k_{eff} rises and positive reactivity is inserted as the moderator temperature and vacancies grow. A core that is over-moderated would provide void feedback and a positive temperature. Unless the positive impact is overridden by another negative feedback mechanism (like Doppler broadening), it will lead to an unstable system.

To guarantee system stability under all operating circumstances, reactor engineers must balance the combined impacts of moderator density, fuel temperature, and other phenomena. The neutron flux spectrum is somewhat harsher (the moderation is somewhat insufficient) than in an ideal situation since the majority of LWRs are built to be so-called under-moderated. However, its design offers a crucial safety element. The reactor tends to become self-regulating when the moderator temperature rises because it causes negative reactivity. Additionally, the comments must be unfavorable overall. Nevertheless, regions with significant water gaps that are over-moderated, like those close to control rods and guide tubes, have local positive coefficients.

The neutron flux trap effect is another phenomenon linked to an under-moderated core. Because neutrons thermalize more effectively in regions with wide water gaps (between fuel assemblies or when the fuel assembly bow phenomena are present), this impact increases local power generation. Because large thermal neutron fluxes are desired for material irradiation, "flux traps" are a regular feature of the majority of contemporary test reactors. However, PWRs can also experience it.

However, under-moderation has its limitations as well. More fissile material is required to guarantee the criticality of the core since it generally results in a fall in total k_{eff}. Additionally, the MTC has a minimum value limit (most negative). The reason for this is because the moderator temperature decline is likewise resisted by the negative temperature feedback. Take into account the consequences of rapidly lowering the moderator temperature, such as in the case of the main steam line break (MSLB, which is the usual starting event for PWRs). The steam pressure and saturation temperature in the steam generators drop quickly as a result of the steam line break. The moderator temperature will drop quickly as a result of the steam generators' declining saturation temperature. A positive reactivity insertion is brought on by the quick moderator temperature reduction. The magnitude of the MTC also determines how much reactivity is added, thus it must be kept to a minimum. MTC = −80 pcm/°C is the usual number for the lower limit, however, technical specifications restrict this value to a particular plant.

The coolant serves as the moderator in the LWR as well. However, the boiling two-phase flow cools the core in the BWR, and the moderator density varies as the volume of vapor bubbles (voids) varies. Figure 8.12 displays the (moderator) void coefficient of BWRs.

It should be mentioned that from a safety perspective, a high negative reactivity coefficient is not always desirable. For instance, positive reactivity is added, and cold coolant is injected into the core if the pump of a shutdown loop is started incorrectly. This incident is known as a cold coolant intrusion. The added positive reactivity rises when the negative reactivity coefficient is high. In BWRs,

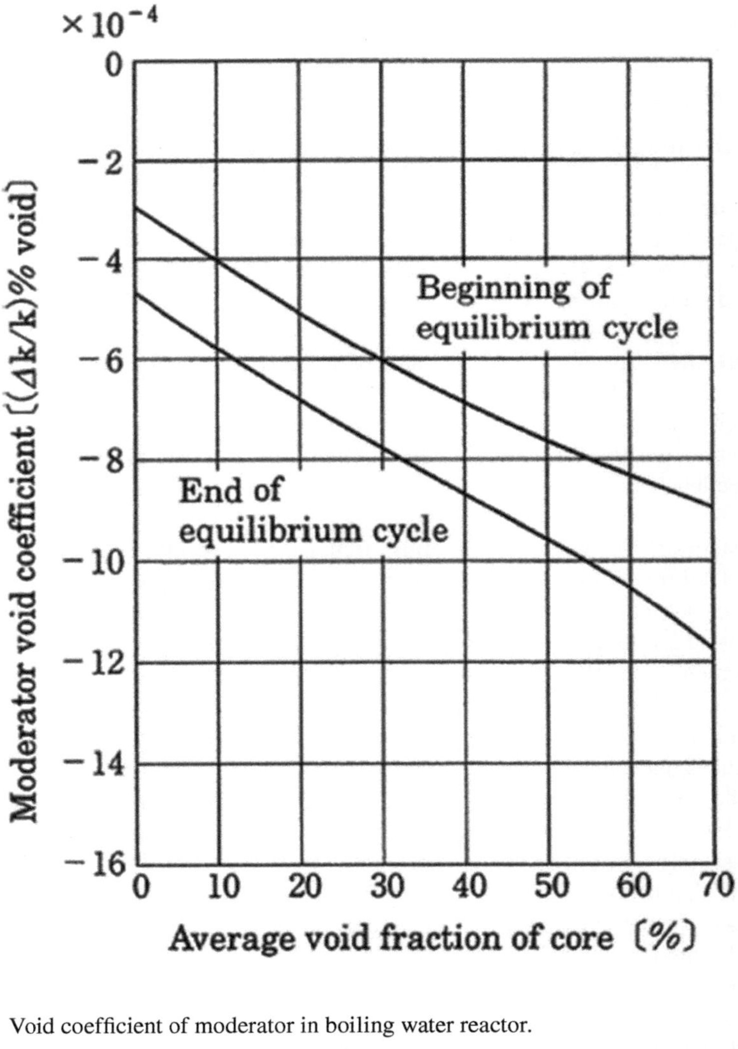

FIGURE 8.12 Void coefficient of moderator in boiling water reactor.

(Source: From the "Application for permission for installation modification of Chubu Electric Power Company's Hamaoka No. 5 nuclear Power plant (April of 1997)")

TABLE 8.1
Reactivity Coefficients of Pressurized Water Reactors (PWRs)

Moderator temperature coefficient*	(+0.8 to −9.4) × 10^{-4}($\Delta k/k$)/°C
Doppler coefficient	(+2.1 to −5.2) × 10^{-5}($\Delta k/k$)/°C
Void coefficient	(+0.7 to −3.1) × 10^{-3}($\Delta k/k$)/%Void fraction
Pressure coefficient	(+0.8 to −0.5) × 10^{-5}($\Delta k/k$)/(kg/cm^2)
Moderator density coefficient	(+0.51 to 0) × ($\Delta k/k$)/(kg/cm^2)

*The moderator temperature coefficient becomes a negative value at high temperature and high power.

TABLE 8.2
Comparison of the Thermal Neutron Cross-Section of Light Water, Heavy Water, and Graphite

	Density	Micro Absorption Cross-Section (Barn)	Macro Absorption Cross-Section (10^{-5}, cm^{-1})	Moderation Ratio
H_2O	1.0	0.66 (hydrogen)	2,200	1.36
D_2O	1.1	0.001 (deuterium)	3.3	0.18
Graphite	1.6	0.0034 (carbon)	27	0.16

the turbine control valve, which is situated on the main steam line from the reactor to the turbine, is rapidly closed to safeguard the turbine in the event that the power transmission line fails and the steam turbine load is removed. This lowers the volume of vacancies in the core coolant and raises the reactor pressure. It introduces positive reactivity. Designing the core to have the proper negative value of the void reactivity coefficient is crucial, even though there are other steps that may be done. Reactor power may be automatically reduced (setback), steam can be released to the condenser by releasing the turbine bypass valve, recirculation pump speed can be decreased, thereby increasing the number of voids, and core coolant flow can be decreased. The PWR reactivity coefficients are shown in Table 8.1.

The moderator and coolant in the graphite-moderated boiling water-cooled pressurized tube reactor (RBMK) and the Canadian Deuterium Uranium (CANDU) reactor are kept apart by pressure tubes. The neutron spectrum does not harden or the moderator density decreases in these reactors when the coolant density decreases. The coolant's neutron absorption rate falls and its reactivity rises as the coolant density falls.

It is possible for the coolant temperature coefficient to turn positive. The reactor physics characteristics and thermal neutron cross-sections of graphite, heavy water, and light water are compared in Table 8.2. In contrast to graphite and heavy water, light water has a significant thermal neutron cross-section while having a strong moderating ratio and the ability to efficiently moderate fast neutrons with a small volume. As a result, in pressure tube graphite moderator reactors (like the RBMK) that employ light water coolant, the coolant void coefficient can frequently turn positive. It has led to a nuclear power excursion and a positive power coefficient.

8.6 EXAMPLES: REACTOR STABILITY

The following are the examples of reactor stability.

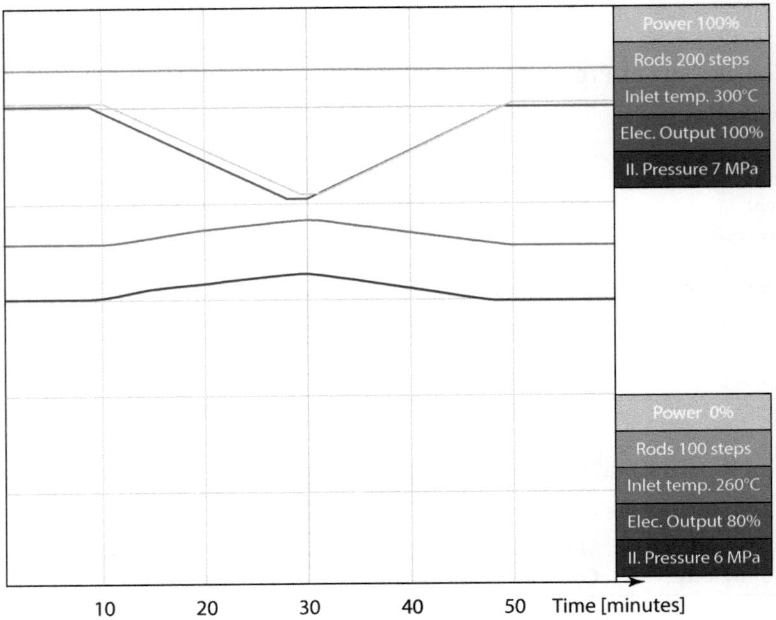

FIGURE 8.13 Illustration of the self-regulating properties of a nuclear reactor.

(Source: U.S. Nuclear Regulatory Commission)

1. **Self-Regulating Properties of Nuclear Reactor**

It is advantageous to operate with a *negative MTC*, even when there are power fluctuations. During regular operation, the primary and secondary circuits have a perfect energy balance. As a result, the steam demand falls when the operator reduces the load on the turbine (for example, because of a grid need). (Note the first electrical output decline in Figure 8.13).

The reactor will now generate more heat than the steam turbine can use. Due to this imbalance, the steam pressure raises the steam generators' saturation temperature (see II. pressure in Figure 8.13). The moderator temperature will only rise when the steam generators' saturation temperature rises (see intake temperature). Reactor power is decreased by adding negative reactivity by raising the moderator's temperature (without any operator interaction). As may be observed, the reactor is somewhat self-regulating, and grid requirements and the steam turbine can both be used to adjust reactor power. Due to the restricted range of permitted intake temperatures, this function has limitations.

2. **Power Increase from 75% to 100%**

Any increase in power causes changes in the void fraction, temperature, or pressure, which in turn affects the core's responsiveness. It is challenging to alter an operational parameter without affecting the core's other characteristics. The power coefficient is defined because it is challenging to distinguish between all of these effects (fuel, moderator, emptiness, etc.). The Doppler, moderator temperature, and void coefficients are combined to form the power coefficient. It is written as $\Delta \rho / \Delta\%$ power, or a change in reactivity per change in percent power. In core life, the power coefficient is always negative. However, the drop in the MTC is the main reason it is more negative near the conclusion of the cycle.

Assume that the plant operator wishes to raise power to 100% of rated power and that the reactor is critical at 75% of rated power. The reactor operator must first introduce a positive reactivity (such as boron dilution or control rod withdrawal) to get the reactor supercritical. The reactor reverts to the critical state when the thermal power rises because the moderator and fuel temperatures rise as well, producing a negative reactivity impact (based on the power coefficient). To keep the power rising, positive reactivity must be added continually (via chemical shims or control rods). The reactor power stabilizes in proportion to the reactivity introduced after each reactivity insertion. The power defect is the entire amount of feedback reactivity that needs to be compensated for during the power increase (from around 1% to 100%) by boron dilution or control rod removal. The power increase scenario is shown in Figure 8.14.

Let us assume

- **The power coefficient: Δ ρ/Δ% = –20 pcm/% of rated power**
- **Differential worth of control rods: Δ ρ/Δstep = 10 pcm/step**
- **Worth of boric acid: –11 pcm/ppm**
- **Desired trend of power decrease: 1% per minute**

75% → ↑ 20 steps or ↓ 18 ppm of boric acid within 10 minutes → 85% → next ↑ 20 steps or ↓ 18 ppm within 10 minutes → 95% → final ↑ 10 steps or ↓ 9 ppm within 5 minutes → 100%

In this scenario, let us imagine that the reactor is critical at 75% of rated power and that the plant operator wants to raise power to 100% of rated power.

3. **Power Decrease from 100% to 75%**

Any drop in power causes changes in the void fraction, temperature, or pressure, which in turn affects the core's responsiveness. It is challenging to alter any operational parameter without affecting the core's other characteristics. The power coefficient is defined because it is challenging to distinguish between all of these effects (fuel, moderator, emptiness, etc.). The Doppler, moderator

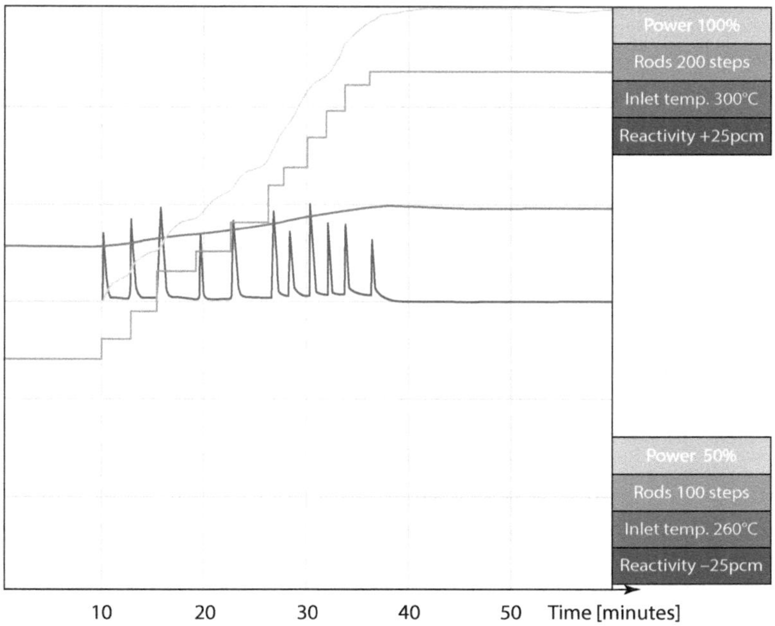

FIGURE 8.14 Power increase situation.

(Source: U.S. Nuclear Regulatory Commission)

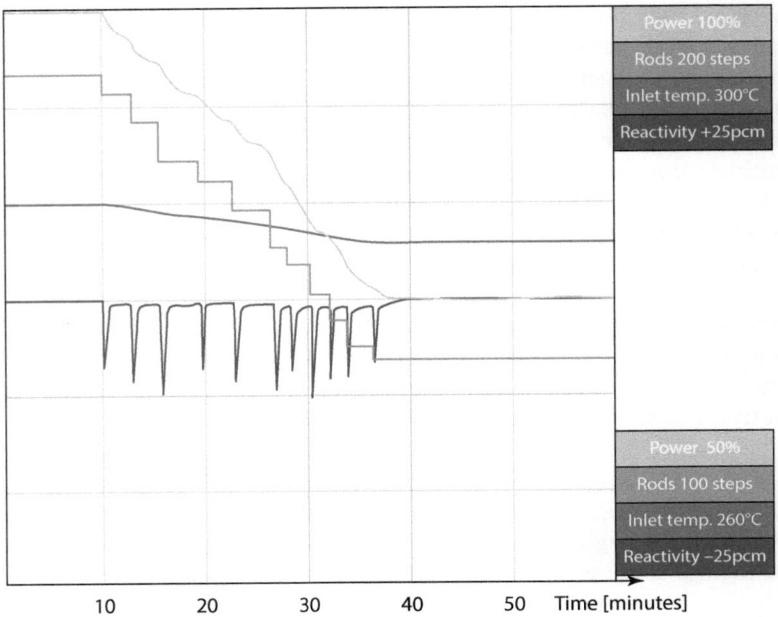

FIGURE 8.15 Power decrease situation.

(Source: U.S. Nuclear Regulatory Commission)

temperature, and void coefficients are combined to form the power coefficient. It is written as $\Delta \rho / \Delta\%$ power, or a change in reactivity per change in percent power. In core life, the power coefficient is always negative. However, the drop in the MTC is the main reason it is more negative near the conclusion of the cycle.

Assume that the plant operator intends to reduce power to 75% of rated power and that the reactor is critical at 100% of rated power. The reactor operator must first provide a negative reactivity (such as by adding boric acid or inserting a control rod) to get the reactor subcritical. The reactor reaches the critical condition again when the thermal power drops because the moderator and fuel temperatures drop as well, producing a positive reactivity effect (based on the power coefficient). To keep the power down, negative reactivity needs to be added continually (via chemical shims or control rods). The reactor power stabilizes in proportion to the reactivity introduced after each reactivity insertion. Figure 8.15 illustrates the circumstances of a power outage.

Let us assume

- **The power coefficient: $\Delta \rho/\Delta\%$ = −20 pcm/% of rated power**
- **Differential worth of control rods: $\Delta \rho/\Delta$step = 10 pcm/step**
- **Worth of boric acid: −11 pcm/ppm**
- **Desired trend of power decrease: 1% per minute**

100% → ↓ 20 steps or ↑ 18 ppm of boric acid within 10 minutes → 90%→ next ↓ 20 steps or ↑ 18 ppm within 10 minutes → 80% → final ↓ 10 steps or ↑ 9 ppm within 5 minutes→ 75%

8.7 FLOW INSTABILITIES CIRCUMSTANCES

Generally speaking, two-phase systems are susceptible to several instabilities. In nuclear engineering, the study of multiphase flow stability is crucial for managing accidents in PWRs and is especially crucial for BWRs under normal and abnormal conditions.

When PWRs are operating normally, the flow is regarded as single-phase. However, there has been a lot of research done on the flow instability in the event of transients and accidents (like the loss-of-coolant accident, or LOCA), or the trip of reactor coolant pumps (RCPs) in the presence of natural circulation, where flow oscillations or reversals may occur.

Variations in flow brought on by void formations are known as flow oscillations, and they are undesirable for a number of reasons.

- Unwanted mechanical load on fuel components, such as spacing grids, might result from flow oscillations. This may cause such components to wear out and fail.
- The properties of local heat transmission are impacted by flow oscillations. Departure from Nucleate Boiling (DNB), which results in the creation of a local vapor layer and a sharp decline in heat transfer capacity, is the major safety concern in the case of PWRs. Testing has shown that while the flow is oscillating, the critical heat flux (CHF) needed for DNB can be reduced by up to 40%. As a result, the power density and temperature limit are significantly decreased along the reactor core's length.

During natural circulation actions such as following RCPs trip, flow oscillations may be an issue. An essential architectural element and the best method of heat dissipation is natural circulation. Coolant boiling might happen due to the low flow rates, which could result in flow oscillations. The steam bubbles created during a flow oscillation may be strong enough to completely reverse the flow in the impacted channel during natural circulation. Refer to Figure 8.16.

Coolant evaporation is a desirable process that happens during normal operation in BWRs. Conversely, depending on the flow rate and thermal power, convective evaporation in the fuel channel alters the flow pattern along the fuel channel. It has been discovered that two-phase flow instabilities can occur in certain instability zones. These two-phase flow instabilities are undesirable because they can impact reactor safety, limit operating parameters, cause mechanical vibrations, and cause issues with system control. It should be mentioned that while flow stability in BWRs is a well-known phenomenon, it has not been a significant problem for many years.

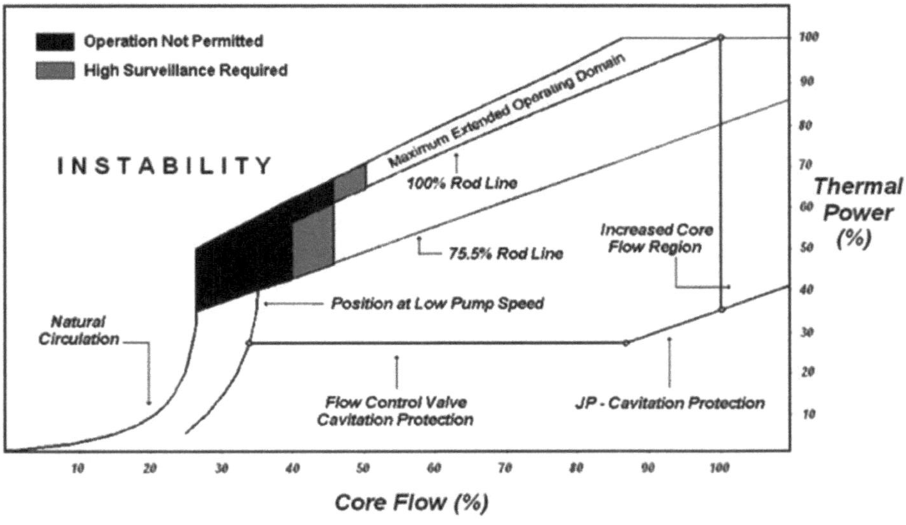

FIGURE 8.16 Instability region in the power-flow map for the Leibstadt Nuclear Power Plant (NPP) [22].

Generally speaking, flow instabilities are classified in a variety of ways. The essential mechanics of thermal-hydraulic systems provide the basis of the following classification.

The static instabilities are

- Flow excursion,
- Boiling crisis, and
- Relaxation types, including flow pattern transition.

The dynamic instabilities are

- Density wave oscillations,
- Pressure drops oscillations, and
- Thermal oscillations.

These have been expanded in the next section.

8.8 DYNAMIC INSTABILITIES

In two-phase flow, the method by which disturbances spread is a complex phenomenon in and of itself. Two types of waves carry the disturbances: vacuum or density waves and pressure or acoustic waves. Both types of waves are present and interact in two-phase systems. One or two orders of magnitude separate their velocities.

Pressure Wave Mechanism

The high frequency and duration of the pressure wave oscillations, often referred to as acoustic instabilities, are comparable to the time it takes for a pressure wave to pass through the system. Subcooled boiling, bulk boiling, and film boiling have all been found to exhibit acoustic oscillations. The frequency of pressure oscillations is quite high, ranging from 10 to 100 Hz, in rather substantially subcooled circumstances. Inlet pressure variations can account for a sizable portion of the pressure during these oscillations, and the pressure drop amplitudes can be significantly larger than the steady state values [13].

Firstenberg, Goldmann, and Hudson [14] used the phrase "boiling songs" to describe the high-frequency oscillations of 1,000 to 10,000 Hz that take place when boiling conditions are met. When the input temperature was lowered and the flow rate was raised, the whistle's intensity seemed to rise. During subcooled boiling, similar noises were also heard at subcritical pressures. The collapse of bubbles, which is influenced by the subcooling temperature and heat flow, is primarily responsible for these noises. According to Ruspini et al., [15] a phenomena that occurs at lower frequencies, between 5 and 100 Hz, is referred to as an acoustic oscillation or a thermo–acoustic oscillation. The acoustic connection of the vapor–liquid media with the presence of a forced oscillation source is the primary mechanism that causes this type of instability [16]. Depending on the pressure, flow patterns, channel shape, and operating circumstances, the resonance happens when the frequency of the nucleation sites matches the inherent acoustic frequencies of the vapor–liquid flows.

Density Wave Mechanism

The coolant conditions at the intake core of a BWR are subcooled, meaning that the fluid's enthalpy at the inlet is lower than the saturation enthalpy at the reactor pressure ($h_{in} < h_f$). The fluid is heated along the core length, and the boiling process begins at a certain point in the core. Thus, two flow areas may be used to visualize the reactor core: (1) a single-phase zone that stretches from the core input to the boiling boundary, often referred to as the boiling length, which is the point at which boiling begins, and (2) a two-phase flow region. In a single liquid phase, the wave pressure

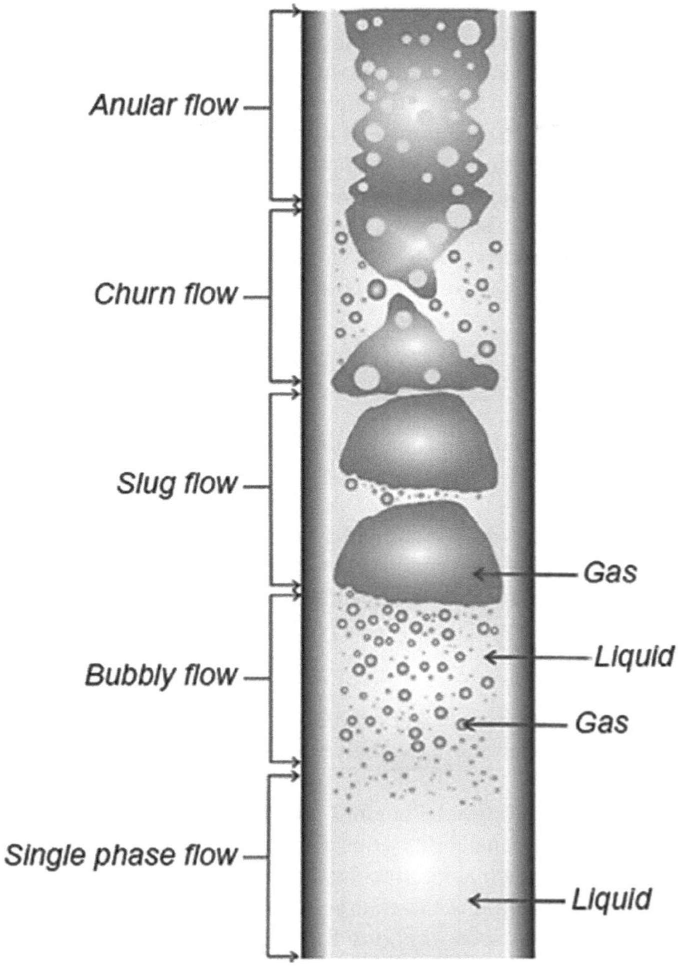

FIGURE 8.17 Two-phase flow regime patterns in vertical flows.

propagation velocity is lower than in a two-phase flow, and the void fraction propagation velocity is significantly lower.

Heat transfer and hydrodynamic processes are strongly coupled, especially in the two-phase flow zone. The topology of the gas phase inside the liquid phase of the two-phase flow patterns is influenced by changes in the void generation and, consequently, the distribution of each phase, which are caused by the addition and removal of heat from a two-phase flow, as shown in Figure 8.17. This leads to changes in heat transport mechanisms, which creates a feedback loop between hydrodynamic and thermal phenomena.

BWR instabilities often arise during transients related to startup and shutdown, which can result in significant fluctuations in the nuclear fuel's wall temperature that could cause mishaps.

When there is enough interaction and delayed feedback between the compressibility of the two-phase mixture and the flow's inertia, or when there are several feedback between the flow rate, pressure drop, and density change due to the rate of void generation in a boiling channel, a flow is considered to be subject to dynamic instability [17, 18].

The density wave is the fundamental mechanism responsible for flow and power instabilities in BWRs. Coolant is routed by the passages encircling the fuel rods and runs upward through the core. A rise in feedwater temperature, a reduction in input flow at constant power, or an increase in power at constant inlet flow can all raise the local vacancy percentage in a fuel bundle. The concentration of voids that results will move higher, creating a propagating density wave that, as it moves upward, alters the localized pressure drop at each axial position. The density wave propagation (DWP) time is the effective time for the voids to ascend through the core.

Density wave oscillation (DWO) is a prevalent kind of dynamic instability that affects a number of two-phase flow systems in addition to BWRs. The phenomena of dynamic instability is caused by the delayed feedback and significant interaction between the compressibility of two-phase flow and flow inertia. As a result, during boiling, there is numerous feedback between the flow rate, pressure drop, density change, and, ultimately, the net vapor production rate.

Two forms of DWOs are used to investigate how a BWR behaves when there is an external flow disturbance.

- Thermal–hydraulic feedback, which refers to instabilities generated by the thermal–hydraulic channel.
- Instabilities resulting from the nuclear–thermal–hydraulic interaction are known as neutronic feedback.

The major mechanism that initiates and maintains instability in BWRs is often described as thermal–hydraulic density wave instability combined with neutronics feedback.

It is crucial to stress that the BWR instability may be divided into three categories based on the occurrences that have been observed.

1. Channel instability
2. In-phase instability (global core power oscillation)
3. Out-of-phase instability (powers in two halves of a core oscillate)

Instabilities Produced by the Channel Thermal–Hydraulic

Unadulterated thermal–hydraulic instability or oscillation is a suitable term for the analysis without neutronic feedback; it is also known as channel instability in the literature.

The coolant mass flow oscillation at the channel intake is caused by the coolant void's damped oscillation in the boiling zone, as seen in Figure 8.18. This oscillation has a substantial impact on the two-phase flow pressure decrease. Lower inlet coolant subcooling, higher channel power and

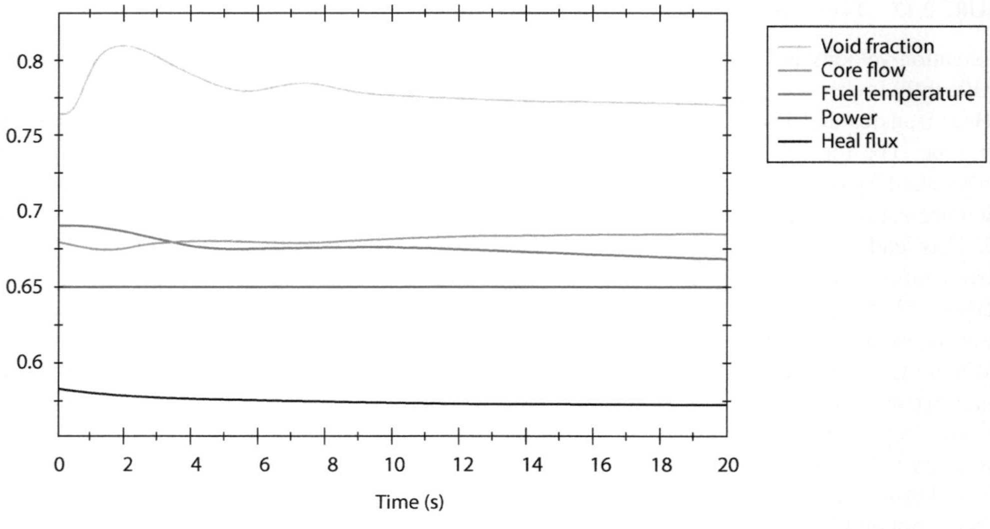

FIGURE 8.18 Thermal-hydraulic feedback in a boiling water reactor (BWR) without neutronic feedback.

lower flow rate, and axial power shape and design in the fuel assembly (number of fuel rods, spacer between them) can all contribute to the DWO because the two-phase flow pressure drop is comparatively larger than the single-phase flow pressure drop. The mechanism has been used to categorize the DWO [15, 19–21].

DWO for Gravity: This kind of instability is particularly significant in natural circulation BWRs and happens under extremely poor steam quality conditions (low power and high intake subcooling). The hydrostatic head (imposed by the core and riser void fraction) is extremely sensitive to changes in flow rate at low pressure and low flow quality. In the heated riser portion, gravity effects are the main cause of these low-frequency oscillations.

Friction DWO: These instabilities are caused by the interplay of the two-phase flow pressure loss, mass flux, and void formation, as well as their propagation in the core. They occur at relatively high power and low inlet subcooling. The two-phase frictional pressure decrease dominates these high-frequency oscillations. In both forced and natural circulation reactors, this phenomenon is significant.

DWO Momentum: Delays and feedback effects in the connection between flow rate, density, and pressure drop result in a momentum DWO.

Instabilities Produced by the Nuclear–Thermal–Hydraulic Coupling

The vacuum and power feedback effects on neutron kinetics and thermal hydraulics, respectively, cause coupled neutronic thermal hydraulic instabilities, also known as reactivity instabilities. While linked neutronic reactivity instability includes power feedback in addition to thermal–hydraulic feedback; thermal–hydraulic feedback solely includes density effects. Through heat transfer from the fission location to the coolant and reactivity effects brought on by temperature and density changes in the coolant involving the void fraction, the thermal–hydraulic and neutronic processes are physically connected.

In summary, types of instabilities observed in BWRs are as follows:

- **Control System Instabilities.** The behavior of controllers that strive to control parts of the reactor's variables via actuators is linked to control system instabilities.
- **Channel Flow Instabilities.** The following is a description of this kind of instability: Assume there is a disturbance in the flow. A "wave" of voids is created by this disturbance and travels upward through the channel, resulting in a two-phase pressure drop that is delayed in relation to the initial perturbation (the pressure drop rises dramatically as the void percentage grows). Instability in the flow rate might result from a rise in the density wave, or channel pressure decrease.
- **Coupled Neutronic–Thermohydraulic Instability.** The coupled neutronic–thermohydraulic instability, often referred to as reactivity instability, is the most common form of instabilities in commercial BWRs. In BWRs, the average void percentage in the core channels through has a high correlation with the fuel neutron flux, which in turn determines the power generation. Reactivity feedback is the term used to describe this phenomenon. As the voids go higher via the fuel channel, the reactivity feedback brought on by variations in the void fraction (void coefficient) is postponed. Sometimes, the void feedback is so severe and the latency is so lengthy that the reactor arrangement becomes unstable. The neutron flux may fluctuate in this situation.

For further information, please refer to [12, 22].

Given our improved knowledge of "flow instability" in the context of the LWR family, which includes BWRs and PWRs generally, we can now swiftly and comprehensively examine a few of the main concerns regarding flow instability situations, including

i. **Neutron Flux – Neutron Intensity,**
ii. **Void Fraction – Two-Phase Flow,**
iii. **Reactivity Coefficient – Reactivity Feedback.**

within the cores described reactors, as they are discussed in the next subsections here.

8.8.1 Neutron Flux Density – Neutron Intensity

The topic of neutron flux has previously been discussed in Chapter 1, Section 1.4 of this book, but as a reminder and new perspective, we examine it from above once again.

> "The number of neutrons moving in all directions through a given cross-sectional unit area in a given amount of time is known as the neutron flux, or $\vec{\phi}$. It is a scalar variable that may be computed by multiplying the neutron velocity ($\vec{v}$) by the neutron density ($\vec{n}$)".

The chance of a neutron traveling through a particular neutron–nuclear reaction was calculated in the section on the neutron cross-section. The material's neutrons' mean free path (MFP) under particular circumstances was ascertained. The reactor core's criticality is influenced by these factors. Stated otherwise, we have no idea what the reactor core's power level is. We need to know how many neutrons are passing through the material in order to determine the reactor core's reaction rate (RR) or thermal power. The distribution of neutron kinetic energies in the thermal reactor and the fast neutron reactor is seen in Figure 8.19. Neutron moderation instantly slows the fission neutrons (fast flux (FF)) in the thermal reactor down to the thermal energy.

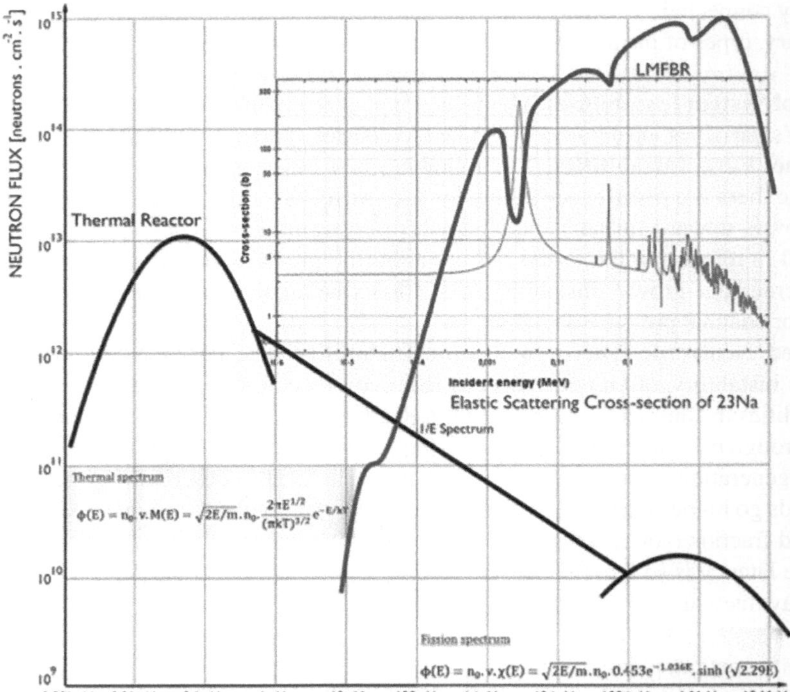

FIGURE 8.19 Distribution of kinetic energies of neutrons in the thermal reactor and the fast neutrons reactor. **(Source: U.S. Nuclear Regulatory Commission)**

It is practical to think about the neutron density, which is the quantity of neutrons in one cubic centimeter. The neutron density, expressed in neutrons/cm^3, is denoted by the sign *n*. Because it more accurately describes the overall route length traveled by all neutrons, the neutron flux is more frequently employed in reactor physics. The overall distance these neutrons may travel in a second is determined by their velocity.

The neutron density ($\vec{n}$) times the neutron velocity ($\vec{\text{v}}$) yields the neutron flux density ($\vec{\phi}$) (i.e., Equation 8.4) value.

$$\vec{\phi} = \vec{n} \cdot \vec{\text{v}} \qquad \text{Eq. 8.4}$$

where

$\vec{\phi}$ = Neutron flux (neutrons·cm^{-2}·s^{-1})
$\vec{n}$ = Neutron density (neutron·cm^{-3})
$\vec{\text{v}}$ = Neutron velocity (cm·s^{-1})

The neutron flux, the number of neutrons crossing through some arbitrary cross-sectional unit area in *all directions* per unit time, is a *scalar quantity*. Therefore, it is also known as the *scalar flux*. The expression $\vec{\phi}(\vec{E}) \cdot d\vec{E}$ is the total distance traveled during 1 second by all neutrons with energies between E and dE located in 1 cm^3.

The number of neutrons that go through a given unit area in all directions in a given amount of time is known as the neutron flux, and it is a scalar quantity. Therefore, it is also known as the scalar flux. The total distance covered in a second by all neutrons with energy between E and dE that are contained inside 1 cm^3 is expressed as $\vec{\phi}(\vec{E}) \cdot d\vec{E}$.

It is clear that the reactor power is related to the RR. We may calculate the rate of interactions (e.g., Rate of Fission Reactions) by knowing the neutron flux, which is the total route length of all the neutrons in a cubic centimeter in a second, and the macroscopic cross-sections, which is the likelihood of having an interaction per centimeter journey length. See Equation 8.5.

They are then multiplied collectively to determine the RR, or the quantity of interactions occurring in that cubic centimeter in a second.

$$RR = \vec{\phi} \cdot \Sigma = \vec{\phi} \cdot N \cdot \sigma$$

or Eq. 8.5

$$\text{R}R = \int_0^\infty \Sigma(\vec{E}) \cdot \phi(\vec{E}) d\vec{E}$$

where

$\vec{\phi}$ = Neutron flux (neutrons·cm^{-2}·s^{-1})
σ = Microscopic cross-section (cm^2)
N = Atomic number density (atoms·cm^{-3})

8.8.1.1 Neutron Flux and Intensity: Examples

It is necessary to differentiate between neutron intensity and neutron flux. Despite this, both physical values are measured in neutrons. Their physical interpretations differ for cm^{-2}·s^{-1}. The number of neutrons moving through a given unit area in a single direction in a unit amount of time is known as the neutron intensity, as opposed to the neutron flux (a surface is perpendicular to the direction of the beam). One vector parameter is the neutron density.

1. **Example – Neutron flux in a typical thermal reactor core**

About 100 tons of uranium with an average enrichment of 2% are found in a typical thermal reactor (do not mistake it with the enrichment of the new fuel). Calculate the average core heat flux and the RR if the reactor has a 3,000 MWth power output.

Solution

The total number of reactions taking place in the reactor core per unit of time is obtained by multiplying the RR per volume unit ($RR = \vec{\phi} \cdot \Sigma$) by the total volume of the core (V). However, we also know that each fission process releases around 200 MeV of energy. The Energy Release Rate (Power) resulting from the fission process may now be calculated. It is given by the following equation:

$$P = \phi \cdot \Sigma_f \cdot E_r \cdot V = \phi \cdot N_{U_{235}} \cdot \sigma_f^{235} \cdot \vec{V}$$

where

P = Reactor power ($MeV{\cdot}s^{-1}$)
ϕ = Neutron flux ($neutron{\cdot}cm^{-2}{\cdot}s^{-1}$)
N = Atomic number density ($atoms{\cdot}cm^{-3}$)
E_r = The average recoverable energy per fission (MeV/fission)
V = Total volume of the core (m^3)
σ = Microscopic section (cm2)

The amount of fissile ^{235}U per the volume of the reactor core.

m_{235} [g/core] = 100 [metric tons] × 0.02 [g of ^{235}U/g of U] × 10^6 [g/metric ton] = 2 × 10^6 grams of ^{235}U per the volume of the reactor core.

The atomic number density of ^{235}U in the volume of the reactor core

$N_{235} \times V = m_{235}.\ N_A/M_{235}$
= 2 × 10^6 [g 235/core] × 6.022 × 10^{23} [atoms/mol]/235 [g/mol] = 5.13 × 10^{27} atoms/core

The microscopic fission cross-section of ^{235}U (for thermal neutrons)

$$\sigma_f^{235} = 585 \text{ barns}$$

The average recoverable energy per ^{235}U fission

$$E_r = 200.7 \text{ MeV/fission}$$

The resulting average neutron flux is

$$\bar{\phi}_{th} = \frac{P}{\left(\sigma_f^{235}\right)\left(E_r\right)\left(N_{235}\right)\left(V\right)}$$

$$\bar{\phi}_{th} = \frac{3000[MW_{th}]\times 10^{6}[W_{th}/MW_{th}]}{585\times 10^{-24}[cm^{2}]\times 200.7[MeV/fission] \times 5.13\times 10^{27}[atoms/core]\times 10^{6}\times 1.6\times 10^{-19}[W_{th}/MeV]}$$

$$\bar{\phi}_{th} = \frac{3\times 10^{9}}{9.64\times 10^{-5}} = 3.11\times 10^{13}\ neutrons\cdot cm^{-2}\cdot s^{-1}$$

The resulting RR entire reactor core [fissions/second] is

$$RR = \Sigma_{f}^{235}\cdot\bar{\phi} = \sigma_{f}^{235}\cdot N_{235}\cdot\bar{\phi}_{th}$$

$$RR = 585\times 10^{-24}[cm^{2}]\times 5.13\times 10^{27}[atoms/core]\times 3.11\times 10^{13}[neutrons\cdot cm^{-2}\cdot s^{-1}]$$

$$RR = 9.33\times 10^{19}\ fission/\sec ond$$

2. **Example: Neutron flux in a MOX-fueled thermal reactor core**

Nuclear fuel that comprises several oxides of fissile material is known as mixed oxide (MOX) fuel, or simply MOX fuel. Typically, natural uranium and plutonium are mixed to create MOX fuels. For the sake of simplicity for this example, it will be assumed that the reactor core is entirely composed of MOX fuel (100% MOX) and that the average proportion of plutonium in MOX fuel is equal to 4% (averaged across all utilized fuel assemblies; not all FAs are fresh FAs). Keep in mind that new MOX fuel has an initial plutonium content of about 7%.

It will also be presumed. Only the fissile ^{239}Pu is present in the recycled plutonium. In fact, the higher isotopes ^{240}Pu, ^{241}Pu, and ^{242}Pu are always present in considerable proportions in MOX fuel. In this case, the existence of these sums will be disregarded.

There are around 100 tons of fuel (HM, or heavy metal) in a conventional thermal reactor. Calculate the average core heat flux and the RR if the reactor has a 3,000 MWth power output.

Solution

The total number of reactions taking place in the reactor core per unit time may be calculated by multiplying the RR per volume unit ($RR = \phi\cdot\Sigma$) by the total volume of the core (V). However, we also know that each fission process releases around 207 MeV of energy. The energy release rate (power) resulting from the fission process may now be calculated. It is given by the following equation:

$$P = \phi\cdot\Sigma_{f}\cdot E_{r}\cdot V = \phi\cdot N_{Pu_{235}}\cdot\sigma_{f}^{239}\cdot\vec{V}$$

where

P = Reactor power ($MeV\cdot s^{-1}$)
ϕ = Neutron Flux ($neutron\cdot cm^{-2}\cdot s^{-1}$)
N = Atomic number density ($atoms\cdot cm^{-3}$)
E_{r} = The average recoverable energy per fission (MeV/fission)

V = Total volume of the core (m^3)
σ = Microscopic section (cm2)

The atomic number density of ^{239}Pu in the volume of the reactor core.

N_{239} x V = m_{239}. NA/M_{239}
= 4 x 10^6 [g 239/core] x 6.022 x 10^{23} [atoms/mol]/239 [g/mol] = 10.1 x 10^{27} atoms/core

For thermal neutrons, the microscopic fission cross-section of ^{235}U is

$$\sigma_f^{235} = 750 \text{ barns}$$

The average recoverable energy per ^{239}Pu fission

$$E_r = 207 \text{ MeV/fission}$$

The resulting average neutron flux is

$$\bar{\phi}_{th} = \frac{P}{\left(\sigma_f^{239}\right)\left(E_r\right)\left(N_{239}\right)\left(V\right)}$$

$$\bar{\phi}_{th} = \frac{3000[MW_{th}]\times 10^6[W_{th}/MW_{th}]}{750\times 10^{-24}[cm^2]\times 207[MeV/fission]\times 10.1\times 10^{27}[atoms/core]\times 10^6\times 1.6\times 10^{-19}[W_{th}/MeV]}$$

$$\bar{\phi}_{th} = \frac{3\times 10^9}{2.51\times 10^{-4}} = 1.20\times 10^{13}\ neutrons\cdot cm^{-2}\cdot s^{-1}$$

The resulting RR entire reactor core [fissions/second] is

$$RR = \Sigma_f^{239}\cdot\bar{\phi} = \sigma_f^{239}\cdot N_{239}\cdot\bar{\phi}_{th}$$

$$RR = 750\times 10^{-24}[cm^2]\times 10.1\times 10^{27}[atoms/core]\times 1.20\times 10^{13}[neutrons\cdot cm^{-2}\cdot s^{-1}]$$

$$RR = 9.09\times 10^{19}\ fission/\sec ond$$

8.8.1.2 Neutron Flux – Uranium versus Metal Oxide

Keep in mind that the neutron fluxes in the cores powered by uranium and metal oxide (MOX) differ from one another. In the first example, which computed the neutron flux in a reactor core filled with uranium, the average neutron flux was 3.11 × 10^{13} neutrons·cm^{-2}·s^{-1}. While the RR is almost constant, the average neutron flux in a core that is 100% MOX fueled is around 2.6 times lower (1.2 × 10^{13} neutrons·cm^{-2}·s^{-1}). This information is crucial for the design of the reactor core and the reactivity control system. It is generally brought on by

- **Higher fission cross-section of ^{239}Pu.** Compared to ^{235}U, which has a cross-section of 585 barns, the fission cross-section is almost 750 barns.
- **Higher energy release per one fission event**. The neutron flux of a uranium-fueled core is not necessary for a MOX core to produce the same amount of energy.
- **Larger fissile loading.** The greater fissile loading is the primary cause. ^{240}Pu and ^{242}Pu accumulation is comparatively high in MOX fuels. A reactivity cost results from a larger concentration of these isotopes, parasitic absorbers, caused by the comparatively lower fission-to-capture ratio. Generally speaking, ^{239}Pu fuel has a lower average regeneration factor (η) than ^{235}U fuel. In order to get the same initial reactivity at the start of the fuel cycle, the MOX fuel needs a higher fissile loading.

The following effects on reactor core design result from MOX cores' comparatively lower average neutron flux.

- The reactivity value of control rods, chemical shims (PWRs), and burnable absorbers (BAs) is lower with MOX fuel due to the lower neutron flux and the bigger thermal absorption cross-section for ^{239}Pu.
- Fuel rods near water gaps or when MOX fuel is loaded with uranium fuel combined experience higher power peaking due to the high fission cross-section of ^{239}Pu and the lower neutron flux.

8.8.1.3 Neutron Flux and Fuel Burnup

The atomic number density of the fuel atoms in a nuclear reactor stays comparatively steady over a brief period of time (days or weeks). Consequently, when a reactor is run at a steady power level over this brief time, the average neutron flux stays constant. However, the fuel burnup causes the atomic number densities of fissile isotopes to drop over months, as well as the macroscopic cross-sections. As a result, the neutron flux increases slowly to maintain the required power level.

For the fuel to reach its design burnup, it must have enough surplus reactivity and be refilled at the right frequency for both the initial cycle of operation when it is first loaded and the fuel's design lifespan. This might be anything from 12 to 24 months for contemporary LWRs.

Excess reactivity is essential to guaranteeing that the PWR core maintains criticality at full power operating conditions throughout the operation cycles, including compensating for fuel depletion, the accumulation of fission product poisons (like xenon and samarium), and loss of reactivity due to variations in fuel, moderator, etc., even though it is neither a design limit nor a requirement directly. Although there is no set limit on the amount of excess reactivity that can be permitted, the amount of excess reactivity can have an impact on other design parameters like shutdown margin or negative reactivity coefficients.

However, there are restrictions on burnups, which are partly caused by fuel vendor warranties, as well as licensing restrictions for the maximum fuel rod average burnup. In both situations, they are vendor-specific and related to fuel design.

Keep in mind that the units are typically megawatt-days per metric ton of heavy metal loaded (MWd/MT_{HM}), and that the burnup of fuel is the quantity of energy recovered per mass of original fuel loaded. For instance, if a 500 megawatt (MW) thermal SMR runs at maximum power for a year and consumes four metric tons (MT) of uranium in the fuel, the fuel's average burnup will be

Average burnup = 500 x 365/4, which is 45,625 MWd/MTHM, or 45.6 GWd/MTHM.

8.8.2 Void Fraction – Two-Phase Flow

The contemporaneous movement of gas (or vapor) and liquid water (found in steam generators and condensers) via a duct of cross-sectional area A will be discussed in this section.

8.8.2.1 Basic Parameters of Two-Phase Fluid Flow

In a wide range of engineering systems, the topic of two or multiphase flow has grown in significance for both safe operations and optimal design. It is by no means restricted to multiphase flow phenomena and contemporary industrial technologies, which both need for deeper comprehension. Below is a list of some of the significant uses.

Two-phase heat transfer devices, such as heat pipes and two-phase closed thermosyphons, have an effective thermal conductivity hundreds of times greater than copper's. When modeling a heat pipe, the following primary physical processes must be taken into account.

1. Energy conduction via the evaporator/condenser wall when there are devices inside the reactor core and internal wall, such as heat pipes, fuel rods, or other instruments.
2. The vapor–gas interface diffusion.
3. The dynamics of ick liquids.
4. The generated vapor moves from the evaporator via the heat pipe vapor space and into the condenser after boiling in the wick of the heat pipe evaporator.
5. The existence of two phases inside the vapor space.

We may say that a phase is only one of the states of matter and can be a gas, liquid, or solid in order to respond to the question, "What is a Two-Phase Flow?"

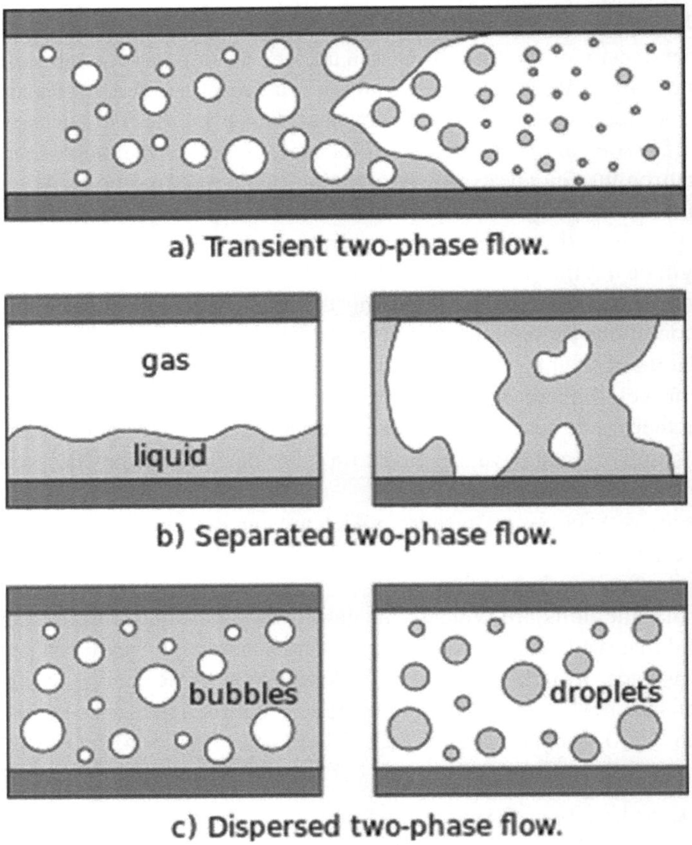

FIGURE 8.20(A) Different modes of two-phase flows.

(Source: www.wikipedia.org)

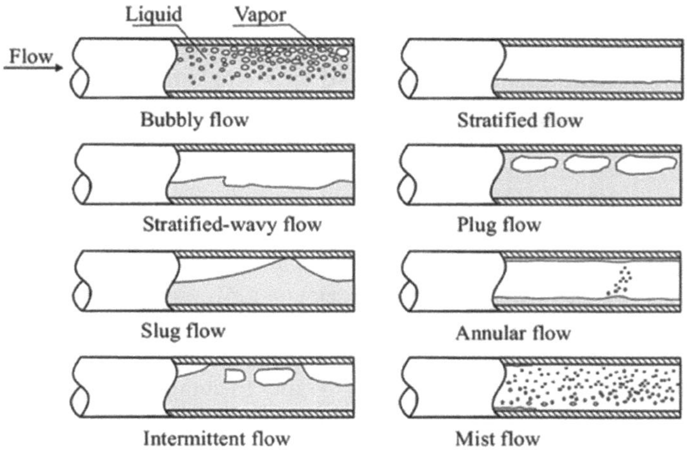

FIGURE 8.20(B) Two-phase flow pattern in horizontal flow.

(Source: www.wikipedia.org)

The simultaneous flow of many phases is known as *multiphase flow*. The most basic type of multiphase flow is a two-phase flow. Refer to Figures 8.20(a) and 8.20(b).

Occasionally, flows when the phases do not contain the same chemical material are referred to as two-component flows. For instance, air-water flows have two components, whereas steam-water flows are two phases. Although they are referred to as two-phase flows, certain two-component flows (usually liquid–liquid) only have one phase, where the phases are distinguished as continuous or discontinuous components. It makes little difference which definitions are used because the mathematics describing two-phase or two-component flows is the same. As a result, in the majority of developments, the two terms will be used interchangeably [23].

There are several typical instances of two-phase flows. Some of these are found in nature and include muck, fast sands, dust storms, fog, smog, smoke, rain, clouds, snow, and icebergs. Other activities, such as brewing tea, boiling water, scrambling eggs, etc., are excellent illustrations of this situation [24].

Two-phase flow, a specific type of multiphase flow in fluid mechanics, is a flow of gas and liquid. Two-phase flows can take many different forms, including separated flows, dispersed two-phase flows, where one phase is present as particles, droplets, or bubbles in a continuous carrier phase (i.e., gas or liquid), and flows that change from pure liquid to vapor due to external heating.

The commonly used approach for classifying two-phase flows involves treating each phase's velocity as though there were no other phases present. The parameter is a fictitious idea known as.

Large-scale power systems have historically likely been the most often researched examples of two-phase flow. Very huge boilers were used in coal and gas-fired power plants to generate steam for turbines. In these situations, heated pipes are used to convey pressured water, which transforms into steam as it passes through. Boiler design necessitates a thorough comprehension of the heat-transfer and pressure-drop behavior of two-phase flow, which differs greatly from the single-phase scenario. More importantly, two-phase flow is used in nuclear reactors to extract heat from the reactor core using water [25].

In order for engineers to plan against potential pipework failures, pressure loss, and other issues (a LOCA), a lot of research has been done on the nature of two-phase flow in such situations [26].

Pump cavitation is another situation where two-phase flow may happen. In this instance, a pump is running near the fluid's vapor pressure. A phase shift may take place and gas will be present in the pump if the pressure continues to drop, which may happen locally close to the pump's vanes, for instance. Marine propellers may also experience similar effects; wherever it happens, it poses a significant challenge to designers. Large pressure spikes may result from the collapse of the vapor bubble, which over time may harm the turbine or propeller.

The two-phase flow examples shown above are for a single fluid that exists alone in two distinct phases, like water and steam. The phrase "two-phase flow" can also refer to combinations of fluids with distinct phases, such as water and air or natural gas and oil. Even three-phase flow is occasionally taken into account, for example, in gas and oil pipelines where a sizable proportion of solids may be present. Oil and water are frequently seen as two-phase flows, even though they are not strictly separate phases because they are both liquids. The flow from an offshore oil well, for example, may be regarded as a three-phase flow when it combines oil, gas, and water.

Water electrolysis [27], climatic systems like clouds [26], and groundwater flow, which studies the flow of air and water through the earth, are other intriguing fields where two-phase flow is investigated.

Additionally, we must discuss "Flow Regime" in order to comprehend two-phase characteristics more fully.

The more accurate the prediction, the more complicated the situation becomes, especially when dealing with a two-phase flow in a heat pipe. The expertise required to do a granule analysis in two-phase flow is frequently shockingly high. For instance, all of the following factors should be taken into consideration while examining the motion of a single gas bubble rising in a stagnant liquid.

- Inertia of the gas and the liquid
- Viscosity of the gas and the liquid
- Density difference and buoyancy
- Surface tension and surface contamination

The last point above is particularly complex on its own since "contamination" might be dirt, dissolved debris, or surface-active chemicals, all of which can significantly affect heat transport in that system. The bubble's motion is also changed by mass and heat transfer. Dividing this hydraulic issue into different regimes, each controlled by dominating geometrical or dynamic characteristics, may be the first step towards making it tractable. A description of the components' morphological arrangement, or flow pattern, is included in the definition of the flow regime.

Figure 8.18 illustrates the intricacy of two-phase flows by showing a series of flow patterns occurring in a heat pipe evaporator as an increasing amount of liquid is turned to vapor. The intricacy of the issue occurs in several evaporator components that need for distinct analytical techniques, and the issue of how one regime evolves from another must also be taken into account.

Numerous writers have mapped out various presentations of flow pattern and flow regime that provide tow-independent coordinates on a graph. As seen in Figures 8.21, Figures 8.22 and Figures 8.23, this is done in terms of the flow rates for a particular apparatus and certain components. However, a two-dimensional graphic is seldom sufficient for general depiction because the flow regime is controlled by around a dozen factors [28, 29].

Bubbles, rain, sea waves, foam, fountains, mousse, cryogenics, and oil slicks are more instances of two-phase flow. The electrical explosion of metal serves as a last illustration.

The vapor and liquid phases are denoted by the subscripts " v " and " ℓ ", respectively. The following basic parameters define this flow.

1. Void Fraction.
2. Static Quality.
3. Flow Quality – Static Quality.
4. Mixture Density.
5. Slip Ratio – Velocity Ratio.
6. Superficial Velocity – Two-Phase Flow

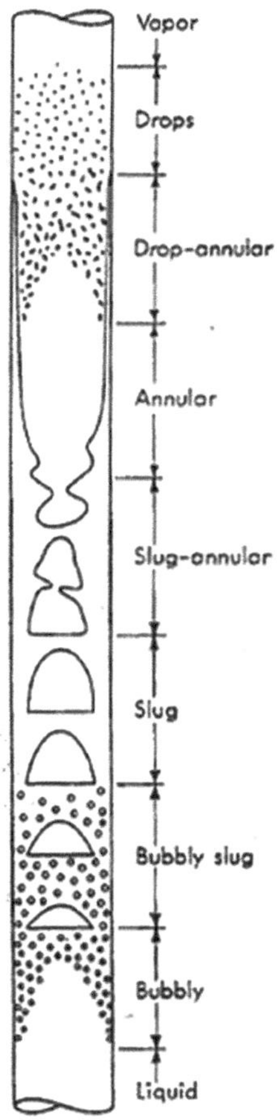

FIGURE 8.21 Approximate sequence of flow patterns in a vertical heat pipe evaporator [28].

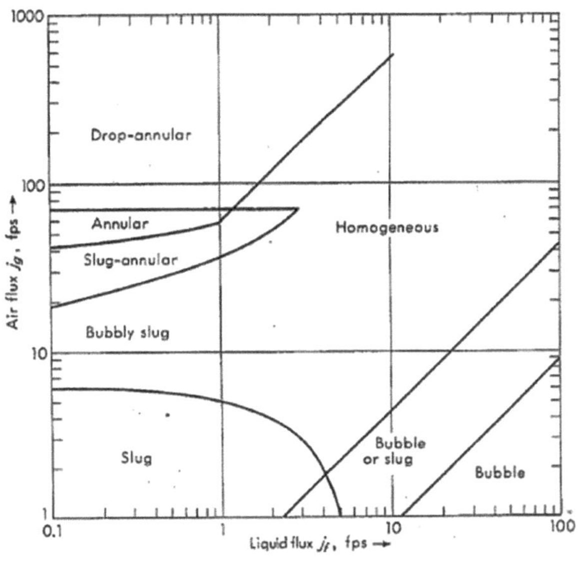

FIGURE 8.22 Flow-Pattern boundaries for vertical up flow of air and water at 15 psi in a 1-Diam tube deduced from equations in Reference [33].

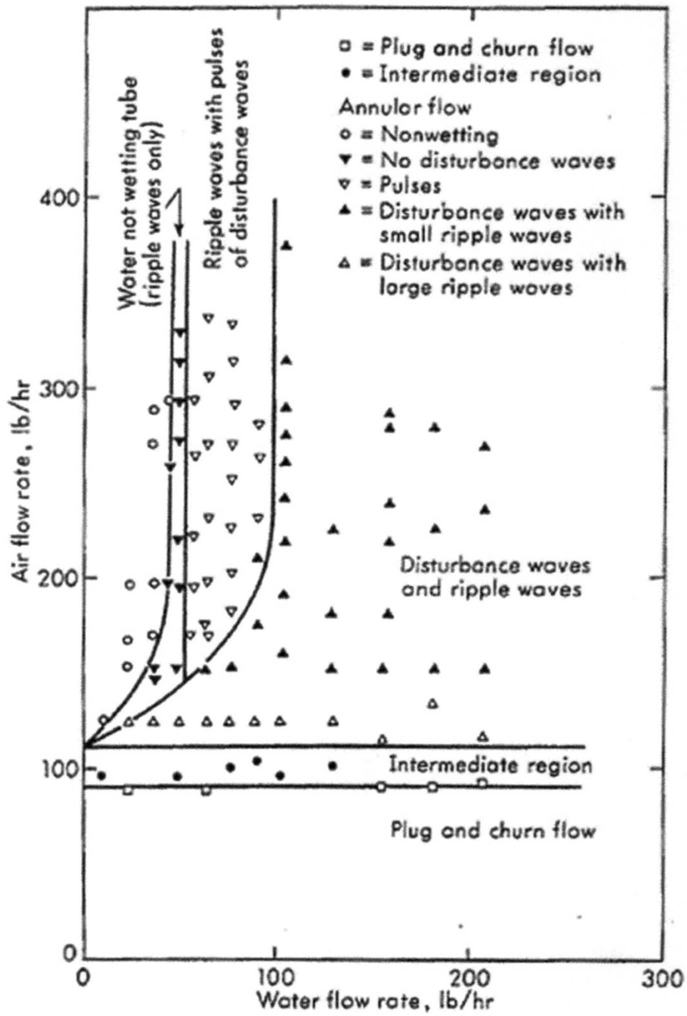

FIGURE 8.23 Various "regime" or subdivisions of the annular flow pattern for current upward flow of air and water in a $1^{1/4}$-in-diam pipe at 15 psi.

(Source: Hall-Taylor and Hewitt AERE-R3952, UKAEA, 1962)

Each of the above six topics are described as follows:

1. **Void Fraction**

One of the most crucial metrics for describing two-phase fluid flow, particularly gas-liquid flow, is the void fraction, or α .

This parameter is specified using a variety of geometric concepts. In a two-phase fluid flow, the void fraction may be described as follows:

I. The percentage of the channel volume that the gas phase occupies. The volumetric void fraction is the name given to this void fraction.
II. The percentage that the gas phase occupies of the channel cross-sectional area. The cross-sectional void fraction is the name given to this void fraction.

III. A single point or extremely tiny volume is referred to as the local empty fraction. As a result, it accepts either 1 or 0.

The proportion of the channel cross-sectional area filled by the gas phase (also known as the cross-sectional void fraction) will be assumed to be a void fraction for the sake of this discussion. This is defined as follows:

$$\alpha \equiv \frac{A_v}{A_v + A_\ell} \qquad \text{Eq. 8.6}$$

In a two-phase flow, the void fraction is significant because it affects crucial physical properties as heat transfer, pressure drop, and viscosity.

$$\alpha = \frac{1}{1+\left(\frac{1-x}{x}\right)\left(\frac{\rho_G}{\rho_L}\right)(S)}$$
$$x = \frac{1}{1+\left(\frac{1-\alpha}{\alpha}\right)\left(\frac{\rho_L}{\rho_G}\right)\left(\frac{1}{S}\right)} \qquad \text{Eq. 8.7}$$
$$S = \left(\frac{1-\alpha}{\alpha}\right)\left(\frac{x}{1-x}\right)\left(\frac{\rho_L}{\rho_G}\right)$$

where

S = Slip ratio, dimensionless parameter
v = Velocity [m/s]
α = Void fraction, dimensionless parameter
ρ = Density of a phase [kg/m^3]
x = Steam quality, dimensionless parameter

Equation 8.7 and flow quality parameters in relation to void percent are shown in Figure 8.24.

2. **Static Quality – Flow Quality**

Vapor quality is often defined as the mass fraction of vapor in a saturated mixture when it comes to static quality. The quality of saturated liquid is 0%, but that of saturated vapor is 100%.

The static quality in a two-phase fluid flow is defined as

$$x_{st} \equiv \frac{M_v}{M_v + M_\ell} = \frac{A_v \rho_v}{A_v \rho_v + A_\ell \rho_\ell} \qquad \text{Eq. 8.8}$$

It is more convenient to use the flow quality rather than the static quality when dealing with two-phase fluid flow. A two-phase fluid flow's flow quality is described as follows:

$$x \equiv \frac{\dot{m}_v}{\dot{m}_v + \dot{m}_\ell} = \frac{v_v A_v \rho_v}{v_v A_v \rho_v + v_\ell A_\ell \rho_\ell} \qquad \text{Eq. 8.9}$$

One may infer the relationships between x, α, and, S and the outcome is shown in sets of Equation 8.7 and Figure 8.24.

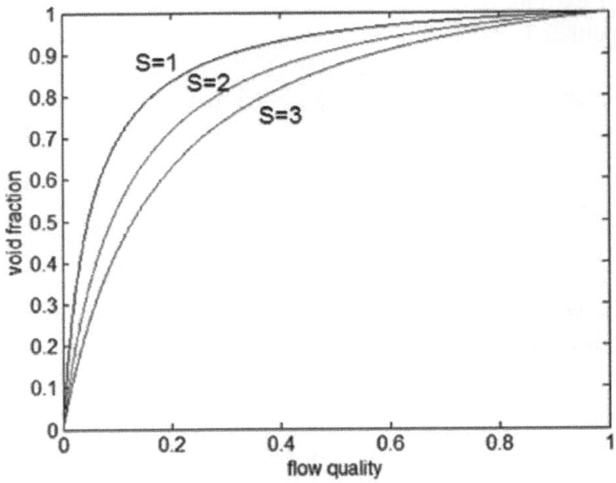

FIGURE 8.24 Effect of S on α versus x for water at 7 MPa.

(Source: Buongiorno Jacopo, MIT Department of Nuclear Science and Engineering, NOTES ON TWO-PHASE FLOW)

3. **Mixture Density – Two-Phase Flow**

Using the Mixture Density in a two-phase fluid flow is convenient. Pressure drop is calculated using the two-phase flow's mixture density. In a two-phase fluid flow, the mixture density is defined as follows [30].

$$\rho_m \equiv \alpha\rho_v + (1-\alpha)\rho_\ell \qquad \text{Eq. 8.10}$$

4. **Slip Ratio – Velocity Ratio**

The slip ratio is a useful tool for fluid flow in two phases. In a two-phase flow, the slip ratio, also known as the velocity ratio, is the ratio of the liquid phase's velocity to the vapor phase's velocity. The definition of the slip ratio in a two-phase fluid flow is:

$$S = \frac{v_v}{v_\ell}$$

$$S = \frac{v_v}{v_\ell} = \frac{V_v(1-\alpha)}{\alpha V_\ell} \qquad \text{Eq. 8.11}$$

Where again

S = Slip ratio, dimensionless parameter
v = Velocity [m/s]
α = Void fraction, dimensionless parameter
ρ = Density of a phase [kg/m^3]
x = Steam quality, dimensionless parameter
V = Superficial velocity [m/s]

One can infer the relationships between x, α, and S, and the outcome is shown in sets of Equation 8.7 and Figure 8.24.

Because they enable the calculation of the two-phase flow's pressure drop, the void fraction and slip ratio are defined. The literature presents a number of correlations for calculating the void percent, S, and slip ratio. The correlations listed below are presented in ascending order of accuracy.

Homogeneous flow model

$$S = 1$$

Zivi's slip correlation

$$S = \sqrt[3]{\left(\frac{\rho_L}{\rho_G}\right)}$$

Chisholm's slip correlation

$$S = \sqrt{1 - x\left(1 - \frac{\rho_L}{\rho_G}\right)}$$

Smith's void correlation

$$\alpha = \left[1 + \left(\frac{\rho_G}{\rho_L}\right)(K)\left(\frac{1-x}{x}\right) + \left(\frac{\rho_G}{\rho_L}\right)(1-K)\left(\frac{1-x}{x}\right)\sqrt{\frac{\frac{\rho_G}{\rho_L} + K\left(\frac{1-x}{x}\right)}{1 + K\left(\frac{1-x}{x}\right)}}\right]^{-1}$$

5. **Superficial Velocity**

A hypothetical flow velocity known as "superficial velocity" is determined by assuming that the specified phase or fluid is the only one present or flowing in a specific cross-sectional region. The provided phase's velocity is computed as though the second phase did not exist.

Because it is an unambiguous metric, superficial velocity (V_{phase} or j_{phase}) is frequently employed in the engineering of multiphase flows and fluxes in porous media. Real velocity, on the other hand, frequently depends on space and is predicated on several assumptions.

Superficial velocity can be expressed as

$$V_{\text{phase}} = \frac{Q}{A} \qquad \text{Eq. 8.12}$$

where

V_{phase} = Superficial velocity [m/s]
Q = Volume flow rate of the phase [m^3/s]
A = Cross-Section of flow area [m2]

To have a better idea, let us look at a pipe with a flow area cross-section of 0.1 m^2. Assume that 1 m^3/s is the flow rate. The true fluid velocity, which is 10 m/s, will be identical to the superficial velocity in a single-phase fluid flow.

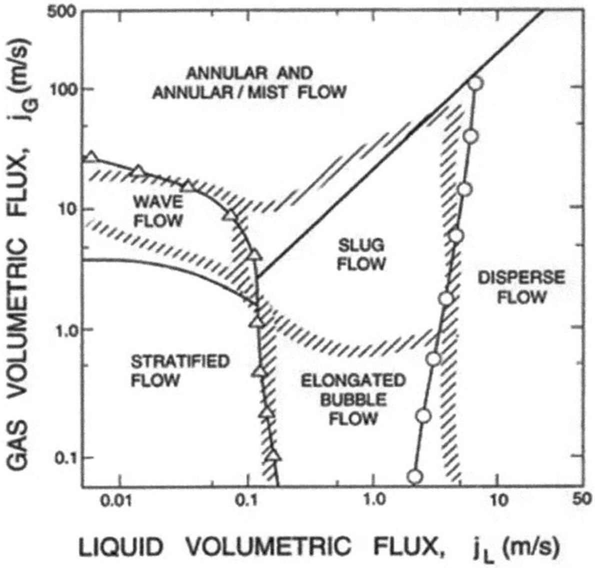

FIGURE 8.25 Gas volume flux versus liquid volumetric flux.

(Source: Mandhane, J.M., Gregory, G.A. and Aziz, K.A. (1974). A flow pattern map for gas-liquid flow in horizontal pipes. Int. J. Multiphase Flow)

The scenario will be different for fluid flow that occurs in two phases, such as vapor–liquid flow. Both phases will have superficial velocities of 5 m/s if the slip ratio is unity and they have been taken independently. At that point, the actual velocity will be 10 m/s. It will be more difficult if the two phases move at different speeds (with slip).

A flow regime map for the flow of an air/water mixture in a horizontal pipe with a diameter of 2.5 cm at 25°C and 1 bar can be found in Figure 8.25. The hatching zones are theoretical predictions, whereas the solid lines and dots are experimental observations of the transition circumstances. Source: Gregory, G.A., Aziz, K.A., and Mandhane, J.M. (1974). A map of the flow patterns in horizontal pipes for gas-liquid flow. Multiphase Flow, Int. J.

8.8.2.2 Characteristics of Two-Phase Fluid Flow

Two-phase flow is an intriguing and difficult area of fluid mechanics because of several characteristics.

- All dynamical problems become nonlinear due to surface tension (see Weber number).
- There is a roughly 1000-fold difference in density between air and water at typical temperature and pressure. Densities of water (liquid) and water vapor often differ in similar ways.
- For materials going through a phase shift, the sound speed varies significantly and can vary by orders of magnitude. Compressible effects are now a part of the issue.
- The liquid vapor system will not always be in phase equilibrium, and the phase transitions are not immediate.
- In contrast to single-phase incompressible flow, where closing a valve would decrease exit velocities, the phase change means that flow-induced pressure drops can cause additional phase-change (for example, water can evaporate through a valve), increasing the relative volume of the gaseous, compressible medium and increasing exit velocities.
- Can result in other counterintuitive, negative resistance-type instabilities, such as Ledinegg instability, geysering, chugging, relaxation instability, and flow maldistribution instabilities as examples of static instabilities, as well as other dynamic instabilities [31].

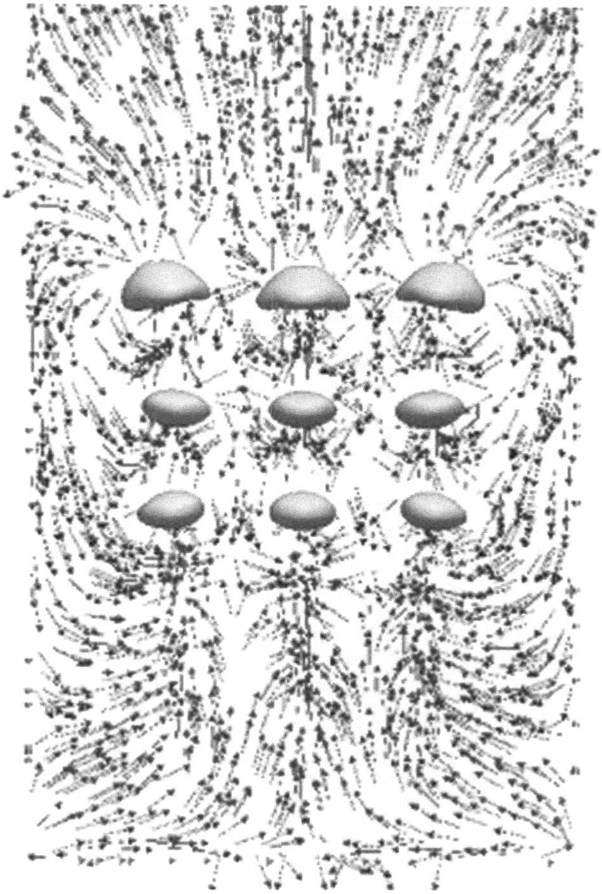

FIGURE 8.26 Simulation of bubble swarm using volume of fluid method.

(Source: www.wikipedia.org)

It should be noted that, according to acoustics physics, gargling is a distinctive sound produced by unstable two-phase fluid flow, such as when liquid is poured from a bottle or while gargling is taking place.

8.8.2.3 Modeling of Two-Phase Fluid Flow

Two-phase flow modeling is still being developed (refer to Figure 8.26). Known techniques include

- Volume of fluid method.
- Level-set method.
- Front tracking by Gretar Tryggvason.
- Lattice Boltzmann methods.
- Smoothed-Particle Hydrodynamics (SPH).

8.8.3 Reactivity Coefficients – Reactivity Feedback

The reactor's stability is determined by reactivity feedback, which is an intrinsic feedback. This feedback is characterized by reactivity coefficients. The amount that the reactivity will vary for a certain change in the parameter is known as the reactivity coefficient. The reactor design must provide negative temperature feedback in all operational situations. This is in accordance with Criterion 11 of 10 CFR Part 50, which is cited below [12].

> "In the power operating range, the net effect of the prompt inherent nuclear feedback characteristics tends to compensate for a rapid increase in reactivity. This is the design for the reactor core and related coolant systems".

The reaction of the neutron population in a nuclear reactor to an external reactivity input has been covered up to this point. It was assumed that the neutron population level had no effect on the system's characteristics, particularly that the neutron power (power from the chain reaction) was low enough that the reactor core's temperature remained constant (i.e., reactivity feedback may be ignored). Because of this, these therapies are commonly called zero-power kinetics.

On the other hand, there is always enough neutrons in a working power reactor to produce heat. Power reactors are designed primarily to produce a lot of heat. Because of the thermal expansion, this results in changes to the system's temperature and material densities.

Because the neutron flux spectrum depends on the density of the moderator, and because macroscopic cross-sections (see Chapter 1 of this book for more information on this topic) are proportional to densities and temperatures, these changes will also result in some variations in reactivity. Reactivity coefficients are used to define these alterations in reactivity, which are typically referred to as reactivity feedback. Because the reactivity feedback affects the reactor's stability, this is a crucial aspect of reactor design. For instance, the design of the reactor must guarantee that, in any operational scenario, the temperature feedback will be negative.

Remember that the reactor kinetics approach of Equation 8.13 shows that the power generated by a reactor is based on the time-dependent multiplication of $k(t)$ or, in turn, the reactivity $\rho(t)$ as follows:

$$\rho(t) = \frac{k(t)-1}{k(t)} \qquad \text{Eq. 8.13}$$

In some situations, it is possible to determine *a priori* the reactivity versus time. Such is often the case, for example, when control rods of known properties are moved in a prescribed manner in a reactor operating at very low neutron flux levels.

As soon as a reactor produces enough power to raise the temperatures in the core above their ambient levels, however, the material densities as well as some of the microscopic cross sections become affected. Because the core multiplication depends on these densities, reactivity feedback loop is established in which $k(t)$, and thus $\rho(t)$ is dependent on temperatures and densities that in turn are determined by the reactor power history.

However, the material densities and certain of the tiny cross-sections are impacted as soon as a reactor generates enough power to increase the core temperatures over their ambient values. A reactivity feedback loop is created in which $k(t)$ and therefore $\rho(t)$ are reliant on temperatures and densities, which are in turn dictated by the reactor power history, since the core multiplication depends on these densities.

The understanding of the nature of mechanisms leading to reactivity feedback is essential to the analysis of power reactor behavior [32].

In conclusion, reactivity – a measure of a reactor's condition relative to where it would be in a critical state – is a frequently used quantity in the nuclear business. Reactivity is zero at criticality, negative at subcritical conditions, and positive at supercritical conditions. There are several techniques to manage reactivity, including increasing the quantity of absorber that competes with the fuel for neutrons, adding or withdrawing fuel, or adjusting the ratio of neutrons that leak out of the system to those that are retained. By altering the absorbers, which are typically movable control rods, the latter technique regulates the reactor's neutron population (although in a less popular design, operators can alter the absorber concentration in the reactor coolant). Conversely, neutron leakage changes are frequently automated. For instance, a reactor's coolant will lose density and could boil as power is increased. By increasing neutron leakage out of the system, this drop in coolant density will lower reactivity – a process called negative-reactivity feedback. Designing a reactor safely requires consideration of neutron leakage and other negative-reactivity feedback processes.

8.8.3.1 Example: Change in the Moderator Temperature

The neutron population is affected by negative feedback as the moderator temperature effect in the following ways. The addition of negative reactivity to the core occurs when the moderator's temperature is raised. The reactor power drops as a result of this negative reactivity. The reactor recovers to the critical state when the power coefficient operates against the decline in thermal power. Reactor power self-stabilizes. Significant variations in the resonance escape probability are the cause of this impact in terms of the multiplication factor (see Chapter 1 of this book) and total neutron leakage, or in the thermal utilization factor (see Chapter 2 of this book), when the chemical shim is used.

8.8.3.2 Chemical Shim – Boric Acid

We imply that boric acid is dissolved in the coolant/moderator when we talk about chemical shim. White and soluble in water, boric acid has the chemical formula H_3BO_3. Boric acid is used as a chemical shim in PWRs to offset the excess reactivity of the reactor core during fuel burnup (long-term reactivity control). Boric acid is most concentrated at the start of the particular fuel cycle (see Figure 8.27(a) and Figure 8.27(b)). Upon completion of this cycle, the boric acid concentration is nearly zero, necessitating the refueling of a reactor.

In certain cases, fine power changes can be also controlled by the chemical shim. If it is desired to increase power, the boric acid concentration must be diluted, removing ^{10}B from the reactor core and decreasing its poisoning effect. Because boric acid is uniformly dissolved in the coolant throughout the reactor core, it avoids the uneven neutron flux density in the core compared to BAs (long-term reactivity control) or control rods (quick reactivity control). However, it is undesirable for high boric acid concentrations to result in a positive MTC. More BAs need to be employed in this situation.

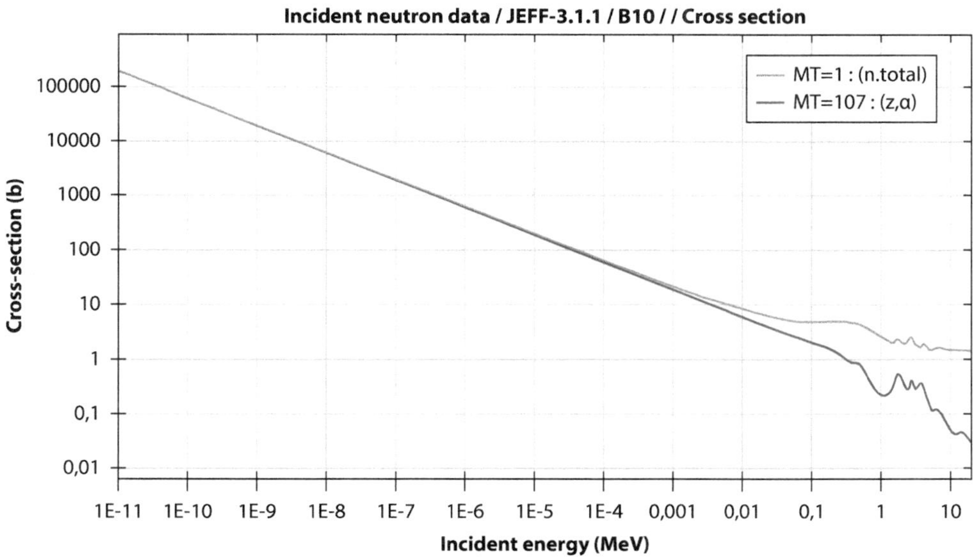

FIGURE 8.27(A) Boron 10. Comparison of total cross-section and cross-section for (n, alpha) reactions.

(Source: JANIS (Java-based Nuclear Data Information Software); The JEFF-3.1.1 Nuclear Data Library)

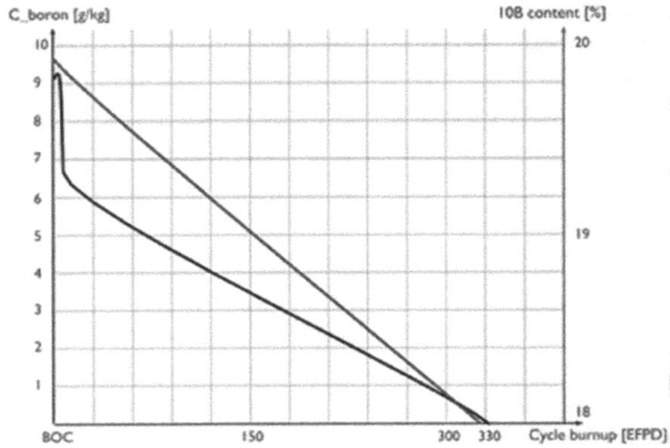

FIGURE 8.27(B) Boron letdown curve (chemical shim) and boron 10 depletion during a 12-month fuel cycle. **(Source: JANIS (Java-based Nuclear Data Information Software); The JEFF-3.1.1 Nuclear Data Library)**

Keep in mind that the concentration of boric acid is greatest at the start of the particular fuel cycle in Figure 8.24(b). The amount of boric acid is nearly nil by the end of this cycle. Since no positive reactivity can be added to offset the negative reactivity of fuel burnup (increased reactor slagging), a reactor needs to be refueled.

8.8.4 Reactivity Coefficient of the Fast Reactor

The Doppler coefficient of the fast reactor is lower than that of the thermal neutron reactor due to its hard neutron spectrum. Furthermore, no moderator is used by the rapid reactor. In general, it has a low reactivity temperature coefficient. Fuel rod bending from differential thermal expansion may be the cause of the reactivity effect in a fast reactor. At the core, where the temperature is high, thermal expansion is significant. When the fuel rods are unable to travel horizontally at the top or bottom of the core, they are bent toward the center of the core. In the early tiny fast reactors, this was an issue since it increased the reactivity. Finding the best location and technique to hold and support the fuel rods and core can help to overcome this issue [Oka, Y. (2013). Temperature Effect of Reactivity].

The sodium void coefficient is often positive in the core of a large liquid sodium-cooled fast reactor. When the sodium coolant boils, it is a problem that is addressed in fictitious incidents. The sodium void coefficient causes a rivalry between the negative reactivity effect, which results from increased neutron leakage, and the positive reactivity impact, which promotes nuclear fission because of the toughened spectrum. The local sodium void coefficient becomes positive toward the center of a big core because the neutron leakage impact is minimal there. One can take countermeasures

1) To increase the neutron leakage by designing a flat core.
2) To decrease the void coefficient by softening the neutron spectrum in the core and to increase the Doppler reactivity; or
3) To form a thin hydride layer between the blanket and seed fuel in order to moderate the fast neutrons (produced in the seed at voiding) in this layer and to absorb the moderated neutrons by the blanket fuel (^{238}U).

8.9 INCORE NUCLEAR INSTRUMENTATION – DETECTORS IN NUCLEAR REACTORS

Reliable measuring, diagnostic, and control equipment is necessary for the safe and efficient use of nuclear technology. Numerous complex applications, including remote sensors for environmental safety, precision imaging systems for medical diagnostics, and the probing and production of the most cutting-edge materials, rely on advanced nuclear technology.

Reactor core temperatures and neutron flux distribution are measured using the Incore nuclear instrumentation (NI) system [33].

The incore instrumentation system includes

1. **Incore neutron flux monitoring system**

The incore instrumentation system's functions include providing data on the distribution of neutron flux and the temperatures of the fuel assembly exit at specific core locations. The incore instrumentation system does not carry out any protective or plant operational control tasks; it only facilitates data acquisition. Both the incore temperature monitoring system and the movable incore neutron flux monitoring system are part of the incore instrumentation system. Whether the plant is a two-, three-, or four-loop Westinghouse design affects the number of incore temperature monitoring thermocouples and the number of flux thimble pathways accessible within the core for the movable detectors. Figure 8.28 shows where the thermocouples and incore flux thimbles are located in the reactor core of a four-loop Westinghouse facility.

The incore neutron monitoring system is composed of up of incore detectors that are adequately sensitive to detect changes in the local neutron flux distribution inside the reactor core. It should be mentioned that a variety of factors greatly affect the distribution of power and flux in power reactor cores. As a result, the temperature in a reactor that is in operation fluctuates throughout the system.

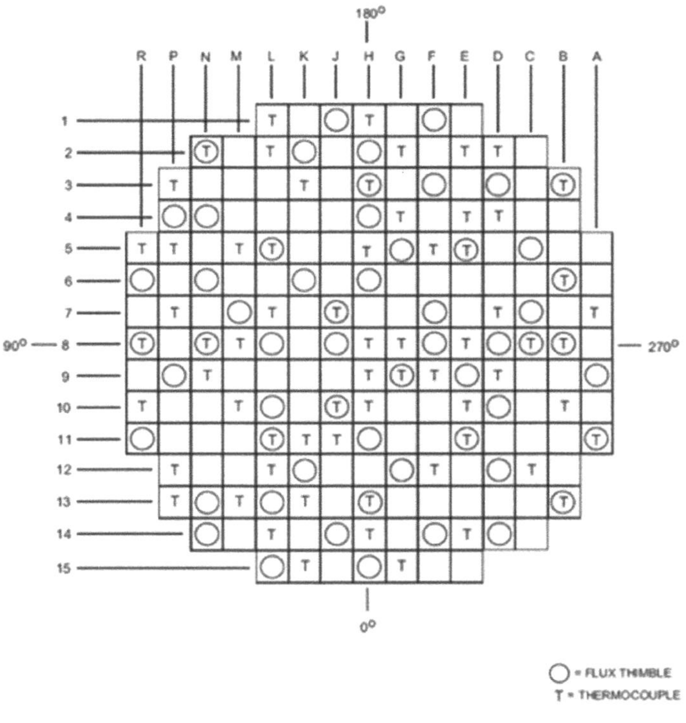

FIGURE 8.28 Thermocouples and flux thimble location (4 Loop Plant Westinghouse Design). [36]

As a result, one local volume and one fuel rod are always hotter than the others. To control these hot spots, peak power limitations must be implemented. A boiling crisis and the circumstances that might lead to fuel pellet melt are linked to the peak power constraints. The neutron flux distribution and, consequently, the margins to these peak power restrictions are described by the incore neutron flux monitoring system.

The incore neutron flux monitoring system usually utilizes

- Miniature fission chambers (MFCs).
- Self-powered neutron detectors (SPNDs).

These portable flux detectors, which are typically installed in a fuel assembly's instrumentation tube, can track a chosen fuel assembly's whole length to provide a very precise, three-dimensional picture of the neutron flux distribution. These data may also be used to rebuild neutron flux in the remaining reactor core. Depending on the reactor architecture, the information gathered from the incore neutron flux monitoring device is often utilized to

1. At any point throughout the fuel cycle, the power distribution in the core may be ascertained using these data. The following power distribution hot channel factors are checked for compliance with technical specification limitations using the monitored power distribution.
   a. The Heat Flux Hot Channel Factor – *FQ(z)* is defined as the ratio of the average local linear power density in the core to the highest local linear power density, when there is little room for restricting fuel temperature (during anticipated operational occurrences (AOOs))
   b. The Nuclear Enthalpy Rise Hot Channel Factor – $F^N_{\Delta H}$ is defined as the ratio of the average fuel rod power in the core to the integral of linear power along the fuel rod when the least amount of deviation from the nucleate boiling ratio occurs (during AOOs) [34].

   AOOs are off-normal events, usually plant transients, which can be coped with by the plant protection systems and normal plant systems but which could have the potential to damage the reactor if some additional malfunction should happen.

2. Fuel burnup and isotopic fuel stocks in the core at any point throughout the fuel cycle may be ascertained using these data.
3. The excore power range nuclear instruments for Axial Flux Difference (AFD) may be calibrated using these data.
4. The Quadrant Power Tilt Ratio (QPTR) may be checked using these data to make sure it satisfies the technical standard limit.
5. In order to take remedial action before a condition gets out of control, the data will also show patterns of core conditions.

2. **Incore temperature monitoring system**

In order to monitor the coolant enthalpy distribution and core radial power sharing, the incore neutron temperature monitoring system uses incore thermocouples placed at certain positions to detect the fuel assembly coolant outlet temperature. Lateral flow mixing largely determines coolant outlet temperatures, and in certain reactor designs, this system serves other purposes including safety function monitoring. Depending on the reactor type, these data (coolant exit temperatures) might be used to as follows [33]

1) During crises, alert the operators to signs of insufficient core cooling conditions (such as core overtemperature).
2) Give details on the fuel assembly's temperature increase. This should be looked at since it could point to a major core problem (such as channel blockage).
3) Supply the subcooling margin monitors with inputs.
4) Supply inputs for plant computational computer programs that calculate fuel assembly enthalpy increases and restricted power distribution data using core-exit temperatures.

As previously mentioned, the incore neutron monitoring system is made up of tiny fission chambers that are sensitive enough to record changes in the local neutron flux distribution inside the reactor core.

8.9.1 Self-Powered Neutron Detector

Because they can adapt to harsh in-core conditions, SPNDs, are frequently employed in reactors to monitor neutron flux. The incore neutron flux monitoring system, which offers comprehensive data on the distribution of neutron flux and, consequently, the margins to these peak power restrictions, may include SPNDs. These detectors provide an output signal by using the fundamental radioactive decay mechanism of their neutron activation material.

As the name suggests, SPNDs generate a voltage potential in the detector without the need for an external voltage source. Rather, neutron activation and the detector's subsequent beta decay result in the production of a current in the detector. These beta particles, or electrons, cause the wire to become increasingly positively charged. A current flows through the resistor, R, as a result of the wire's positive potential. An amperemeter may be used to directly measure the electron current from beta decay.

There are two main advantages of the SPND.

- Very little equipment is needed, typically only an amperemeter or a millivoltmeter.
- The lifespan of the emitter material is far longer than that of the uranium-235 or boron liner used in fission chambers.

However, there are drawbacks as well, one of which is related to the extremely low currents even while operating at full power. As a result, while operating at low power (10% and below), SPNDs are unable to offer information on flux distribution. The primary drawback of the SPND is that the response time of the detector is determined by the characteristic half-life of the emitter material. These detectors are often divided into the following categories based on reaction time

- **Prompt response detectors.** Reactor protection and regulation applications employ rapid response detectors like cobalt and Inconel.
- **Delayed response detectors.** For Flux Mapping Systems (FMSs), delayed response detectors such as vanadium and rhodium are frequently utilized.

The typical SPND is a coaxial cable consisting of

- **Emitter.** The substance used to make the inner electrode absorbs a neutron and then decays radioactively, releasing an electron (beta decay). The emitter, which creates electrons, is often composed of rhodium.
- **Insulation.** The emitter is surrounded by insulation, usually made of aluminum oxide.
- **Collector.** These components are enclosed by the metal detector walls, which also act as a collector for the generated electrons. Ground potential is connected to the collector.

Typically, SPNDs are installed in a fuel assembly's instrumentation tube. They can generate a very precise, three-dimensional image of the neutron flux distribution by tracking the full length of certain fuel assemblies. These data may also be used to rebuild neutron flux in the remaining reactor core.

The emitter is often made of cobalt, cadmium, rhodium, and vanadium. These materials should be utilized because they are compatible with the production process of SPNDs and have relatively high melting points and cross-sections to thermal neutrons.

In general, neutron detectors are employed for nuclear applications such as high-energy physics experiments, special nuclear material detection [35], monitoring of nuclear reactor core [36], fuel reprocessing plants, fast neutron radiation therapy [37], and several other applications, including space [38–42]. In most of the occasions, the generated neutrons are of high energy, for example, in nuclear reactors, neutrons generated through fission are born as fast neutrons with high kinetic energy [43, 44]. Also, neutron interaction with matter is strongly dependent on the energy, as the energy increases, the probability of interaction goes down due to the rapid reduction in cross-section [45, 46]. Knoll [46] has quoted that, in principle, almost all techniques of slow neutron detection can be applied for fast neutron detection as well. But the detection efficiency will be very poor due to the cross-section, and such a low efficiency limits the practical application. Thus, fast neutron devices have to incorporate a revamped or different detection scheme to achieve acceptable detection efficiency. Contrary to slow neutron detection, the detection of fast neutrons could provide additional information associated with the energy of the incident neutron, vital in some applications [47, 48]. Information on neutron kinetic energy is lost in conversion for slow neutrons, as the energy liberated in the conversion reaction itself (Q-value) is extremely large compared to the neutron energy. However, the energy of fast neutrons is not negligible compared to the reaction Q-value. In principle, subtracting the reaction Q-value from the reaction product energies can yield the incident neutron energy.

Various methods have been developed for fast neutron spectroscopy and detection [38, 46].

8.9.2 Rhodium Emitter – Rhodium-Based SPND

Rhodium (Rh) is one potential substance that might be the emitter. An SPND with a rhodium emitter has a (two-fold) delayed signal, perturbs the local power density, and has a reasonably high sensitivity and high burnup rate. The beta-current kind of self-powered detector, known as a rhodium-based detector, generates a measurable current by the following activation process.

$$^{1}n + {}^{103}Rh \rightarrow {}^{104}Rh \rightarrow {}^{104}Pd + \beta$$

As is evident, a rhodium-103 atom becomes a radioactive rhodium-104 atom when it absorbs a neutron. After that, the rhodium-104 decays to produce palladium-104 along with an electron or beta particle. With sufficient energy, the beta particle can travel through the insulator and arrive at the collection. Activated rhodium-104 has a half-life of 42.3 seconds, which postpones the charged particle's emission. This creation of beta particles (electrons) is used by the rhodium-based detector to generate a current proportional to the number of neutrons that the emitter captures, which is also proportional to the power density of the local reactor. Gamma rays are responsible for a percentage of the detector's current flow. To counteract for this false signal, a background correction is carried out using a background detector that has the identical parts as the detector, with the exception that the rhodium is eliminated.

Rhodium-103 has a resonance at 1.25 eV and a capture cross-section of 133 barns for thermal neutrons. Beta radioactive ^{104}Rh with $T_{1/2}$ = 42 seconds is produced as a result of this process. Note that about 11 barns are part of a reaction that yields an isomer of 104mRh (with $T_{1/2}$ = 4.4 min).

When utilized in thermal power reactors (such as PWRs), the following traits are common.

- The rhodium burn-up rate is 0.39% per month in a thermal neutron flux of 10^{13} n/cm^2/s.
- 92% of the signal has a half-life of 42 seconds.
- 8% of the signal has a half-life of 4.4 minutes.
- The beta emission has an energy of 2.44 MeV.

8.9.3 Vanadium Emitter – Vanadium-Based SPND

Vanadium emitters have a very long delayed signal, a low burn-up rate, negligible disturbance of the local power density, and very low sensitivity. A vanadium-based detector is a self-powered beta-current detector that generates a measurable current by the following activation process.

$$^{1}n + {}^{51}V \rightarrow {}^{52}V \rightarrow {}^{52}Cr + \beta$$

The capture cross-section of vanadium-51 for thermal neutrons without resonances is 4.9 barns. ^{52}V, which is beta radioactive, is produced as a result of this process with $T_{1/2} = 3.74$ minutes.

The following characteristics are typical when used in thermal power reactors (e.g., PWRs).

- The vanadium burnup rate is 0.012% per month in a thermal neutron flux of 10^{13} n/cm^2/s.
- 99% of the signal has a half-life of 3.8 minutes.
- 1% of the signal is prompt.
- The subsequent beta emission has an energy of 2.6 MeV.

In summary of this section, from "Human Factors Engineering (HEF) Aspects of Instrumentation and Control System Design" perspective, we may be able to state that

NPP personnel play a vital role in the safe, efficient and productive generation of electrical power. Operators monitor and control the plant to ensure it is functioning properly. Test and maintenance personnel help ensure the equipment is functioning properly and restore components when malfunctions occur. Personnel performance and the resulting plant performance are influenced by many aspects of plant design, including the Instrumentation and Control (I&C) systems and the Human–System Interfaces (HSIs) provided for personnel to interact with the plant. Information presented to plant personnel through I&C systems and HSIs has to be accurate, sufficient, operationally relevant, timely and dependable.

HSIs can have different characteristics depending on the technologies installed in the plant. The general operational and maintenance environment for nuclear facilities is becoming more computer based, incorporating features such as soft controls, computerized procedures, mobile interfaces, and touch screen interfaces. These interfaces affect the ways that humans operate and maintain the plant.

While HSI design and digitalization can greatly improve personnel and plant performance, it is important to recognize that, if poorly planned, designed, or implemented, there is the potential to negatively impact performance and safety as well as to reduce human reliability, resulting in a detrimental effect on safety and cost-effective power production. Human Factors Engineering (HFE) is needed to ensure that the benefits of the technologies are realized and problems with their implementation, operation, and maintenance are minimized. The I&C systems can consequently affect HSI design and, likewise, HFE can drive some aspects of I&C systems.

In final, bear in mind that, digital technology is rapidly replacing analogue technology in many areas of NI. This transition is enabling numerous benefits and opportunities in nuclear spectrometry that were previously not possible, or very difficult to implement using traditional analogue technology. Digital Pulse Processing (DPP), which replaces the traditional analogue pulse processing, provides technical improvements in spectroscopic performance and reliability, and instruments with small physical size, low weight, and battery-powered operations. These features deliver improvements in the quantity, quality, and reliability of data for laboratory and research-based nuclear spectroscopy systems, providing end users with data with a higher degree of accuracy and confidence to support their findings and conclusions, and the development and deployment of hand-held instruments and field-based instruments for remote and unattended operations.

Therefore, the reactor core's temperature and neutron flux distribution are measured using the incore NI equipment. The incore instrumentation system's objectives are to give precise data on the distribution of neutron flux and the temperatures of the fuel assembly exit at specific core regions.

8.10 NUCLEAR INSTRUMENTATION

The thermal power generated by nuclear fissions in nuclear reactors is directly proportional to the level of neutron flux. Therefore, accurately measuring and controlling the neutron flux and its spatial distribution within the reactor using the right instruments is crucial from the perspective of reactor safety. Numerous nuclear instruments are placed for this reason. Although certain measurements are taken from inside the reactor core, they are typically taken from outside. As a result, nuclear instruments are typically divided into:

- Nuclear instrumentation that is exceptional. Excore neutron detectors placed outside the core are often used to measure the neutron flux. The Nuclear Instrumentation System (NIS) excore includes several detectors. Three categories of instruments are employed to acquire precise flux level measurements since the neutron flux varies greatly (by around 12 decades).
- Source Range Detectors
- Intermediate-Range Detectors
- Power Range Detectors
- Nuclear Instrumentation Core. The reactor core's temperature and neutron flux distribution are measured via the incore NI equipment. The incore instrumentation system's objectives are to give precise data on the distribution of neutron flux and the temperatures of the fuel assembly exit at specific core regions. Among the components of the incore instrumentation system are
- Incore neutron flux monitoring system
- Incore temperature monitoring system

Neutron detection is the foundation of both systems. Excore neutron detectors placed outside the core are often used to measure the neutron flux. The so-called excore NIS is comprised of these detectors. Typically, an incore equipment put within the reactor measures the neutron flux and its distribution throughout the core. The NIS has to be calibrated even though it is an irreplaceable system that responds quickly to changes in neutron flux. Only techniques based on the energy balance of the primary circuit or the energy balance of the secondary circuit may accurately assess the reactor's thermal power.

Although these techniques yield precise reactor power, they fall short for safety systems. The hot leg temperature or the feedwater system flow rate are two examples of signal inputs used in these computations, and they vary extremely slowly in response to variations in neutron power. Stated differently, colorimetric techniques provide precise measurements of thermal power. On the other hand, the only technology that can identify reactivity excursions quickly is the nuclear power measured by excore neutron detectors.

8.10.1 Reaction Rate – Proportionally between Neutron Flux and Thermal Power

We may calculate the rate of interactions (e.g., rate of fission reactions) by knowing the neutron flux (the total route length of all the neutrons in a cubic centimeter in a second) and the macroscopic cross-sections (the likelihood of having an interaction per centimeter route length). They are then multiplied collectively to determine the RR, or the quantity of interactions occurring in that cubic centimeter in a second:

$$\begin{cases} RR = \Phi \cdot \Sigma = \Phi \cdot N \cdot \sigma \\ \text{or} \\ RR = \int_0^{\infty} \Sigma(E) \cdot \Phi(E) dE \end{cases} \qquad \text{Eq. 8.14}$$

Where, in Equation 8.14, the parameters are identified as follows:

Φ = Neutron flux (neutrons·cm^{-2}·s^{-1})
σ = Microscopic cross-section (cm^2)
N = Atomic number density (atoms·cm^{-3})

The relevant cross-section type is used to determine the response rate for different kinds of interactions.

- $\Sigma_t \cdot \Phi$ = Total reaction rate
- $\Sigma_a \cdot \Phi$ = Absorption reaction rate
- $\Sigma_c \cdot \Phi$ = Radiative capture reaction rate
- $\Sigma_f \cdot \Phi$ = Fission reaction rate

To calculate the thermal power, we must concentrate on the fission RR. Assume for the sake of simplicity that the reactor's fissionable material is evenly distributed. In this instance, location has no bearing on the macroscopic cross-sections. The total number of reactions taking place in the reactor core per unit time is obtained by multiplying the fission RR per unit volume ($RR = \Phi \cdot \Sigma$) by the total volume of the core (V). However, we also know that each fission process releases around 200 MeV of energy. The following equation may be used to calculate the rate of energy release (Power P) caused by the fission reaction.

$$P = RR \cdot E_r \cdot V = \Phi \cdot \Sigma_f \cdot E_r \cdot V = \Phi \cdot N_{^{235}U} \cdot \sigma_f^{235} \cdot E_r \cdot V \qquad \text{Eq. 8.16}$$

Where

P = Reactor power (MeV·s^{-1})
Φ = Neutron flux (neutron·cm^{-2}·s^{-1})
σ = Microscopic cross-section (cm^2)
N = Atomic number density (atoms·cm^{-3})
E_r = The average recoverable energy per fission (MeV/fission)
V = Total volume of the core (m3)

Example
The number of fissions per second required to generate one watt of electricity in a reactor with an average energy per fission of 200 MeV may be calculated using the conversion factors shown below.

1 fission = 200 MeV
1 MeV = 1.602 x 10^6 ergs
1 erg = 10^7 Watt·s

$$1watt(\frac{1erg}{1\times10^{(-7)}watt-sec})(\frac{1MeV}{1.602\times10^{6}erg})(\frac{1fission}{200MeV})=3.12\times10^{10}\frac{fission}{second}$$

This is equivalent to stating that 3.12*10[10] fissions release 1 Watt-second of energy.

Equation 8.14 may be used to compute the power released in a reactor. Consequently, the total fission rate for the reactor as a whole would be obtained by multiplying the RR by the reactor volume. The power released by fission in the reactor would be expressed in watts, which would be calculated by dividing by the number of fissions per watt-sec. Equation 8.15 provides a mathematical illustration of this connection.

$$P=\frac{\varphi_{th}\Sigma_f V}{3.12\times10^{10}\frac{fission}{wat-\text{sec}}} \qquad \text{Eq. 8.15}$$

where

P = Power (Watts)
φ_{th} = Thermal neutron flux (neutrons/cm^2·s)
Σ_f = Macroscopic cross-section for fission (cm^{-1})
V = Volume of core (cm3)

The reactor's volume remains constant while it is functioning. The numerical density of the fuel atoms is likewise rather stable throughout a short time span (days or weeks). The macroscopic cross-section must be constant since the microscopic cross-section and atom density are both constant. It is clear from looking at Equation 8.15 that the reactor power and the neutron flux are exactly related if the reactor volume and macroscopic cross-section are both constant. In terms of daily operations, this is accurate. Because of the fuel burnup and the ensuing drop in atom density and macroscopic cross-section, the neutron flux for a given power level will grow extremely slowly over several months.

8.11 NUCLEAR PRIMARY ENERGY, PRESENT STATUS, AND FUTURE PERSPECTIVE

Given that the availability of enough primary energy is essential to the future of the national and global economies, it is easy to understand the great public interest in the energy issue among experts as well as active and concerned individuals. The issues that arise are: What are the major kinds of fundamental energy? How can they be put to good use, and how much will it cost? and What potential part may they play in the energy situation of the present and the future?

The fact that the current conscious or subconscious fears arising from confrontation with scientific and technological advancements – to which even the educated layman finds intellectual access difficult – have been sublimated onto the energy problem, and particularly onto the problem of nuclear energy, is another reason that may not be as evident but may ultimately prove to be of great importance as far as public acceptance of energy technologies is concerned. Unlike earlier breakthroughs, the rise of nuclear energy has brought to our awareness the ambiguity of increasing science and technology, which may either be exploited peacefully or misused violently.

Nuclear energy has the potential to assist the world solve its growing energy needs, but it also has the potential to wipe out human existence as we know it. Humanity is increasingly faced with the opportunities and dangers of new discoveries. New developments in science and technology are emerging more often. Compared to the time scales of biological evolution, they have shrunk to minuscule proportions since the start of the industrial revolution. However, compared to a generation, they are currently even getting shorter. Because of this, it is challenging to stay up with the environmental changes that we are all exposed to. Along with the hazards associated with these advances, their effects on both individuals and communities are growing. Will the capacity for self-organization be enough for humanity to adjust to the circumstances of rapid advancement and preserve the balance of social, political, and economic systems?

The use of nuclear energy through nuclear reactors has become increasingly significant globally during the past 20 years. Notwithstanding varying degrees of antinuclear sentiment, the European Community's nuclear power generating share has steadily grown to 14% (in France, 1981: 37%).

The exceptionally high energy content of the nuclear fuel uranium and its comparatively low cost provide the economic justification for using nuclear energy: In 1980, the cost of 1 kilogram of natural uranium in a conventional pressurized "Hater" reactor (PWR) was around 220 DM, or \$40/lb U_3O_8; the unit cost was 4.5×10^{-7} DM/kJ [49]. It produces approximately 4.9×10^8 kJ.

Currently, hard coal is 15 times more expensive per unit of energy than uranium. For other fossil fuels, this is comparable. Although the costs of nuclear fuel enrichment and processing, NPP capital costs, and waste management must still be taken into account, there are unquestionably cost advantages over conventional power plants for a wide range of power and load levels. There are also benefits in terms of pollution emissions and area requirements.

Existing and planned coal-fired generating is under intense pressure to "clean up its act" in order to continue being a viable source of electricity. According to the most recent U.S. Energy Information Association (EIA) forecasts, between 2020 and 2050, the proportion of electricity produced in the United States by coal burning is expected to decrease by 11%, from 24% to 13% (Figure 8.29).

However, there are additional sorts of cost related to energy use. These include, for example, the costs to society of environmental pollution and hidden or explicit government subsidies. If one takes a very broad view, one could also include the costs of providing energy security through a strategic oil reserve or military action. These are sometimes called "social costs", or "hidden costs", but a common present usage is to refer to them as external costs or externalities.

In some models of a desirable approach, a system of taxes and fees would be created in which the external costs are all recovered and are reflected in the price of electricity. However, determining these costs would be fraught with ambiguity. What, for example, is the cost of carbon dioxide emissions? To answer this, it is necessary to know the climate changes that are produced by the emissions and the price to attach to the impacts of these changes.

Extensive debates surround these matters, and we have neither the scientific knowledge nor the accounting principles needed to resolve the disagreements.

One of the more ambitious attempts to address external costs in a comprehensive fashion is the ExternE project, undertaken by the European Commission with the collaboration of the U.S. DOE.1 Although the ExternE study represents an interesting effort toward assessing external costs in a rational fashion, it does not – and probably cannot – resolve the crucial uncertainties.

For example, the largest cost component for the coal fuel cycle is listed as "global warming". For Germany, this cost is indicated as lying between 3 and 111 mECU/kWh – a range that is too wide to be very helpful. The estimated range for the full costs of the coal fuel cycle is 17 to 138 mECU/kWh.

For nuclear energy in Germany, the total external cost is indicated to be between 4.4 and 7.0 mECU/kWh. The largest component (3.5 mECU/kWh) is attributed to the global effects of long-lived radionuclides, primarily ^{222}Rn emissions from milling tailings (see Section 9.2.1). The assumed harm comes from small radiation doses received by many people, calculated over 10,000 years – raising unanswerable questions as to the effects of low radiation doses and the discount, if any, that should be applied to injuries that occur thousands of years from now.2 A further large part (0.1 to 2.7 mECU/kWh) of the total cost is attributed to the global warming consequences of fossil fuel use in the construction of nuclear facilities and in the production of nuclear fuel. The magnitude of this "cost" will depend on the fossil fuel share of the energy economy.

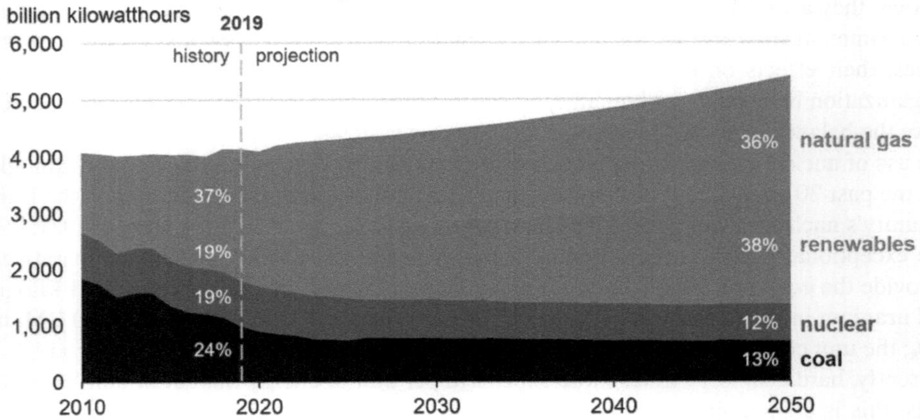

FIGURE 8.29 Electricity generation by fuel, 2010–2050.

(Source: U.S. Energy Information Association)

Overall, the ExternE studies are useful in focusing attention on the importance of externalities. However, in view of the uncertainties in the numerical estimates and in the appropriateness of including some of the cost components, they do not provide decisive guidance on energy choices.

In practice, some of the external costs (only a small fraction) are addressed by assessing charges against utilities or by compelling utilities to take steps to reduce undesirable impacts. Thus, producers of nuclear power are charged a fee of 0.1/kWh to cover the costs of nuclear waste disposal. Similarly, under the Price–Anderson Act, utilities are required to contribute to a fund maintained to reimburse victims of a hypothesized nuclear accident. Coal-burning utilities have been required to install equipment to limit effluents. There also have been recurring suggestions that a "carbon dioxide tax" or a "BTU tax" be imposed on fossil fuel use.

1. Results of the ExternE studies can be found through its homepage (http://externe.jrc.es). The most concise results of these studies are all-Europe overview tables of the external cost in mECU/kWh for each country and type of electricity source. [The European Currency Unit (ECU) was approximately equal to one U.S. dollar, and therefore 1 mECU ~0.1 cent. The Euro has now replaced the ECU].
2. The ExternE results quoted here are for a 0% discount rate. The study also considers a 3% discount rate, which reduces these costs from 3.5 mECU/kWh to 0.01 mECU/kWh.

Such fees, equipment expenditures, and taxes, when paid by the energy producer, automatically become a part of the producer's expenses and thus are internalized in the narrow definition of cost. The issue of the appropriate magnitude to assign to an external cost is thereby resolved by law or regulation, even if the decision is arbitrary. When an external cost is not defined by law or regulation, attempts to assign an appropriate cost carry the discussion into regions of great uncertainty, as discussed earlier.

Normally, when the costs of different options for electricity generation are compared, only generation costs are considered, and these costs dominate most of the active process of choosing among alternative technologies. However, external costs, even if not quantified, play a major part in the broad decision process. When a country bars the construction of NPPs, as a number have done, that is tantamount to attaching an infinite external cost to the perceived dangers of nuclear power. Similarly, when the construction of new natural gas- or oil-fired baseload power plants was barred, as was done briefly following the 1974 oil embargo, this amounted to assigning an infinite external cost to excessive dependence on limited resources. On the reverse side of the coin, the incentives given to renewable energy, in the form of tax credits, are tantamount to assigning a negative external cost.

Keep in mind that coal-fired power plants are a particular kind of power plant that uses the burning of coal to produce energy. Approximately 37% of the power produced worldwide comes from their utilization. By 2040, coal will still account for 22% of global electricity production. Therefore, many consumers are concerned about the cost calculation of coal-fired power plants.

A power plant's cost is closely correlated with how much electricity it produces. A 1 MW power plant, for instance, costs significantly more than a 10 MW power plant. Therefore, the amount of power plant required must be ascertained before estimating the cost of a power plant. Now that the power plant has been defined, let us examine its cost. The startup and operating costs of a coal-fired power plant are included in the initial cost, which also includes the cost of establishing the power plant's civil structure, equipment transportation, and other expenses. Fuel costs, coal transportation costs from the source to the power plant, and other expenses are included in the operational cost.

The boiler, generator, and turbine are the main pieces of equipment needed by the power plant. Auxiliary equipment includes blowers, draft fans, water supply pump shafts, cold water pumps, and certain big pumps. Coal-fired power stations must be outfitted with ash disposal equipment in addition to coal handling and transportation facilities. As a result, various equipment is selected, and the price varies. Estimates of construction costs for coal-fired power stations are very speculative and have gone up dramatically in recent years. Without finance fees, the anticipated cost of constructing new coal plants has already risen to $3,500 per kW and is still predicted to rise.

We live in a time when we cannot afford to use harmful and inefficient methods of producing power, which pollutes the air we breathe. An effective substitute with almost no negative environmental impacts is nuclear energy. Despite this, it is still highly expensive and might not be the best option from an economic standpoint. Coal is and will probably continue to be the most economically appealing commodity in nations like the United States, China, and Australia. This is due to the fact that the nations indicated above have readily available and substantial fossil fuel sources, but they will be actively looking for a suitable substitute when these run out or diminish.

There are two categories of expenses associated with nuclear power production: operational costs and waste management costs. Fuel expenditures, operations, and maintenance are among the components in the first category that add up to the production cost. Among coal, natural gas, and oil-fired power plants, nuclear has had the lowest production costs since 2001 (Nuclear Energy Institute). Since 1995, output using coal and nuclear technologies has maintained extremely consistent trends and is not significantly impacted by fluctuations in fuel prices, as seen in the figure below, which shows the most commonly used fuel sources expressed in cents per kilowatt hour. As you can see, fluctuations in fuel costs have a significant impact on gas and oil.

The expenses associated with waste disposal should also be taken into account when calculating the cost of producing nuclear energy. Nuclear plant waste is volatile and highly contentious, in contrast to many other fuel sources. The Nuclear Waste Fund was implied as one method of accounting for these expenses. In order to have a consolidated fund for reducing the expenses related to trash treatment, this fund was created in 1983. $10.8 billion has been spent and $35.8 billion has been contributed to the Nuclear Waste Fund since its inception. Additionally, a power plant undergoes a decommissioning procedure once it is no longer usable, which includes expenses for site restoration, radiological investigation, and the disposal of spent fuel. Additionally, a power plant undergoes a decommissioning procedure once it is no longer usable, which can cost anywhere from $300 to $500 million and involve radiological analysis, the disposal of spent fuel, and site restoration.

Before any power can be produced, any type of electrical producing facility will require a substantial initial investment. Site preparation, building, manufacturing, commissioning, and project finance are all included in this capital cost. Nuclear reactors have higher capital costs than coal-fired facilities and significantly higher costs than gas-fired ones. Very low fuel expenses, which make up a very tiny percentage of overall output costs, offset this greater capital cost.

An Organization for Economic Co-operation and Development (OECD) study was carried out in 2010 to compare 17 OECD countries and some data from developing nations. The study weighed all of the costs associated with nuclear energy and compared them on a cents per kilo-watt hour metric with other technologies like coal, coal with Carbon Capture and Storage (CCS), gas with Combined Cycle Gas Turbine (CCGT), and onshore wind. About 190 power plants from all around the world were included in this analysis, and a 5% and 10% discount rate were applied for future value comparison. They discovered that, at $30 per ton CO_2, nuclear power was highly competitive. In every country, nuclear energy is more cost-effective than coal and gas when discounted by 5%. Although the differences close when a 10% discount rate is used, nuclear is still less expensive than coal.

CCGT is a form of highly efficient energy generation technology that combines a gas-fired turbine with steam turbine. The same principle is also used for marine propulsion, where it is called a Combined Gas and Steam (COGAS) plant. Combining two or more thermodynamics cycles improves overall efficiency, which reduces fuel cost.

Carbon Capture and Storage (CCS) is an emission reduction process designed to prevent large amounts of carbon dioxide (CO_2) from being released into the atmosphere. It is considered a key and necessary technology to actively reduce industry-driven greenhouse gas (GHG) emissions.
From a levelized cost of power/electricity (LCOE), we can also examine how power is generated using a variety of techniques. The competitiveness of various generating technologies is assessed using this parameter. The U.S. Energy Information Administration used finance costs, plant life and capacity, constant and variable operation and maintenance expenses, construction expenditures, and fuel prices in its assessment. In order to properly incorporate a meter for future expenses, such as carbon taxes and initiatives aimed at decreasing greenhouse gas emissions, percentage points are also added at the conclusion of the computations based on GHG intensity.

Table 8.3 shows varying technology and how it compares to all of the production choices available today.

Table 8.3 shows that nuclear is competitive in almost every category. Compared to technologies that use non-renewable fuel sources, it has the lowest variable O&M (including fuel) and the best capacity factor of all the present technologies.

The generation of nuclear energy has several advantages and disadvantages, making it challenging to reach a judgment just on the basis of cost. One of the largest economic disadvantages of new nuclear technology is the extremely high capital costs of constructing new facilities, as well as the time and expense required to have new development approved by the federal government. Initial investors and those wishing to enter the nuclear energy sector are discouraged by the high capital expenses. On the other hand, compared to most current technologies, it has cheaper fuel, operating, and maintenance expenses, making it a wise long-term investment.

It is difficult to evaluate the relative costs of new producing facilities using various technologies, and location has a significant impact on the outcome. As long as carbon emissions are free or not properly costed, coal is and likely will continue to be economically appealing in nations like China and Australia. In many locations, gas is also competitive for base-load electricity, especially when combined-cycle facilities are used.

Although they are costly to construct, NPPs are reasonably priced to operate. When it comes to producing power, nuclear energy and fossil fuels are competitive in many locations. Typically, operational costs include all expenses related to waste disposal and decommissioning. Nuclear power becomes more competitive when the environmental, social, and health costs of fossil fuels are also included.

The LCOE is the fundamental economic statistic for every producing unit. Typically, the cost per megawatt-hour is calculated by dividing the entire cost of building and operating a power plant during its lifetime by the total amount of energy it dispatches during that time. It considers not just the "overnight" cost but also the financing expenses of the capital component.

Combining the benefits of security, dependability, and extremely low greenhouse gas emissions, nuclear power is a cost-effective way to generate energy on a levelized (i.e., lifetime) basis. Existing

TABLE 8.3
Average U.S. Levelized Costs (2012 $/MWh) for Plants Entering Service in 2019

Plant Type	Capacity Factor (%)	Levelized Capital Cost	Fixed O&M	Variable O&M (Including Fuel)	Transmission Investment	Total System LCOE	Subsidy[1]	Total LCOE Including Subsidy
Dispatchable Technologies								
Conventional Coal	85	60	4.2	30.3	1.2	95.6		
Integrated Coal-Gasification Combined Cycle (IGCC)	85	76.1	6.9	31.7	1.2	115.9		
IGCC with CCS	85	97.8	9.8	38.6	1.2	147.4		
Conventional Combined Cycle	87	14.3	1.7	49.1	1.2	66.3		
Advanced Combined Cycle	87	15.7	2	45.5	1.2	64.4		
Advanced CC with CCS	87	30.3	4.2	55.6	1.2	91.3		
Conventional Combustion Turbine	30	40.2	2.8	82	3.4	128.4		
Advanced Combustion Turbine	30	27.3	2.7	70.3	3.4	103.8		
Advanced Nuclear	90	71.4	11.8	11.8	1.1	96.1	−10	86.1
Geothermal	92	34.2	12.2	0	1.4	47.9	−3.4	44.5
Biomass	83	47.4	14.5	39.5	1.2	102.6		
Non-Dispatchable Technologies								
Wind	35	64.1	13	0	3.2	80.3		
Wind-Offshore	37	175.4	22.8	0	5.8	204.1		
Solar PV[2]	25	114.5	11.4	0	4.1	130	−11.5	118.6
Solar Thermal	20	195	42.1	0	6	243.1	−19.5	223.6
Hydro[3]	53	72	4.1	6.4	2	84.5		

(**Source: Unites States Energy Information Administration,** *Annual Energy Outlook 2014*)

1. http://www.nei.org/Knowledge-Center/Nuclear-Statistics/Costs-Fuel,-Operation,-Waste-Disposal-Life-Cycle
2. http://www.world-nuclear.org/info/Economic-Aspects/Economics-of-Nuclear-Power/
3. http://www.eia.gov/forecasts/aeo/electricity_generation.cfm

plants are highly predictable and perform well. With a relatively minimal risk of operational cost inflation, these facilities have lower operating costs than nearly all of their fossil fuel counterparts. It is currently anticipated that plants will continue to function for 60 years or more. The effects of low-cost gas-fired generating and subsidized intermittent renewable energy are the primary economic threats to current plants. These dangers are increased by the political risk of greater taxes, particularly nuclear taxes.

Early in 2017, Nuclear Power Economics and Project Structuring was released by the World Nuclear Association. According to the analysis, the capital cost of new nuclear facilities, which makes up at least 60% of their LCOE, has a significant impact on their economics. The construction length and interest rates are crucial factors in figuring out the total cost of capital. According to the International Energy Agency (IEA), the rise in nuclear capital costs in some nations has peaked and is more noticeable than actual, given the lack of new reactor building in OECD nations and the introduction of new designs. Capital expenses have been kept in check and, in the case of South Korea, even decreased in nations that have maintained ongoing development initiatives. In the past 15 years, worldwide. Global median construction timeframes have decreased within the past 15 years. After an NPP is built, energy production costs are predictable and inexpensive.

Due to the aging of existing facilities, the economic rationale for capital investments has been eroding in deregulated wholesale power markets, while the real demand has been growing. According to the IEA, two-thirds of electricity investment moved into regulated markets with some guarantee of return on capital around the turn of the century, while one-third went into deregulated markets subject to wholesale price volatility. Just 10% of investments were going into deregulated areas by 2014. Governments worried about medium-term energy security are now conducting urgent studies as a result. Since long-term income and cost recovery were almost guaranteed, governments or regulated utilities constructed all currently functioning nuclear power reactors. Some of these plants are currently operating in a deregulated market, particularly in the United Kingdom and the United States.

Government and regulated utilities invest in generation assets, pay for the fuel and operation of power plants, and decide whether to retire existing facilities. These choices are supported by long-term planning procedures meant to guarantee dependable operation while lowering overall expenses in the long run. A merchant generator's revenue in a deregulated market is reliant on the market, which is essentially short-term and frequently unpredictable. This puts the operator at risk, and the developer of a new plant confronts significant uncertainty because of the increased completion risk. In order to reduce these risks and make new initiatives financially feasible, government assistance is required.

The system cost of ensuring that supply from any source satisfies real grid demand is another economic consideration. With dispatchable sources like nuclear, the system cost is negligible; however, intermittent renewables, whose production is dependent on sporadic wind or solar inputs, become a consideration. System expenses rise sharply and easily surpass the real generating cost from those sources if the share of such renewables rises over a nominal percentage of the total. In addition to the LCOE, this is a crucial factor to take into account when comparing sources, as demonstrated in Germany and predicted in 2019 OECD Nuclear Energy Agency research. Refer to Figure 8.30.

There is not a lot of information available on the costs of constructing new nuclear plants because there have not been many nuclear reactors built in North America and Western Europe in the last 20 years. Uncertainty has increased with the switch to Generation III reactors. Depending on the location, large infrastructure projects like roads and bridges, as well as other non-nuclear power technologies, also exhibit diversity. In contrast to roads, bridges, and dams, which are often less complicated, the variation is especially important for electricity generating because its economics heavily rely on minimizing capital investment costs that must be passed on to customers [50].

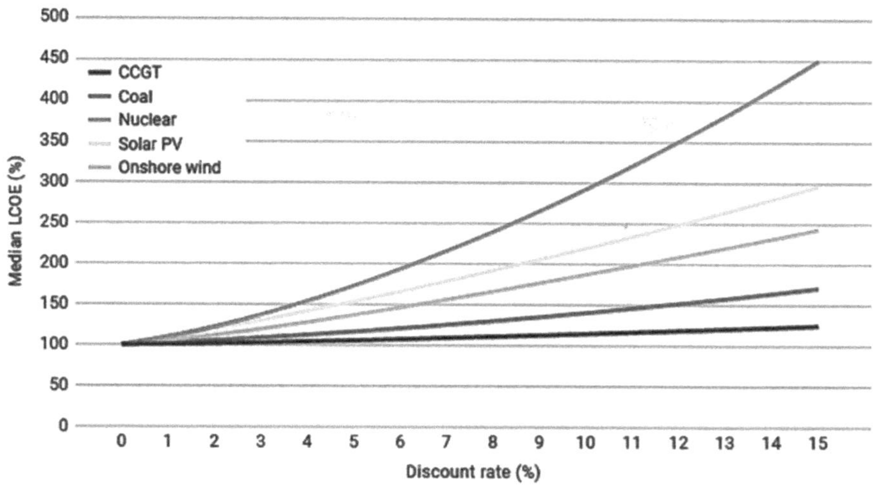

FIGURE 8.30 Effect of discount rate on levelized cost of electricity (LCOE) for different technologies.

(Source: OECD Nuclear Energy Agency)

8.12 SMALL MODULAR REACTORS

The creation of nuclear energy that is safe, dependable, and clean is essential to meeting America's energy needs. There will always be a need for consistent, base load output in the system, even while renewable energy sources are significantly replacing fossil fuel-based electrical generation. Since the 1979 Three Mile Island catastrophe, no nuclear power facility has been constructed in the United States. Another setback to the future sustainability of nuclear energy in the United States was the deactivation of the San Onofre Nuclear Generating Station in California in June 2013. However, the creation of Small Modular Reactors (SMRs) might play a key role in reviving the nuclear energy sector, which has halted [51].

Recently, SMRs have attracted worldwide attention owing to their advanced design features, such as structural safety improvement, simplification, and component modularization to achieve highly enhanced safety and improved economics. To increase their competitiveness in the global SMR market, their design issues related to further safety enhancement and cost reduction have been widely investigated [1–4].

The most outstanding feature of the SMR is that all the main components such as the pressurizer, steam generators, and RCPs are installed inside of the reactor pressure vessel [5] as shown in Figure 8.30. The integral design of the SMR is possible to enhance the safety of nuclear reactor by removing possibility of the severe accident such as the LOCA.

In order to maximize the heat exchange area in a limited space, a once through helical type steam generator is widely adopted for the design optimization of SMRs [6–10]. However, it requires a more complicated design and manufacturing process compared with the conventional PWR with U-bending type tubes [52].

8.12.1 The Need for Small Modular Reactors

The world's population has grown by over 12% in the past ten years, and by 2050, it is predicted to exceed 9 billion people. Population growth has resulted in a 20% rise in primary energy consumption and a 31.5% increase in electricity usage. Up to 2030, the global demand for power is expected to increase by 2.5%. Many countries have been creating varied energy mixes in attempt to fulfill this need. For many nations, SMRs might be a powerful non-fossil alternative [53].

Small reactors are defined by the International Atomic Energy Agency (IAEA) as having an electrical output of less than 300 MWe, however, SMRs generally agree that anything with an output of less than 700 MWe qualifies [54]. Mass-produced, factory-fabricated reactors with a smaller, more compact design may be delivered to a nuclear power facility by rail or truck. The creation of Grid-Appropriate Reactors was recognized as essential to the expansion of safe, peaceful nuclear electricity globally in 2007 by the U.S. DOE-led Global Nuclear Energy Partnership (GNEP) program.

Of the 442 commercial power plants operating globally, 139 are SMRs, according to the IAEA. However, these reactors are not what are referred to as intentionally tiny reactors; rather, they are just scaled-down copies of huge reactors. SMRs are reactors that make use of their compact size to accomplish certain performance goals. Compared to conventional large-scale reactors with a capacity of more than 1,000 MWe, small modular reactors have several advantages. They can go to places that are hard to get to or lack infrastructure for fuel transportation. Because the design is modular, less work needs to be done on-site, which speeds up, simplifies, and lowers construction costs. SMRs require less refueling and have a longer lifespan. All things considered, they are less expensive to start, run, and maintain, have a less environmental impact, and are resistant to proliferation [55].

8.12.2 Scalability and Flexibility of Small Modular Reactors

Due to their purported lack of economies of scale, SMRs are commonly referred to as being uneconomical. However, because many more tiny plants are needed than big plants for a given installed power, mass manufacturing efficiencies do result in cost reductions. This enables more consistent procurement procedures and economies of scale in mass manufacturing for small businesses. SMRs' intrinsic modularity results in more flexibility and quicker fabrication timeframes. Investors can closely and swiftly respond to early indicators of shifting market conditions because plant capacity can be easily modified to changing market conditions. Big, conventional reactors have to handle fluctuations in demand and price. As investors ponder whether or not to invest, large sums of money might be wasted [56].

8.12.3 Plant-Grid Matching

The stability of nuclear reactors and their connectivity to the electrical grid are crucial technological requirements. Large power plants may be supported by well-connected grids in certain industrialized nations and regions, such as the western European Union. A huge, concentrated power plant that produces 1,000 MWe or more cannot be connected to other, less developed locations. By eliminating grid instability issues, fossil fuel consumption, and greenhouse gas emissions, the development of SMRs might offer energy solutions to a sizable market segment with smaller electrical grids [55].

8.12.4 Cogeneration for Small Modular Reactors

SMRs are capable of producing more than just energy. The heat produced by the nuclear reactor can be utilized for desalination or urban heating. Resource and income are maximized by using the heat to be rejected to the environment, as needed by the Rankine thermal cycle in the heat-to-electricity conversion process. The cogeneration value of big nuclear reactors is limited by the technical necessity that heat or desalination units be located close to end-user locations. However, SMRs may be deployed strategically at a reasonable distance from metropolitan areas because to their smaller size, enhanced safety, and lower radiation. Urban areas' heating demands can be met by the thermal energy produced [56].

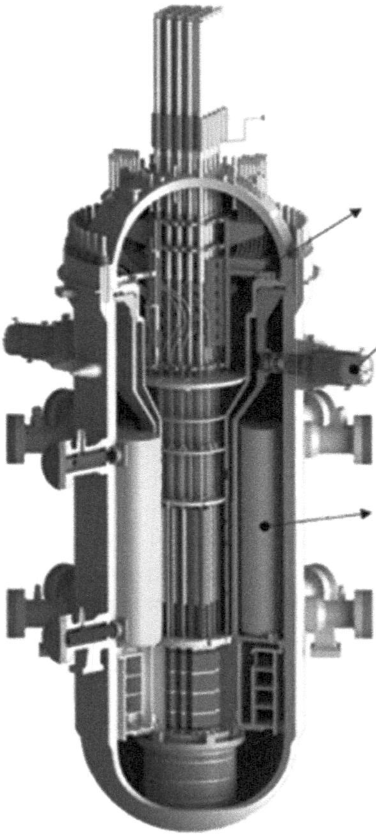

FIGURE 8.31 Schematic representation of small modular reactor (SMR).

(Source: U.S. Department of Energy)

8.13 THE FUTURE OF NUCLEAR ENERGY MINIATURIZATION

The three biggest issues facing nuclear energy today are waste management, safety, and cost. There are now less than 100 reactors in the United States, the lowest number in the industry's history [57]. In addition, Westinghouse, the leading commercial nuclear designer, filed for bankruptcy the previous year [58]. Following a sharp increase in expected costs and an extension of the estimated construction period, its final two reactors, each weighing 1,100 MW, were abandoned.

Reactors of a new generation will begin to generate electricity in the next years. Despite their relatively small size, they might play a significant role in helping us achieve our climate objectives without carbon emissions or greenhouse impacts [59].

The future of nuclear power has been waiting for the past 20 years in a high bay laboratory hidden away on the campus of Oregon State University in the western portion of the state. A new chapter in the politically troubled and conflict-ridden history of NPPs is represented by this prototype reactor, which is being operated by NuScale Power as SMRs, an energy company located in Oregon. Or even historical firms like Westinghouse, which has decades of experience with NPPs in the form of Generation III, and the recent launch of the transportable Nuclear Micro Reactor eVinci, which has military use for a mobile brigade for a quick deployment process in addition to space exploration into the terrestrial domain [59].

There will be no need for expansive evacuation zones or enormous cooling towers with NuScale's reactor. It is based on ancient and traditional understanding of LWR technology and can be manufactured in a factory and sent to any area, no matter how remote, thanks to its modulation technical approach. Numerous simulations indicate that it is capable of managing nearly any situation without experiencing a meltdown. Its little consumption of nuclear fuel, at least as compared to current reactors, is one explanation for this.

The cooling system of the eVinci Micro Reactor is based on Advanced Heat Pipe technology, which is a very dynamic passive cooling system that is the safest method of preventing meltdown disasters from either natural or manmade threats.

Another use for these tiny reactors with large energy sources that enables space travel beyond Earth is NASA's strategy of using heat pipe-cooled kilopower reactors for space exploration and the upcoming Mars mission.

For a globe facing a climatic crisis, this is excellent news. Although some environmentalists have a negative opinion of nuclear energy, many energy professionals and politicians concur that breaking atoms will be a crucial component of decarbonizing the world's power. About two-thirds of all clean electricity in the United States comes from nuclear power, but the regulatory lifespans of the current reactors are quickly coming to an end. The United States is only building two nuclear reactors, but they are years behind schedule and billions of dollars over budget [59].

Here comes the compact modular reactor, which combines many reactors into a single unit. Do you require a little amount of energy? Just a few modules should be installed. Do you want to power a large city? Add a few more. This makes it considerably simpler to design a power plant that works well in a variety of scenarios. These reactors are tiny enough to be mass-produced and transported in a few parts to any site. Most significantly, compact modular reactors may benefit from a number of safety and cooling features that their larger counterparts cannot, almost ensuring they will not experience another Chernobyl or Fukushima.

The industry is seeing some significant potential as a result of the shrinking size of nuclear reactors. Within the next 10 years, a few micro reactor designs that are now in development in the United States may be prepared for deployment. These plug-and-play reactors might assist address energy issues in a variety of locales, from military posts to isolated business or residential regions. They will be small enough to be transported by truck.

Countries all over the globe are being forced to search for zero-emissions options for producing power due to the catastrophic effects of climate change brought on by the use of fossil fuels.

The International Energy Agency, a body that works with 32 member nations on energy security, development, and environmental sustainability, argues that without nuclear energy, the shift to a cleaner energy system will be much more difficult.

As Canada transitions to a low-carbon future, the government seems to be on board, stating that nuclear innovation plays a "critical role" in lowering greenhouse gas emissions.

Although certain Canadian towns have been fueled for decades by Canadian-designed Canada Deuterium Uranium (CANDU) reactors, the government is now looking at technologies of a different scale. According to the federal government, SMRs provide a "important technology opportunity for Canada" and the "next wave of innovation" in nuclear energy technology [59].

It gets more difficult to rationalize the usage of nuclear reactors when you combine this economic inefficiency with growing worries about the frequency of accidents like Fukushima and the out-of-sight, out-of-mind proposal to simply bury the poisonous waste underground. Many see this as unmistakable proof that the sector has to reform quickly. Some people believe that miniaturization holds the key to the future of nuclear energy, mirroring the trend that Moore's Law is bringing about in the electronics industry.

8.13.1 The Transportable Small Modular Reactor

Reactors that produce less than 300 MWe of power are known as small modular reactors, or SMRs [60]. After being created at the production facility, they are fully assembled and installed on-site (see Figure 8.32). Three advantages are available to the consumer: SMRs are less expensive, more

FIGURE 8.32 Transportable small modular reactors (SMRs) are made onsite and then transported fully constructed [61].

(Source: Courtesy of the Department of Energy (DOE))

portable, and use less uranium fuel. They take less time and money to construct and generate less energy. Nonetheless, some people worry that because SMRs lack economies of scale, they are not economically feasible. Although this is true, economies that rely on mass production might offset this. This happens because about three times as many tiny plants are needed for every necessary electricity installation as large plants. The reactors' compact, modular design also greatly reduces production time and boosts adaptability [62].

In previously unthinkable locations, SMRs may also be able to supply electricity without the use of fossil fuels. For larger nuclear power reactors to be stable, they must be connected to larger power networks. For more economically developed nations (MEDCs) worldwide, this might not be a problem, but many people reside in smaller, more remote villages that lack the stability required for a huge power plant. In addition to the economies of scale associated with mass manufacturing, SMRs provide a significant potential benefit in the form of a sizable unexplored market for greener energy delivery to smaller towns [63].

While the SMR do show promise, not all issues have been addressed, and waste management continues to be a major problem facing the future of all nuclear energy.

A developed country (or industrialized country, high-income country, MEDC, advanced country) is a sovereign state that has a high quality of life, developed economy and advanced technological infrastructure relative to other less industrialized nations.

8.14 A TECHNOLOGY ROADMAP FOR GENERATION IV NUCLEAR ENERGY SYSTEMS

Meeting the demand for power caused by the 10% worldwide population growth is a crucial function of nuclear energy. By 2050, there will likely be 10 billion people on the planet, up from the current 7.9 billion, all of whom will be vying for a higher standard of living. The need for energy and its advantages – such as higher living standards, longer life expectancy and better health, more literacy and opportunity, and many more – will increase along with the world's population.

However, even increasing energy usage with the current mix of production alternatives will continue to have negative effects on the environment and may have long-term repercussions due to worldwide changing climate conditions. We need to use more clean, safe, and affordable energy sources if we want to sustain the planet's population. Nuclear energy is a prominent source of these resources.

Currently, 438 nuclear power stations are operating worldwide, contributing 10% of global energy production, the second-highest percentage of any non-GHG source. As a result, today's electric generating has a far less environmental impact. As plants retire, new systems will be required to maintain this benefit. The environmental advantages of nuclear energy may grow in the later half of this century and perhaps spread to other energy sources than electricity. For instance, nuclear energy may be used to desalinate water in places with limited freshwater supplies and to produce hydrogen for use in petroleum processing and as a fuel for vehicles to lessen reliance on oil. It will be necessary to build nuclear reactors in the near future and conduct extensive research and development (R&D) on next-generation technologies in order to provide this advantage.

Many countries throughout the world, both developed and developing, think that in order to attain energy security, more nuclear energy will need to be used. They have faith that nuclear energy can be utilized not just as a decarbonized energy source now but also in the future to safely and affordably fulfill their expanding energy needs, with a long-term supply guaranteed and no negative environmental effects.

This technological roadmap identifies and outlines the required R&D to enable a generation of cutting-edge nuclear energy systems known as Generation IV (GEN-IV), which will increase the future role of nuclear energy systems. Nuclear reactors and their energy conversion systems, along with the infrastructure required for the whole fuel cycle from ore extraction to ultimate waste disposal, make up GEN-IV nuclear energy systems [64].

8.14.1 The Long-Term Benefits from Nuclear Energy's Essential Role

This roadmap outlines challenging technical targets for Generation IV nuclear energy systems in four domains, namely in the form of advanced Small Modular Reactors (aSMR) and innovative SMRs [65–67].

1. Sustainability,
2. Economics,
3. Safety and Reliability, and
4. Proliferation Resistance and Physical Protection.

New nuclear systems can attain certain long-term advantages that will contribute to nuclear energy's crucial role globally by working to satisfy the technical goals [67].

8.14.1.1 Sustainable Nuclear Energy

The ability to satisfy current requirements while improving the capacity of future generations to satisfy societal demands endlessly into the future is known as sustainability. This roadmap outlines sustainability objectives with an emphasis on resource use and waste management. To emphasize their significance, other elements that are frequently linked to sustainability – like the environment and economy – are taken into account independently in the technological roadmap. According to the conclusions of this roadmap, achieving sustainability goals has the following advantages.

- By recycling spent fuel to recover its energy content and turning ^{238}U into new fuel, the nuclear fuel supply may be extended into future millennia.
- Benefiting the environment by using nuclear power generation and nuclear-produced hydrogen to replace energy and transportation sources that emit pollutants [68, 69].

- By significantly lowering waste volumes and associated decay heat, geologic waste repositories will be able to accommodate the waste of many more plant-years of nuclear plant operation.
- By significantly lowering the lifespan and toxicity of the leftover radioactive wastes transported to repositories for eventual geologic disposal, safe repository performance for extended periods of time (beyond 1,000 years) may be shown, and scientific analysis made much easier.

8.14.1.2 Competitive Nuclear Energy

In general, economic objectives take into account the financial risks and competitive costs of nuclear energy systems.

In the future, achieving economic objectives will have the following advantages:

- Through a variety of creative developments in plant and fuel cycle efficiency, design simplifications, and plant sizes, it is possible to achieve economic life-cycle and energy production costs.
- Lowering the financial risk associated with nuclear projects by creating reactors that use cutting-edge fabrication and construction methods, as well as potentially modular designs.
- Permitting district heating, fresh water, hydrogen, and other energy products to be generated locally where they are required.

8.14.1.3 Safe and Reliable Systems

When developing next-generation systems, maintaining and improving their safe and dependable functioning must be a top focus. In general, safety and reliability goals take into account reduced requirement for off-site emergency response, enhanced accident management and consequence reduction, safe and dependable operation, and investment protection. In the future, achieving these objectives will have the following advantages:

- Making more use of sturdy designs, built-in safety measures, and transparent safety elements that even non-experts can understand.
- Enhancing public confidence in the safety of nuclear energy.

8.14.1.4 Proliferations Resistance and Physical Protection

Physical protection and proliferation resistance take into account ways to manage and safeguard nuclear materials and installations. In the future, achieving these objectives will have the following advantages.

- Increasing physical security against terrorism by making new facilities more resilient;
- Providing ongoing effective proliferation resistance of nuclear energy systems through enhanced design features and other measures.

8.14.2 Meeting the Challenges of Nuclear Energy's Essential Role

For future nuclear energy systems to be crucial, they must supply

1) Reduced radioactive waste, efficient fuel use, and enhanced environmental advantages,
2) Competitive economics,
3) Recognized safety performance, and
4) Safe nuclear materials and nuclear energy sources.

This roadmap's aims for next-generation nuclear energy systems are based on these issues, which are explained below.

Most nations prefer to dispose of released fuel or other high-level radioactive leftovers in a geological repository, and significant technological advancements are being made in this area. Assessments are also being conducted for long-term retrievable surface or subsurface repositories. The success of repositories as a solution is demonstrated by the advancements made in the United States at Yucca Mountain and in other nations like Finland and Sweden. Future widespread nuclear energy usage, however, necessitates making the best use of repository space and taking fuel cycle closure into account.

While some nations end the fuel cycle through recycling, the majority of nations now employ the once-through fuel cycle. By recovering uranium and plutonium from spent fuel through recycling (either one or several passes), more power may be produced and less uranium mining and enrichment is required. Proliferation hazards can be further reduced by recycling without producing separated plutonium. Today, however, recycling has proven to be unfeasible due to the abundance of uranium available at steady, cheap costs. Eventually, this will alter, and when the cost of keeping an open cycle higher than that of a closed cycle, shutting the fuel cycle will be preferred. Recycling has other advantages, such as reducing the volume of high-level radioactive leftovers, making them less poisonous, and processing them into a form that is more suited for disposal. Additionally, problematic long-lived heavy metals can be transformed by reactor design. However, substantial fuel cycle technology R&D will be necessary to realize these advantages.

Nuclear power's economic success has been inconsistent. Positively, in many nations, the cost of generating energy from nuclear power is equal to or less expensive than that of generating electricity from natural gas, coal, or oil. However, building sophisticated nuclear energy systems requires overcoming their historically high construction costs as well as addressing their economics in a number of shifting markets. Even if the present generation of reactors produces power at competitive prices, building costs are too high, and more reliable licensing is required to encourage general interest in future nuclear development. For new facilities, significant R&D is required to lower capital costs and construction timelines.

Overall, nuclear power has a very good safety and environmental record. Nevertheless, there is a need to boost public trust in nuclear power's safety. This need should be met by new systems using visible and unambiguous safety techniques that come from advanced systems R&D.

An efficient international framework effectively protects fissile materials used in civilian nuclear power operations. Plants of the current generation are built with strong designs and extra security measures against terrorist attacks. However, future nuclear fuel cycles and nuclear materials safeguards should be designed with an even greater level of resistance to undeclared manufacture or diversion of nuclear materials from the beginning. Concerns have also been raised regarding nuclear reactors' susceptibility to terrorist attacks. Future nuclear energy systems will respond by offering better physical defense against terrorist attacks.

Numerous specialists from nations with expertise building and running nuclear reactors and infrastructure have contributed to the preparation of this plan. These professionals offered a wide-ranging, global viewpoint on the prospects and requirements for nuclear energy in the twenty-first century. Gaining public trust, which can be improved by making the process of creating and implementing GEN-IV technologies more transparent, will also be necessary to advance these systems.

Figure 8.33 presents the historical progress of NPP energy development from past to present and going forward with the future of GEN-IV, utilizing SMRs approaches.

In summary, global nuclear energy research initiatives have been formulating ideas that may serve as the foundation for GEN-IV systems. The Generation IV International Forum (GIF) countries will work together more on R&D projects, which will accelerate the development of such systems.

With international commitment and resolve, the world can begin to realize the benefits of GEN-IV nuclear energy systems within the next few decades.

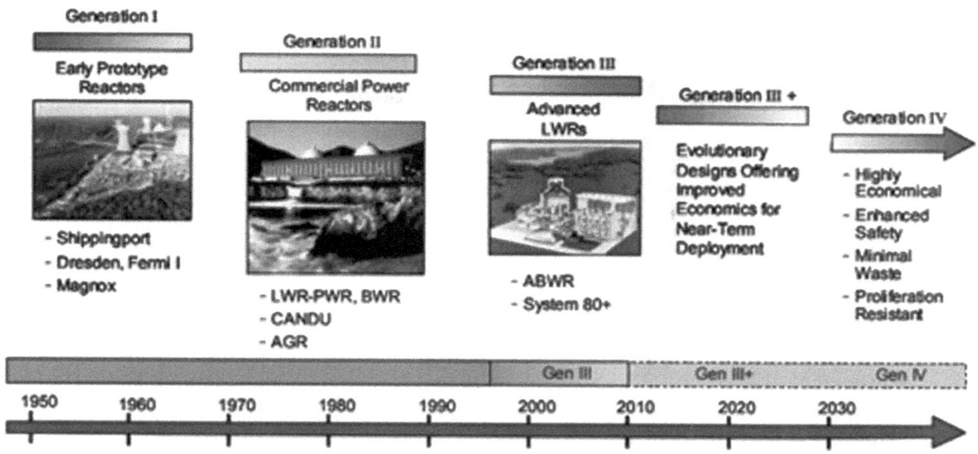

FIGURE 8.33 Historical generation of nuclear power plants.

(Source: U.S. Department of Energy)

REFERENCES

1. PRIS - Home. (n.d.). https://pris.iaea.org/pris/
2. RRDB. (n.d.). https://nucleus.iaea.org/rrdb/#/home
3. W. Oldekop, Electricity and Heat from Thermal Nuclear Reactors. In *Primary Energy*, Springer Berlin Heidelberg, 1982, pp. 66–91, https://doi.org/10.1007/978-3-642-68444-9_5, retrieved 2 February 2021
4. "DOE Fundamentals Handbook: Nuclear Physics and Reactor Theory" (PDF). U.S. Department of Energy. Archived from the Original (PDF) on 23 April 2008. Retrieved 24 September 2008.
5. "Reactor Protection & Engineered Safety Feature Systems". *The Nuclear Tourist.* Retrieved 25 September 2008.
6. "Bioenergy Conversion Factors". Bioenergy.ornl.gov. Archived from the original on 27 September 2011. Retrieved 18 March 2011.
7. J. Bernstein, *Nuclear Weapons: What You Need to Know*, Cambridge University Press, 2018, p. 312. Retrieved 17 March 2011.
8. "How Nuclear Power Works". HowStuffWorks.com. 9 October 2000. Retrieved 25 September 2008.
9. "Reactor Protection & Engineered Safety Feature Systems". *The Nuclear Tourist.* Retrieved 25 September 2008.
10. "Chernobyl: What Happened and Why? by CM Meyer, Technical Journalist" (PDF). Archived from the original (PDF) on 11 December 2013.
11. P. Tsetkov, S. Usman, Nuclear Energy Encyclopedia: Science, Technology, and Applications. In S. Krivit (Ed.), *Nuclear Energy Encyclopedia*, Wiley, 2011, pp. 48, 85.
12. Appendix A to Part 50–General Design Criteria for Nuclear Power Plants. (n.d.). NRC Web. https://www.nrc.gov/reading-rm/doc-collections/cfr/part050/part050-appa.
13. R. A. Fleck, Oscillations in Two-phase Flow Systems, *Nuclear Science and Engineering*, 9, no. 2, 140–146 (1961). https://doi.org/10.13182/NSE61
14. H. Firstenberg, H. Goldmann, J. C. Hudson, *An Investigation of Acoustic Phenomena in Boiling Systems*, General Electric Report, GEAP-3866, 1960.
15. L. C. Ruspini, G. F.Hewitt, A. Manera, Two-phase Flow Instabilities: A Review, *International Journal of Heat and Mass Transfer,* 71, 521–548 (2014). https://doi.org/10.1016/j.ijheatmasstransfer.2013.12.038
16. A. V. Smirnov, A. V. Zrodnikov, M. M. Boshkova, Thermo-acoustic Oscillations in Boiling Channels and their Influence on the Characteristics of Heat Transfer, *Thermal Engineering,* 44, no. 1, 59–63 (1997).
17. S. Kakaç, B. Bon, A Review of Two-phase Flow Dynamic Instabilities in Tube Boiling Systems, *International Journal of Heat and Mass Transfer,* 51, no. 3–4, 399–433 (2008).
18. M. C. Paul, S. N. Singh, Two-phase Flow Instabilities: A Review, *Renewable and Sustainable Energy Reviews,* 29, 406–415 (2014). https://doi.org/10.1016/j.rser.2013.08.102

19. J. A. Bouré, A. E. Bergles, L. S. Tong, Review of Two-phase Flow Instability, *Nuclear Engineering and Design,* 25, no. 2, 165–192 (1973). https://doi.org/10.1016/0029-5493(73)90025-3
20. K. Fukuda, T. Kobori, Classification of Two-phase Flow Instability by Using Density Wave Oscillation Model, *Journal of Nuclear Science and Technology,* 16, no. 1, 19–31 (1979). https://doi.org/10.1080/18811248.1979.9731903
21. R. Prasad, S. K. Das, P. Saha, A Review of the Density Wave Oscillation Phenomenon, *Journal of Fluids Engineering,* 129, no. 12, 1533–1545 (2007). https://doi.org/10.1115/1.2805675
22. D'Auria, F. The BWR Stability Issue. In Seminar on the transfer of competence, knowledge and experience gained through CSNI activities in the field of thermal-hydraulics (pp. 423–454). (2008). Pisa, Italy. INIS-IT-008. Retrieved from INIS: https://inis.iaea.org/records/6z6db-dar14
23. J. March-Leuba, *Density-Wave Instabilities in Boiling Water Reactors.* NUREG/CR-6003, ORNL, 1992. https://www.osti.gov/servlets/purl/10183139.
24. B. Zohuri, *Heat Pipe Design and Technology, Modern Applications for Practical Thermal Management*, 2nd edition, Springer Publication Company, 2016.
25. R. C. Bueno, P. H. F. Masotti, J. F. Justo, D. A. Andrade, M. S. Rocha, W. M. Torres, R. N. de Mesquita, Two-Phase Flow Bubble Detection Method Applied to Natural Circulation System Using Fuzzy Image Processing, *Nucl. Eng. Des.* 335, 255–264 (2018).
26. S. Levy, *Two-Phase Flow in Complex Systems*, Wiley, 1999.
27. J. M. Bisang, A. N. Colli, Current and Potential Distribution in Two-Phase (Gas Evolving) Electrochemical Reactors by the Finite Volume Method, *J. Electrochem. Soc.* 169, no. 3, 034524 (2022). Bibcode:2022JElS.
28. C. A. Busse, Theory of the Ultimate Heat Transfer of Cylindrical Heat Pipes, *Int. J. Heat Mass Transf.* 16, 169–186 (1973).
29. G. B. Wallis, *One-Dimensional Two-phase Flow*, McGraw-Hill, 1969.
30. B. Zohuri, P. McDaniel, *Thermodynamics in Nuclear Power Plant Systems*, Springer, 2015.
31. S. M. Ghiaasiaan, *Two-Phase Flow, Boiling, and Condensation: In Conventional and Miniature Systems*, Cambridge University Press, 2008, page 362.
32. E. E. Lewis, *Fundamentals of Nuclear Reactor Physics*, 1st edition, Academic Press, 2008.
33. Westinghouse Technology Systems Manual, Section 9.2. Incore Instrumentation System, https://www.nrc.gov/docs/ML1122/ML11223A264.pdf.
34. K. O. Ott, R. J. Neuhold, Introductory Nuclear Reactor Dynamics, American Nuclear Society, 1985, ISBN: 0-894-48029-4.
35. R. C. Runkle, A. Bernstein, P. E. Vanier, Securing Special Nuclear Material: Recent Advances in Neutron Detection and Their Role in Nonproliferation, *J. Appl. Phys.* 108, no. 11, 111101 (2010).
36. M. Sivaramakrishna, C. P. Nagaraj, K. Madhusoodanan, P. Chellapandi, Neutron Flux Monitoring System in Prototype Fast Breeder Reactor, *Int. J. Eng. Innov. Technol. IJEIT* 3, 10 (2014).
37. L. Cohen, M. Awschalom, Fast Neutron Radiation Therapy, *Ann. Rev. Biophys, Bioeng.* 11, no. 1, 359 (1982).
38. N. Tsoulfanidis, *Measurement and Detection of Radiation*, Hemisphere Publishing Corporation, 1983, 600 pp.
39. J. H. Agarwal, Electronic Aids for Enhancement of Agricultural Production, *IETE Tech. Rev.* 10, no. 1, 57 (1993).
40. N. Kalra, D. K. Das, T. K. Sarkar, Electronics in Water Management, *IETE Tech. Rev.* 8, no. 6, 348 (1991).
41. W. C. Feldman, G. F. Auchampaugh, R. C. Byrd, A Novel Fast-Neutron Detector for Space Applications, *Nucl. Instrum. Methods Phys. Res. Sect. Accel. Spectrometers Detect. Assoc. Equip.* 306, no. 1, 350 (1991).
42. G. Chen, R. C. Lanza, Fast Neutron Resonance Radiography for Elemental Imaging: Theory and Applications, *IEEE Trans. Nucl. Sci.* 49, no. 4, 1919 (2002).
43. V. V. Verbinski, M. S. Bokhari, A Fast-Neutron Spectrometer for Reactor Environments, *Nucl. Instrum. Methods* 46, no. 2, 309 (1967).
44. M. Cortesi et al., Concept of a Novel Fast Neutron Imaging Detector Based on THGEM for Fan-Beam Tomography Applications, *J. Instrum.* 7, no. 2, C02056 (2012).
45. S. Glasstone, *Nuclear Reactor Engineering: Reactor Design Basics*, 4th edition, Vol. 1, CBSPD, 2004, 495 pp.
46. G. F. Knoll, *Radiation Detection and Measurement*, 4th edition, John Wiley and Sons, 2010, 860 pp.

47. D. C. Stromswold, R. Hansen, P. Reed, A. Peurrung, Direct Fast-Neutron Detection PNNL-13068, *Pacific Northwest National Laboratory*, Richland, 2000, https://www.osti.gov/servlets/purl/750259).
48. S. Tripathi et al., Geant4 Simulations of Semiconductor Detectors (SiC) for Fast Neutron Spectroscopy, 2015 Annual IEEE India Conference (INDICON), IEEE, New Delhi, 2015, 1–6.
49. W. Oldekop (Ed.), *Druckwasserreaktoren fur Kernkraftwerke*, Thiemig, 1979.
50. J. Lovering, A. Yip, Ted Nordhaus, Historical construction costs of global nuclear power reactors, *Energy Policy*, 91, 371–382 (April 2016).
51. D. T. Ingersoll, Deliberately Small Reactors and the Second Nuclear Era, *Prog. Nucl. Energy* 51, 589e603 (2009).
52. K. Ahn et al., *Analytic Springback Prediction in Cylindrical Tube Bending for Helical Tube Steam Generator*, Nuclear Engineering and Technology, https://doi.org/10.1016/j.net.2020.02.004
53. Z. Liu, J. Fan, Technology Readiness Assessment of Small Modular Reactor (SMR) Designs, *Prog. Nucl. Energy* 70, 20e28 (2014).
54. G. Locatelli, C. Bingham, M. Mancini, Small Modular Reactors: A Comprehensive Overview of Their Economics and Strategic Aspects, *Prog. Nucl. Energy* 73, 75e85 (2014).
55. M. K. Rowinski, T. J. White, J. Zhao, Small and Medium Sized Reactors (SMR): A Review of Technology, *Renew. Sustain. Energy Rev.* 44, 643e656 (2015).
56. J. Vujic, R. M. Bergmann, R. Skoda, M. Miletic, Small Modular Reactors: Simpler, Safer, Cheaper? *Energy* 45, 288e295 (2012).
57. International Atomic Energy Agency, *Nuclear Power Reactors in the World*, International Atomic Energy Agency, 2017.
58. D. Cardwell, J. Soble, Westinghouse Files for Bankruptcy, in Blow to Nuclear Power, *New York Times*, March 29, 2017.
59. B. Zohuri, *Nuclear Miro Reactors*, 1st edition, Springer Publishing Company, 2020.
60. A. Glaser et al., *Small Modular Reactors*, Andlinger Center, Princeton University, June 2015.
61. S. E. Geene et al., *Pre-Conceptual Design of a Fluoride-Salt-Cooled Small Modular Advanced High-Temperature Reactor (SmAHTR)*, Oak Ridge National Laboratory, ORNL/TM-210/199, December 2010.
62. D. Handoko, *Small Modular Reactors, Physics 241*, Stanford University, Winter 2014.
63. G. Locatelli, M. Mancini, Small - Medium Sized Nuclear Coal and Gas Powerplant: A Probabilistic Analysis of Their Financial Performances and Influence of CO2 Cost, *Energy Policy* 38, 5072 (2010).
64. U.S. Department of Energy (DOE). (2002). *A Technology Roadmap for Generation IV Nuclear Energy Systems.*GIF-002-00. Generation IV International Forum (GIF). https://www.gen-4.org/gif/upload/docs /application/pdf/2013-09/gen_iv_roadmap.pdf
65. B. Zohuri, *Advanced Smaller Modular Reactors: An Innovative Approach to Nuclear Power*, 1st edition, Published by Springer Publishing Company, 2019.
66. B. Zohuri, *Small Modular Reactors as Renewable Energy Sources*, 1st edition, Published by Springer Publishing Company, 2018.
67. B. Zohuri, P. J. McDaniel, *Combined Cycle Driven Efficiency for Next Generation Nuclear Power Plants: An Innovative Design Approach*, 2nd edition, Published by Springer Publishing Company, 2019.
68. B. Zohuri, *Hydrogen Energy: Challenges and Solutions for a Cleaner Future*, 1st edition, Published by Springer Publishing Company, 2018.
69. B. Zohuri, *Nuclear Energy for Hydrogen Generation through Intermediate Heat Exchangers: A Renewable Source of Energy*, 1st Edition, Published by Springer Publishing Company, 2016.

9 Dynamics and Control of Nuclear Reactors

9.1 INTRODUCTION

Human operators typically oversee complicated systems known as Nuclear Power Plants (NNPs). An operator must keep an eye on a lot of data from sensors that indicate a certain kind of event when a transient happens. Faults and failures, which are associated with the system's deviance from the intended state and its inability to carry out the intended function, respectively, cause the transitory in NPPs (Isermann and Balle, 1997) [1]. It is more likely that transients will occur on older NPPs [2].

Failures or defects that cause plants to deviate from their usual state make it difficult for operators to comprehend the trends of interacting variables because the changes are either too subtle or occur too quickly. In NPPs, transient identification refers to the categorization of transitory kinds by the interpretation of the primary plant factors. Accordingly, the operator may benefit from accurate transient identification (Jeong et al., 1996a, Jeong et al., 1996b) [3].

In order for the heat generated in fission to be transferred to the coolant and pushed out of the core, such transitions necessitate that the reactor be on for a duration that is significantly longer. As the reactor moves through a number of distinct states, where the extra reactivity is almost completely eliminated by the control poisons, we may then use steady state formulas for heat transfer.

On the other hand, there are situations where faster neutron population changes are necessary. For example, starting a reactor from cold critical reduces its initial power by orders of magnitude compared to operating it. At higher power levels, when temperature feedback becomes important, changes in operating levels to follow power demand or for other reasons frequently need to happen faster than a quasi-static model can account for. Last but not least, the study of hypothetical reactor accidents frequently makes use of transients that take place over time spans comparable to or shorter than those required to transfer heat from fuel to coolant. Point kinetics equations are commonly used by engineers to analyze the kinetic behavior of a system. The rationale for the name of point kinetics.

The study of reactor dynamics examines how the neutron flux changes over time when the macroscopic cross-sections are permitted to rely on the neutron flux level. Reactor kinetics with feedback and spatial effects are other names for it.

Among other things, we will briefly explain how the reactivity of a multiplication system that is not equal to zero affects the neutron flux, or reactor power, in this section. We shall investigate how nuclear reactors behave over time. It is crucial for nuclear reactor safety to comprehend how the neutron population in a reactor behaves over time in response to either an unanticipated and abnormal situation or a planned adjustment in the reactor's reactivity.

9.2 REACTOR DYNAMICS MODEL

We use a straightforward model to study reactor transients, where a positive reactivity is introduced and the feedback is derived from the fuel and coolant temperature coefficients, respectively, which can be taken to be negative constants. During transient initiation, if the reactor is critical and operating in steady state, the total reactivity becomes

DOI: 10.1201/9781003590194-9

$$\rho(t) = \rho_i(t) - |\alpha_f|\left[\bar{T}_f(t) - \bar{T}_f(0)\right] - |\alpha_{fc}|\left[\bar{T}_c(t) - \bar{T}_c(0)\right] \quad \text{Eq. 9.1}$$

where we impose absolute value signs to indicate negative temperature coefficients. To analyze the transient, we substitute this expression into the kinetics equations.

The point kinetics equations are a series of linked ordinary differential equations that are solved to calculate the neutron density and delayed neutron precursor concentrations in a strongly connected reactor as a function of time. These equations are often solved using a model reactor that has at least six delayed precursor groups, resulting in a system of seven interconnected differential equations. Because analytical solutions are generally unachievable due to the intricacy of the issue, numerical techniques are required. Accurate results are frequently difficult to obtain due to the stiffness of the equations and the need for very tiny time steps in many procedures. The point kinetics equations provide a good description of the time-dependent behavior of small, tightly coupled reactors. When expressed in terms of reactor power and the external source is set to zero, they turn into [1]

$$\frac{d}{dt}P(t) = \frac{\left[\rho(t) - \beta\right]}{\Lambda}P(t) - \Sigma_i \lambda_i \tilde{C}_i(t) \quad \text{Eq. 9.2}$$

and

$$\frac{d}{dt}\tilde{C}_i(t) = \frac{\beta_i}{\Lambda}P(t) - \Sigma_i \lambda_i \tilde{C}_i(t) \quad i = 1, 2, \cdots, 6 \quad \text{Eq. 9.3}$$

We approximate the temperatures using the transient heat transfer model.

$$\frac{d}{dt}\bar{T}_f(t) = \frac{1}{M_f C_f}P(t) - \frac{1}{\tilde{\tau}}\left[\tilde{T}_f(t) - T_i\right] \quad \text{Eq. 9.4}$$

and

$$\bar{T}_c(t) = T_i + \frac{1}{2R_f W C_f}T_f(t) \quad \text{Eq. 9.5}$$

Where

M_f = Mass of fuel
W = Coolant mass flow rate
C_f = Specific heat capacity of fuel
C_i = Concentration of the ith delayed neutron precursor
T_f = Fuel temperature
$R_f = \frac{1}{UA}$ = where U represents overall fuel-to-coolant heat transfer coefficient and A is fuel cylinder surface area (i.e., fuel-to-coolant heat transfer area)
P = Power released in the fuel node

Keep in mind that the word "$M_f C_f / UA$" contains time units. It is the fuel-to-coolant heat transfer time constant. Light Water Reactors (LWRs) and CANDU reactors typically have values between 4 and 5 seconds.

Equation 9.5 permits the coolant temperature to be removed from Equation 9.1, as long as we estimate the thermal time constant as $\bar{\tau} \approx \tau = M_f C_f R_f$.

$$\rho(t) = \rho_i(t) - |\alpha|\left[\bar{T}_f(t) - \bar{T}_f(0)\right] \quad \text{Eq. 9.6}$$

Where

$$|\alpha| = |\alpha_f| + \frac{1}{2RWC_p}|\alpha_c| \quad \text{Eq. 9.7}$$

The steady state circumstances provide $C_i(0) = \left(\frac{\beta_i}{\lambda_i \Lambda}\right)P(0)$ and the starting temperatures if a reactor is critical prior to the transient being started.

$$\bar{T}_f(0) = \left(R_f + \frac{1}{2WC_p}\right)P(0) + T_i \quad \text{Eq. 9.8}$$

and

$$\bar{T}_c(0) = \frac{1}{2WC_p}P(0) + T_i \quad \text{Eq. 9.9}$$

9.2.1 Transient AnalysisFirst, we apply the above model to very slow transients, that is, those where the added reactivity increment $\rho_i(t) = \rho_0 \ll \beta$ is so small that the reactor period in the absence of feedback would be significantly longer than the fuel time constant, τ. In these cases, we can use the resulting quasi-steady state heat transfer equation and ignore the temperature derivative on the left of Equation 9.4.

$$\bar{T}_f(t) = \left(R_f + \frac{1}{2WC_p}\right)P(t) + T_i \quad \text{Eq. 9.10}$$

This, in conjunction with Equation 9.5, will produce

$$\bar{T}_c(t) = \frac{1}{2WC_p}P(t) + T_i \quad \text{Eq. 9.11}$$

The estimated reactivity obtained by substituting Equations 9.10 and 9.11 into Equation 9.1 is

$$\rho(t) \approx \rho_0 - \left[R_f|\alpha_f| + \frac{1}{2WC_p}\left(|\alpha_f|\right)\right]\left[P(t) - P(0)\right] \quad \text{Eq. 9.12}$$

However, take note that the quantity in brackets is only equal to the power coefficient's set magnitude.
Hence,

$$\rho(t) \approx \rho_0 - |\alpha_f|\left[P(t) - P(0)\right] \quad \text{Eq. 9.13}$$

There will be a slight initial prompt boost in power in suitably slow transients, but it will stabilize when the increase in core temperatures offsets ρ_0, transmitting $\rho(t) \to 0$. The following will be the new equilibrium power.

$$P(\infty) = \frac{\rho_0}{|\alpha_p|} + P(0) \quad \text{Eq. 9.14}$$

We may substitute ρ_0 with $\dot{\rho}t$, where $\dot{\rho}$ is known as the ramp rate, and get a reactor power that rises linearly with time if tiny increments of reactivity are added one after the other at a very slow pace.

$$P(t) = \frac{\dot{\rho}t}{|\alpha_p|} + P(0) \qquad \text{Eq. 9.15}$$

It cannot be stressed enough that the preceding quasi-static study is only relevant to extremely small step reactivity insertions or very moderate ramp rates of reactivity addition. If not, the transient must be described using the whole set of Equations 9.1 through 9.7. Equations 9.14 and 9.15 become relevant over long periods of time only when balance can be re-established.

We now go on to find that the basic quasi-static model is unable to forecast. They need equations and the full dynamics mode, as shown by Equations 9.1 to 9.7. Figures 9.1 and 9.2 show results based on settings typical of a large pressurized water reactor (PWR).

Step reactivities of 0.20 and 0.95 dollar, or ..., are introduced into a reactor running at maximum power in Figure 9.1. A transient-approaching, quick critical, which is typical of those considered for hypothetical accidents, and what one may observe during a typical operating transient are both illustrated by them. It should be noted that curve *a*, which is the outcome of the 0.20-dollar transient, experiences a brief, quick jump before gradually declining to an equilibrium value as established by Equation 9.14. For the 0.95-dollar injection in curve *b*, the reactor continues for a very short time and shows a significant power surge before the fuel temperature rises sufficiently for negative feedback to cancel out the initial reactivity insertion. Heat cannot exit the fuel for such short periods of time. Therefore, by setting the final part in Equation 9.4 to zero and integrating it to obtain the fuel temperature for short durations, we get

$$\bar{T}_f(t) \approx \frac{1}{M_f C_f} \int_0^t P(t')dt' + \bar{T}_f(0) \qquad \text{Eq. 9.16}$$

where $\bar{T}(t) \sim \bar{T}_c(0)$. Inserting this expression into Equation 9.1 gives

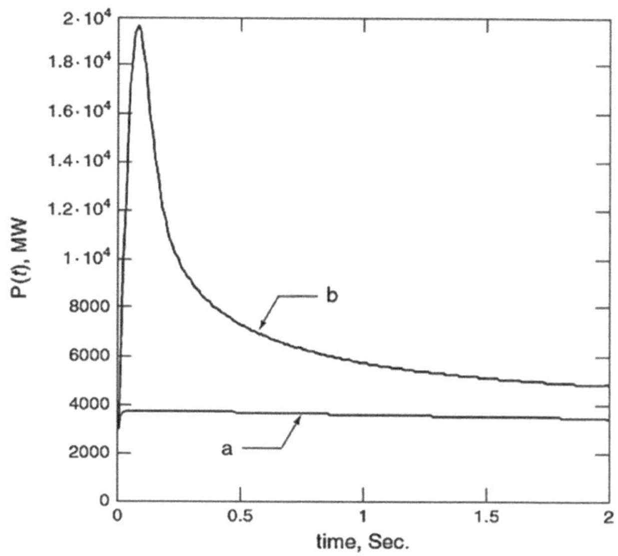

FIGURE 9.1 Power transients versus time initiated from full power for step reactivity insertions. (a) 0.20 dollars insertion, (b) 0.95 dollars insertion [8].

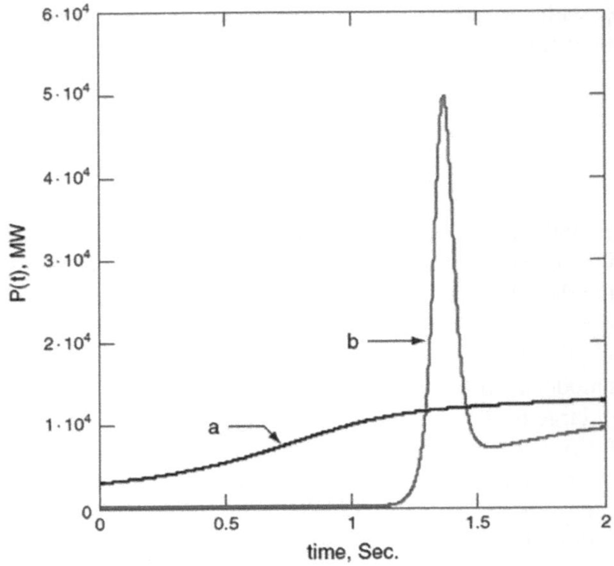

FIGURE 9.2 Power transients vs time for a reactivity insertion rate of 1.0 dollars. (a) Initiated from full power, (b) Initiated from low power [8].

$$\rho(t) \approx \rho_0 - \frac{|\alpha_t|}{M_f C_f} \int_0^t P(t')dt' \qquad \text{Eq. 9.17}$$

where the prompt temperature coefficient is simply equal to α_f. The feedback is therefore proportional to the energy produced by the transient across small time spans in relation to the thermal time constant. Only if the transient produces no boiling or other disturbances that render the straightforward model put out here incorrect will the power eventually drift to a value established by Equation 9.14.

Whether the rate is due to the planned removal of a control rod bank or a hypothetical accident scenario where poison is abruptly ejected from the core, in practice, the rate at which reactivity is injected is frequently just as significant as the overall quantity of reactivity accessible. In fact, more realistic research often focuses on the rate at which reactivity may rise since it is challenging to speculate about physical mechanisms that might inject substantial amounts of reactivity "instantaneously" on time scales shorter than a prompt neutron lifespan.

Two power transients are shown in Figure 9.2, both of which are caused by introducing reactivity at the extremely fast rate of $\dot{\rho} = \beta$, or 1.0 dollar/s, but with distinct beginning circumstances. We use the same model, just swapping out ρ_0 for $\dot{\rho}t = 0$ in the equations above. Both transients would go through the quick critical in a single second if there was no temperature feedback.

The severities of the transients, however, are very dependent on the beginning conditions because feedback does occur in both. When the short-time approximation provided by Equation 9.17 is modified to apply to the ramp insertion of reactivity, the changes become evident.

$$\rho(t) \approx \dot{\rho}t - \frac{|\alpha_f|}{M_f C_f} \int_0^t P(t')dt' \qquad \text{Eq. 9.18}$$

Hence, given that we are employing a fuel thermal time constant of τ = 4.5 s, it is logically relevant throughout the first 1 or 2 seconds. With $P(0) = 3{,}000$ MW(t), the transient is started at full power in curve *a*, allowing negative feedback to build very quickly. Hence, ρ (*t*) never gets close to the prompt critical, and as a result, the power rise rate is quite low. On the other hand, curve *b* is the outcome of starting the transient at hot zero power with $P(0) = 1.0$ MW(t). As a result, the integral on the right side of Equation 9.18 increases somewhat more slowly, enabling the reactivity to surpass the prompt critical before being slowed down by feedback from negative reactivity. Thus, a power surge occurs prior to the fuel receiving enough energy to cause the reactivity to turn negative and end the transient.

There are significant ramifications for reactor operations from the variations between curves *a* and *b* in Figure 9.2. Reactor trips typically occur when the power increases by a few percent while the reactor is operating at full capacity.

Curve *a* reaches this point in a tiny fraction of a second, as shown in Figure 9.2. Therefore, with only a modest increase in power, the reactor is shut down if control rod banks are inserted within the next few seconds. However, trip levels cannot be kept too near the beginning power at lower powers since there would not be any space to adjust power to accommodate rising demand. Even if there were, curve *b* shows that the chance of producing a trip signal is minimal until the transient is far over 1 second. Control rods cannot be placed quickly enough to restrict the effects of the power surge since it lasts for much less than a second. To avoid the risk of a so-called starting disaster, more caution is needed when manipulating reactors at extremely low power, when there is little or no negative temperature feedback, for this and several other reasons [1].

9.3 REACTOR CONTROL

The two main functions of all control systems are to transfer a process from one desired state to another and to maintain a process at a desired condition (also known as a set point) in the face of external disruptions. The control system's initial task is to ensure that the supplied power matches the intended power because a power reactor's main function is to generate power, often electricity but occasionally process heat. Additionally, the control system makes a number of plant variables (temperature, pressure, liquid level, etc.) reach the appropriate levels and does so quickly. The extra duties of these plant controllers include creating economic settings, preventing unfavorable plant conditions, maximizing component longevity, and creating beginning conditions that guarantee a safe and controlled reaction to anomalous events.

The independent variable that affects control in a nuclear reactor is excess reactivity. It establishes the relative rate of neutron concentration rise and, consequently, the rate of power level increases. The time integral of the excess reactivity that causes the shift is therefore connected to a change in power level. The temporal integral of the input, excess reactivity, is the output power. Thus, the power increases endlessly over time for a step of excess reactivity. The reactor is dynamically unstable in this regard. To make it a steady proportional device, feedback control – internal or external – is required. Reactivity can be changed in a variety of ways, including by adding or removing fuel or absorbing material in solid (rod) or liquid (concentration) form, altering the reflector's or moderating element's aspect, or changing one of the reactor's physical parameters, like temperature or density. Reactivity in a thermal reactor has traditionally been controlled by the location of neutron-absorber (boron, hafnium, or cadmium) control rods [3].

Positioning these rods to maintain a constant (or otherwise defined) neutron concentration at one or more neutron flux detectors within the reactor has been standard procedure in both research and industrial reactors. The reactor's working conditions of a central station power plant, where it is tightly connected to a turbo generator, are dictated by the turbine's power requirement. The location of control rods or other methods of modifying the reactivity are generally accomplished by the intermediary impact of the steam pressure, a dependent variable.

Regulating the reactor steady power level against arbitrary reactivity changes introduced from external sources, determining the nature of the transient response of the reactor temperature and power level to unintentional or intentional reactivity disturbances, and controlling the power level in alignment with power demand are the main issues that reactor control engineers have focused on the most.

By using the device's output to offer feedback to the input, the regulation approach makes sure that changes in the output tend to counteract changes in the input. Consequently, the system is self-regulating. Using a device like an ionization chamber to monitor neutron flux and using the electric signals that are produced to place a neutron-absorbing control rod are popular techniques for power level stabilization. The power level is kept constant by maintaining the neutron flow at the neutron detector. Although this technique has been widely employed, it necessitates a comparatively fast reaction speed and significant rod-driving power for efficient control of reactors with short mean neutron lifetimes. As a result, the equipment is often unreliable and might pose a risk to the facility when it malfunctions.

The required property of intrinsic self-regulation is present in reactors with negative temperature coefficients of reactivity. As seen in the heavy rectangle-labeled reactor in the top left side of Figure 9.3, the negative temperature coefficient may be thought of as a component of an internal feedback loop within the reactor that tends to hold the average reactor temperature constant. Below the rectangle, a dashed line represents a loop representing separate additional ways for controlling the neutron flux and power output using neutron flux instruments.

One idea of a complete power plant, including the two control loops that regulate the plant in the operational or power range (10%–100% full power), is depicted in the block diagram in Figure 9.3. By no means are the blocks limited to linear elements. The diagram's goal is to depict the intricate plant using the most straightforward explanation possible, effectively illustrating the information and energy transfer between key elements that affect both the plant's static and dynamic activities. Since they only activate when anomalies arise that result in any of the monitored variables exceeding predetermined limits, the supervisory or safety circuits are not shown. Many nuclear-powered central station systems may be accurately shown by the diagram with only minimal modifications or deviations.

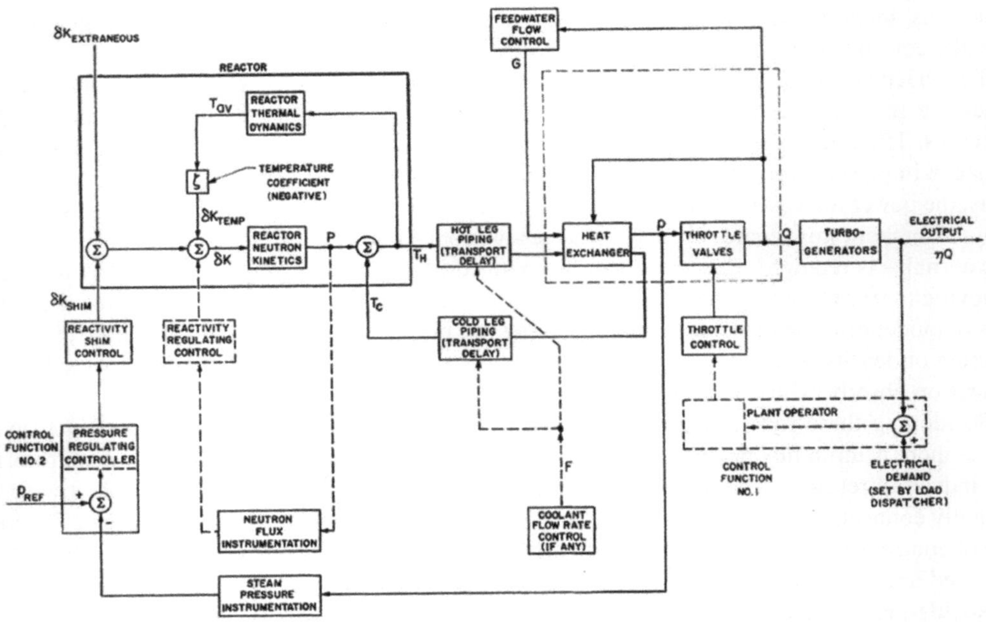

FIGURE 9.3 Block diagram of power plant including feedback control loops.

(Source: U.S. Department of Energy)

A reactor with a negative temperature coefficient of reactivity is only critical at a specific temperature if all other reactivity-affecting factors, such as fuel, poison content, and control rod position, remain fixed. The power generated fluctuates rather than being constant at any other temperature. The excess or lack of power produced results in a brief increase or decrease in reactor temperature in response to a change in power demand. The reactivity first decreases or increases in accordance with this temperature change. When the power created equals the power requested, the temperature returns to the value that corresponds to criticality because the power generated falls or rises above the newly demanded value. At the increased necessary power level, any such power transient must eventually revert to the same temperature. A different temperature at which criticality is once more reached will be established as a result of a change in reactivity brought on by another factor [3].

9.3.1 Reactor Control Elements

Through the use of a device that alters reactivity, the reactor may be locally controlled to enhance its natural process behavior. Only reactor control from a process perspective is covered here.

Other parts of the plant, such as the turbine's throttle or the reactor and heat exchanger's primary circuit coolant circulation rate, may also use automatic control systems. Nonetheless, we must be explicit about the many degrees of control that are required, which might be summed up as follows:

(a) To always have a sufficient reactivity control margin to ensure that the fission process is shut down when necessary and on a time scale suitable for safety, within the range of anticipated operation, startup, and all potential malfunction scenarios.
(b) To supply surplus reactivity that is controlled in order to satisfy specifications like [4]
    i. Reactivity loss on heating up from a cold start;
    ii. Reactivity loss in xenon and other poisoning;
    iii. Reactivity loss (occasionally gain) in fuel burnup and isotope production.
(c) To ensure the reactor can quickly adjust its reactivity (i.e., how many neutrons are available to sustain the nuclear chain reaction) in certain situations. The list that follows gives examples of when this fast control of reactivity is needed:
    i. Startup and shutdown;
    ii. Localized modifications to regulate the balance and temperature distributions;
    iii. Between fuel elements channel;
    iv. Process control of fluctuations in load demand.

Although NPPs' economics now mandate that they must operate at full capacity to fulfill base load demand, this is not always the case (for example, naval plants), and the plant may experience unavoidable local oscillations. As a result, the user may specify that the plant must be able to react in only 1 or 2 minutes to a 10% change in demand, for example. Naturally, the plant has to be safeguarded against more drastic alterations or disruptions.

Magnox reactor's potential control margin needs are shown in Table 9.1.

It is observed that a significant margin of excess reactivity is required to hold it down in the cold, shutdown condition throughout the transition from cold start to hot running. In the event of an unplanned shutdown, this thermal reactor type has a limited capacity for xenon override and a comparatively low xenon poisoning risk.

9.4 CONTROL METHODS

Since elements like cadmium, boron, hafnium, etc., have high capture probabilities at these neutron energies, thermal reactors are often regulated by adding or removing neutron-absorbing materials. If there is a liquid coolant, such poisons can also be supplied in a soluble form in addition to the control rod form, which involves the mechanical insertion of an absorber. In an appropriate process

TABLE 9.1
Typical Magnox Control Margins

Control Requirement	Reactivity
Bulk control	
Cold to hot margin	0.015
Xenon poisoning	0.020
Error allowance	0.005
Safety rods	0.010
Operational control	0.010
Long-term fuel changes	0.020
Total margin	0.080

plant, boron and its compounds may be introduced to the coolant and removed from it. Although soluble poisons are more widely disseminated and do not disrupt local flux, which tends to encourage localized temperature fluctuations, they alter levels more slowly than rods.

As a result, weak control rods are typically added to them for process control. In light water reactors, coolant circulation and coolant pressure variations are also helpful control mediums. Once more, they are unable to offer the localized control required to make comparable modifications to temperature distributions, etc. The amount of a liquid moderator (like SGHWR) or absorber, known as "liquid control rods", can change the gross reactivity of some water reactors.

Fast reactors struggle to provide adequate control because moderator effects in the coolant are missing and typical resonance absorption cross-sections are tiny in the high-energy spectrum. In the first case, the Doppler effect makes it crucial to provide a good negative fuel coefficient. Reactivity has been attempted to be controlled by altering the efficacy of reflectors or by withdrawing gasoline (so-called driving rods, in which fuel is withdrawn when the control absorber is inserted).

However, the techniques make it mechanically difficult to provide sufficient cooling in a power reactor, thus semiconventional rods are employed, but they must be operated carefully in a sodium environment. Because reactivity fluctuates less due to xenon and temperature effects, both during power ascent and throughout core life, than in thermal reactors, fast reactors fortunately require less active reactivity management.

Some designs employ burnable poisons, which may be modified to lose efficacy at the same rate that the fuel charge loses reactivity, in order to address the unique requirements of regulating reactivity against fuel burn-up. Clearly, they are slow-acting gadgets. Once more, we will continue the investigation using the traditional control rod motion since soluble poisons are sluggish to respond.

Beginning at the right-hand side of Figure 9.4, the control rod effect block diagram, it is possible to assume that the rod's tip is shifted from a low flux to a high flux zone, causing more neutrons to be absorbed and a consequent increase in the rod's related reactivity loss.

Standard reactor physics textbooks demonstrate that, when adjoint weighting is taken into account, the change in reactivity per unit length of travel may be roughly represented as a constant, say Kc reactivity/m.

The kind of reactor – light water, gas-cooled, etc. – has a major impact on how the reactor and its equipment, including turbines and coolant circulators, are controlled. However, before being shown in the context of particular systems, it is worthwhile to identify a few generic reactor control principles. In addition to comprehending the function of power reactors, or the type of load they are applied to, these broad considerations also encompass what may be referred to as

a. Part-load specification;
b. Load coupling and decoupling;

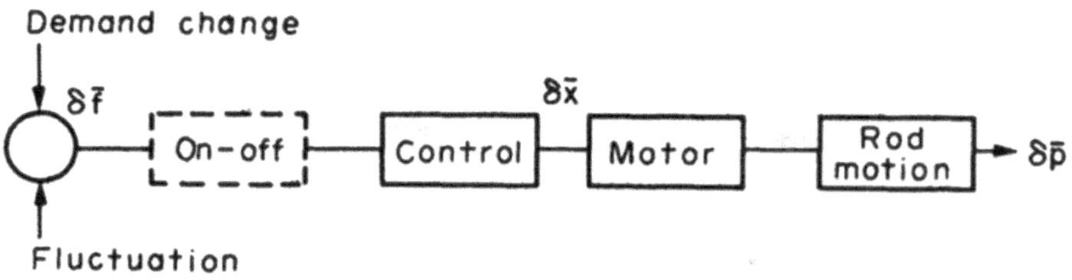

FIGURE 9.4 Block diagram for control rod effects [8].

c. Load following;
d. System stability.

Applications for power reactors to generate work range from a tightly integrated central energy supply (utility) to an extreme, such as a nuclear marine propulsion plant operating independently. In the former case, the captain effectively determines the load demand, and control is necessary to maintain a consistent output at the shaft as desired and to adjust this based on bridge demand [4].

9.5 GENERAL CONTROL CONSIDERATIONS

As part of general control consideration, we look at Control Action as described below.

9.5.1 Control Action

A generic reactor plant schematic is shown in Figure 9.5 to highlight certain general characteristics of any control system, especially locations where signals or information may be received and locations where control action can be implemented. Both locations are equally important in a control system.

Principle signals available on which to base automatic control are

(a) Reactor power P from neutron or flux measurements (less conveniently from combined temperature increments and coolant flow measurements);
(b) Pressure P_c and temperature T_c measurements on the primary coolant at core exits and boiler entrance/exit; primary flow rates;
(c) Turbine generator shaft speed, f, perhaps measured or at least expressed as generated frequency;
(d) Secondary pressure, particularly at the boiler exit P_s and at the high and intermediate entrances to the turbines; secondary water level in the boiler.

Principle points for the application of control signals

(a) Control rods (or similar) vary in reactivity and hence power;
(b) Primary coolant circulation, that is, pump speed, to vary coolant temperature and perhaps, indirectly, reactivity;
(c) Secondary steam flow rate via turbine throttle opening to either high or intermediate pressure stages and bypass valve dumping steam flow; boiler feed water flow (secondary pump) F_W.

With the support of controllers, appropriate measured control signals may be compared to predetermined signals, shaped and amplified, and then applied to the designated application sites. In many

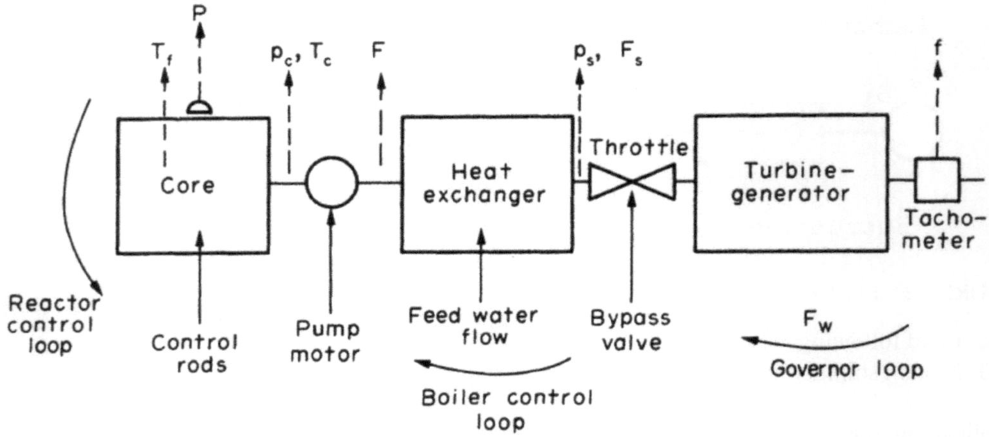

FIGURE 9.5 A reactor power plant control schematic [8].

cases, these controllers are traditional three-term analog process controllers that can easily provide proportional, derivative, and integral term control within the time scale we are considering. The parameters can be changed to achieve the desired dynamic response while meeting various restrictive limits on the system's pressures and temperatures.

As an alternative, Direct Digital Control (DDC) can be implemented using a computer that models the system's behavior and computes the necessary control signals once again while taking efficiency and state constraints into account. The dynamic stability of a suggested control system may be impacted by the typical significant delay in measuring a control signal, which may be indicated by an appropriate simple lag in the controller response function [4].

9.6 BOILING WATER REACTOR CONTROL STRATEGY

After an increase in steam flow, a Boiling Water Reactor (BWR) without any control action reacts first incorrectly. Therefore, the fundamental principle of BWR control is to boost reactor power before increasing steam output to the turbine in response to a rise in demand. This type of reactor is known as a "turbine following boiler". In other words, the reactor power is changed first after a power demand maneuver. The turbine does not experience a change in steam flow until the reactor's power level changes.

The reactor power controller, feedwater controller, and pressure controller are the three primary control systems of a BWR. Reactivity is managed by the reactor controller by core flow modification and control rod motion. As seen previously, the reactor state determines which control action – control rods or core flow – is used. Permitted flows at all power levels are shown on the power-flow map.

The so-called three-element controller is the feedwater controller. The downstream level, feedwater flow rate, and steam flow rate are all determined by measurements. In order to eliminate a level departure from its fixed point and a mismatch between the feedwater flow rate and the steam flow rate, the feedwater flow rate is changed. Similar to U-tube steam generators, BWRs experience shrink and swell issues, necessitating this kind of regulation.

The electrohydraulic controller, also known as the pressure controller, controls the steam valve to maintain a steady steam pressure.

Think about how you would react if the demand for power increased. An increase in core flow is the initial action. Reactivity, power level, and steam output all rise as a result. The pressure controller opens the steam valve in response to the ensuing pressure increase, supplying the steam flow required to meet the rising power demand. To keep the downcomer level at a certain level, the feedwater flow is controlled. The "turbine following boiler" method employed in BWRs is demonstrated in this example [5].

9.6.1 Boiling Water Reactor Monitoring

Neutron-monitoring channels, positioned inside the reactor core, monitor reactor power from the source range up through the power operational range. The in-core neutron flux monitor offers efficient monitoring in the intermediate and power levels and the highest sensitivity for controlling rod movement during starting [6]. Depending on the power level, three different types of monitors are used to do power monitoring in the reactor core. Three of them are the Local Power Range Monitor (LPRM), Monitoring in the Intermediate Range (IRM), and Source Range Monitoring (SRM). With fission counters ranging from source level to 109 nv (where nv = n/cm^2·s), SRM monitors the neutron flux in the source range. According to Lahey and Moody [7], the IRM that corresponds to the intermediate range is between 108 and 1,013 nv, where a system employing the voltage variance approach monitors the neutron flux.

In the power range when the neutron flux is tracked by fixed in-core ion chambers distributed uniformly around the core, a Local Fission Range Monitor (LFRM) is employed. The rated power of these chambers ranges from 1% to 125%. When a control rod is chosen, the operator's control bench board shows the location of the rod together with the data from the nearby detectors. There are four fission chambers in each detector assembly. The chambers are located in four horizontal planes and are evenly spaced axially.

Four Average Power Range Monitors (APRMs) in the control room of a conventional BWR record the average power level. The bulk power in the core is represented by each APRM. The APRM, which is functionally categorized as safety-related, receives inputs from the LPRMs. A fission chamber and the signal wires that go with it are housed in each LPRM. Each APRM channel may provide a signal proportional to the average neutron flux in the reactor core, thanks to the LPRM assignments. Only functioning LPRM signals are averaged by the circuit, and the output of the averaging circuit for every APRM channel is sent to the process computer. The Reactor Protection System (RPS) trips are controlled by the output signals from these monitors, which are also shown in the control room.

For the estimation of thermal power and other reactor operation-related factors, including the RPS, instrumentation within a BWR's core is essential. LPRMs can be used to identify BWR oscillations associated with instability. As seen in Figure 9.6, these detectors are positioned both axially and radially within the core vessel. From above the reactor vessel, the LPRM assemblies are inserted into the core. An incorporated spring-loaded plunger secures the assembly's top to the upper core grid. The assembly joins with the detector guide tube after passing vertically through the core area and core support plate. A structure made of stainless steel braces the guide tube at various points throughout its length. The assembly's electrical connections go through a flange, a seal assembly, and the in-core detector housing. Because of this configuration, LPRM connections may be accessed from beneath the reactor vessel.

The APRM channel or LPRM group to which each LPRM circuit is assigned provides all of the power for that circuit. Every LPRM detector has a bias supply, which provides a high voltage to the detector. All LPRMs that are assigned to an APRM or LPRM group share all other direct current (DC) power supply. Circuit boards installed on hinged vertical assemblies house the LPRM flux amplifiers. The detector output current signal is transformed into an analog voltage signal by the flux amplifier. As shown in Figure 9.6, there are N radially placed LPRM assemblies (strings) depending on the reactor layout. Each assembly has L detectors (usually four) separated by a distance d, creating four detection levels. A hollow dry tube for the Traversing In-core Probe (TIP) system is also included in each LPRM assembly.

It is important to keep in mind that monitoring systems in NNPs may or may not raise the need for shielding protection. As is well-known, ionizing nuclear radiation poses a risk to nuclear reactor

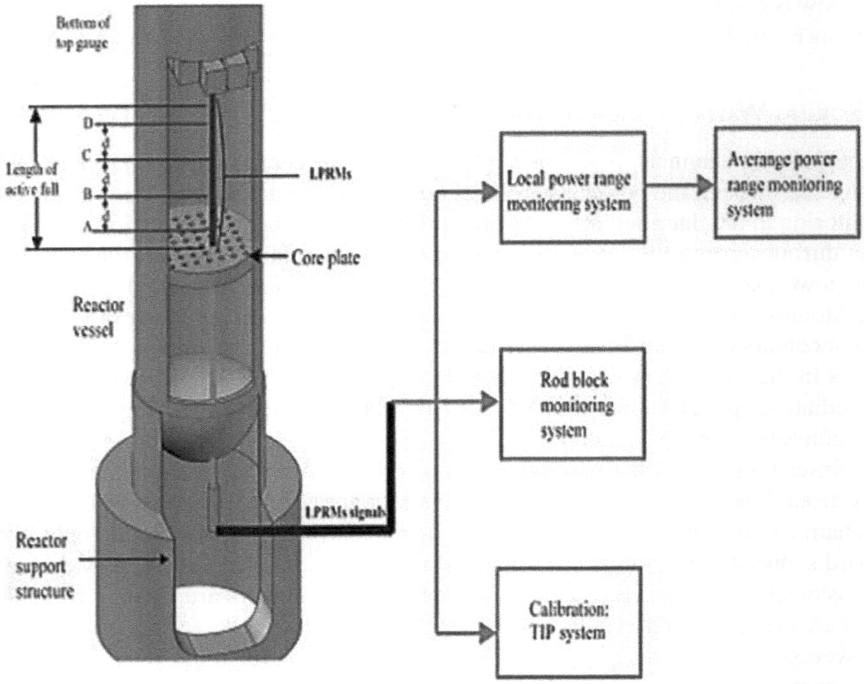

FIGURE 9.6 Scheme of the local power range monitor (LPRM) system. (Source: U.S. NRC)

safety. Therefore, by increasing signal accuracy, precision, and fidelity, putting sophisticated sensors and related electronics closer to a nuclear reactor core will make sense for reactor management and operation and lead to safer, more effective energy production.

However, because nuclear applications involve extremely high temperatures and intense radiation, electronics and electronic materials technologies have been developed to enhance the safety, monitoring, and control of both the current fleet of nuclear reactors and the next generation of reactors, including microreactors. By lowering the noise that is normally introduced into the long cabling between the sensor and electronics, instrumentation employing strategically positioned Radiation-Hardened (Rad-Hard) electronics will increase reactor safety and efficiency [8]. In nuclear reactors of the fourth generation, this is what matters most [9].

9.7 CONTROL OF CANDU REACTOR (NEUTRONIC FEATURES)

In a neutronic sense, a Pressurized Heavy Water Reactor (PHWR) is big. It is hence vulnerable to spatial instability. Both global variables and power shape control are handled by the control system.

Reactors with PHWR are over-moderated. In other words, reactivity rises when moderator or coolant levels are decreased.

Every day, there is a slight reactivity "bump" and a slight shift in the distribution of power because some of the exhausted fuel is replaced by fresh fuel [5].

Properties of the core materials are normally averaged over discrete volumes and over discrete energy intervals and collected in tabular form suitable for interpolation over space and fuel irradiation so as to select appropriate values to represent the specific state of the core (fuel irradiation, temperature, near-neighbor effects) at any location.

The time-dependence of these properties (due to changes in density, temperature, etc.) is also precalculated and stored in interpolation tables for use in dynamic analysis. Most reactors must shut down for refueling. In these systems, extensive computer simulations are done to select the optimum new-fuel loading (enrichment, burnable poison concentrations, spatial core layout, excess reactivity requirements, etc.) for the next "cycle" for about a 1-year duration. Once the final calculation is made, fuel is ordered and the decisions become very difficult to reverse. The CANDU system is fueled at full power, at a rate of 15 to 20 identical fuel bundles per day (2–3 fuel channels). Precalculation is necessary to select the best channels for refueling, to maintain the correct local power distribution, core reactivity, etc. Neither enrichment nor burnable poison is required. The shift supervisor makes the final choice of channels to be fueled and the fueling system operator initiates the automatic fueling system [10].

The addition of new fuel in CANDU initiates a small reactor transient because the new fuel reactivity is higher than that of old fuel. The zone controllers compensate for this reactivity addition automatically, to maintain reactor power and correct local power levels. Removal of any failed fuel elements is essentially identical to refueling. New bundles are added to replace those containing failed elements. Power reactors must be designed to operate for 30 years or more. During such long periods, the high-intensity neutron and gamma irradiation lead to the degradation of materials.

Examples are radiolysis of water, transport of active corrosion products to inhabited areas, swelling and embrittlement of steel, production and transport of tritiated water, and creep of zirconium pressure tubes. Such processes are an intrinsic part of the "dynamics" of any power reactor's operation even though their characteristic time scale is quite long, because the operating life of a power reactor may itself extend for 60 years or more [10].

9.8 REACTIVITY CONTROL MECHANISMS AND CONTROL SYSTEM IN CANDU

There are four reactivity control systems in heavy water reactors. It is possible to add or withdraw light water from the chambers of the calandria. The introduction of light water lowers reactivity because it absorbs neutrons considerably more strongly than heavy water. The primary reactivity control system is made up of the light water chambers. The reactor core's flux shape is likewise controlled by the light water chambers. To affect the neutron flux in the area around a chamber, the water level in each chamber can be changed.

"Adjuster rods" are absorber rods that are typically fully inserted and are used to flatten the flux distribution. When adding light water to the in-core light water chambers is insufficient, they can also offer beneficial reactivity. To aid in xenon override, they can also be eliminated [5].

"Control absorber rods" are often put vertically into the core and placed outside of it. When the light water chambers are insufficient, they might be employed to introduce negative reactivity.

To decrease or boost reactivity, dissolved poison – usually boron or gadolinium – can be added to or taken out of the moderator in the calandria.

Although PHWR control systems are more intricate than LWR control systems, they are also more adaptable, offering a variety of reactivity control choices. Computer control is used in PHWRs. Digital computer systems that are redundant are used to operate and monitor every aspect of nuclear stations. The following tasks are carried out by the digital control system [5]:

- Adaptability to instrumentation failures and continuous safe operation;
- Automated control of normal reactor operation at startup and at any power level;
- Automatic reactor shutdown in the event of a reactor safety issue;
- Maneuverability under both normal and abnormal plant circumstances.

The following describes the five primary control systems. Table 9.2 shows the core CANDU reactor control program.

- **Unit Power Regulator:** To get the electrical power up to the set point power, the unit power regulator modifies the steam flow to the turbine.

TABLE 9.2
Main CANDU Control Program

Program Name	Measured Parameters(s)	Variable(s) Controlled	Variable(s) Manipulated
1. Unit Power Regulating (UPR)	• Electrical output	• Electrical output	• Steam flow
2. Reactor Regulating System (RRS)	• Reactor neutron power • Reactor thermal power	• Neutron flux	• Reactivity mechanisms • Control rod position • Zone water level
3. Heat Transfer Pressure and inventory (HTPand I)	• Reactor Outlet Heater pressure	• Pressurizer pressure • Pressurizer level	• Pressurizer steam bleed and heaters • D_2O feed and bleed
4. Steam Generator Pressure Control (SGPC)	• Steam Generator pressure • Reactor power	• Steam Generator pressure	• Reactor setpoint • Steam flow
5. Steam Generator Level Control (BLC)	• Steam Generator pressure • Reactor power • Feed water flow • Steam flow	• Steam Generator level (inventory)	• Feed water flow

- **Reactor Regulating System:** The reactor power is calculated by the computer using a variety of thermal and neutronic readings. The computer introduces reactivity modifications through the reactivity control devices when the measured power and the power set point diverge. Keep in mind that unlike a PWR, the coolant temperature cannot be directly controlled.
- **Pressure and Inventory Control:** Principal pressure control and principal D_2O coolant inventories are handled by a PHWR pressurizer. Electrical heaters used in PHWR pressurizers can be adjusted to raise or lower pressure. Bleeding steam to a condenser in a PHWR pressurizer can start the pressure reduction process. The pressurizer's D_2O level is monitored. D_2O is added or removed by a feed-and-bleed system when the measured level and the level set point differ.
- **Steam Generator Level Control:** Heat is transferred from the heavy water main coolant inside tubes to secondary light water outside the tubes using U-tube steam generators in PHWRs. The steam generator's internal preheaters bring the feedwater to saturation, after which it boils as it continues to rise. The riser is the term for the heated area.

The heated area releases a combination of around 90% liquid water and 10% steam. Water and steam are separated using steam separators. Less than 0.1% water is present in the steam that enters the turbine. Before entering the riser zone, the separated liquid water first mixes with feedwater in the downcomer region.

PHWRs operate on a three-element feedwater control system. The steam generator level, steam flow rate, and feedwater flow rate are all measured by the three-element controller. The controller uses flow mismatch error and level set point mismatch error to modify the feed flow rate in order to reduce the total mistakes. Similar to U.S. reactors, this method gets over shrink and swell issues.

- **Steam Generator Pressure Control:** Power operation maintains a consistent steam pressure. The pressure of a steam generator can be controlled in two ways. Reactor power is changed to regulate pressure in the "normal" mode. By altering the steam flow to the turbine, pressure is managed in the "alternate" mode.

9.8.1 Reactor Leading Turbine Mode

Reactor power adjustments are the main method of controlling steam pressure during normal operation. The "normal" mode of controlling the secondary side pressure of the steam generator is interrupted if the reactor regulating system is unable to respond to pressure controller demands for any reason, or if a trip, step back, or setback causes a reactor power reduction. In these cases, the reactor setpoint is directly controlled by the corresponding reduction signal. To modify the plant loads, the steam pressure control shifts to the "alternate" mode [11].

The control system adjusts the position of the reactivity control devices and, consequently, the level of the reactor neutron flux if there is a discrepancy between the setpoint and the actual reactor power level. Reaction power changes will also alter the steam generator pressure (in the same direction), and the steam generator pressure control system modifies the steam flow by adjusting the position of the governor valve to maintain the steam generator pressure at its setpoint and the turbine power will fluctuate accordingly. The turbine governor valves are managed by the unit power regulator program while the plant is operating in "normal" mode. This means that the unit power regulator determines the ideal valve setpoint and pulses to that value. When the plant is in the "alternate" mode, the turbine is controlled by the steam generator pressure control system based on the reactor power change rate, steam pressure error, and steam pressure error rate of change. Outside of the control computers is a low steam pressure unloader on the turbine. This overrides directly the turbine governor action, including the steam generator pressure control signal and generates a quick runback of the turbine (Figure 9.7).

9.9 CANDU REACTOR DYNAMICS

PHWR computers allow versatility in maneuvering methods. Operators can opt to operate in a reactor-follow-turbine mode or a turbine-follow-reactor mode. The steam flow to the turbine reacts first to a change in power demand in the reactor-following, or "normal", mode. As a result, the reactor power set point is calculated to vary due to a change in steam pressure. Until the delivered reactor power reaches the power set point, the reactor regulating system then modifies reactivity using the reactivity control mechanisms.

In the "alternate" or turbine-following mode, the reactor power shifts first. The reactor regulating system uses the new power demand as the reactor power set point. Reactivity is altered by activating reactivity control devices until reactor power reaches the predetermined point. Heat transmission in the steam generators is altered by variations in reactor power, which in turn affects steam pressure. After that, the turbine controller will modify the steam flow until the steam pressure reaches its predetermined level again. A block schematic of the primary process and control system is shown in Figure 9.8 [9].

An example of a CANDU reactor modeling and dynamic simulation is presented in this section. For the modeling details, please refer to the references [12, 13].

9.9.1 Modeling Strategy

The creation of a lumped parameter CANDU reactor core dynamics model is presented in the reference [12]. The reactor core is separated into 14 zones due to the configuration of the fuel in the Calandria vessel, the addition of coolant and moderator flows, and the size of the reactor core. A single-point reactor kinetics equation, like the one used for light water reactor modeling, was

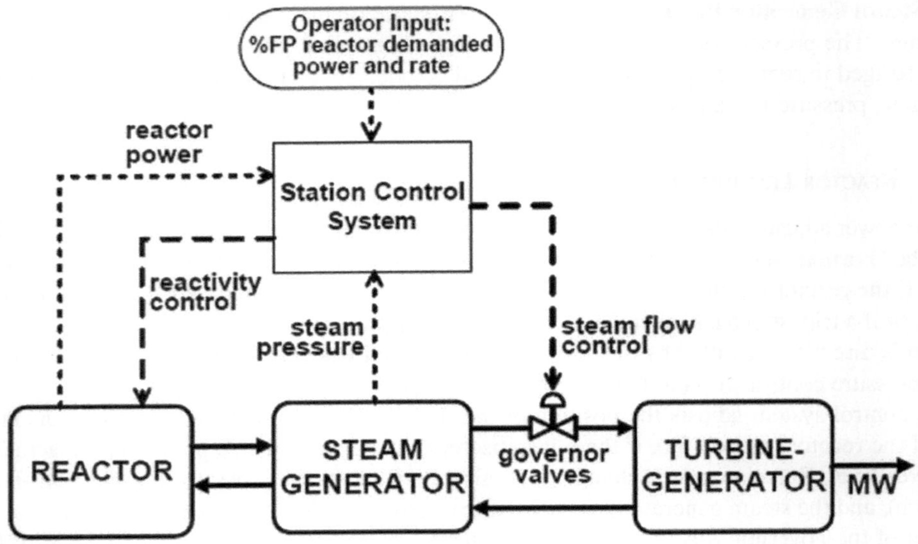

FIGURE 9.7 Simplified reactor-leading-turbine overall unit control system [9].

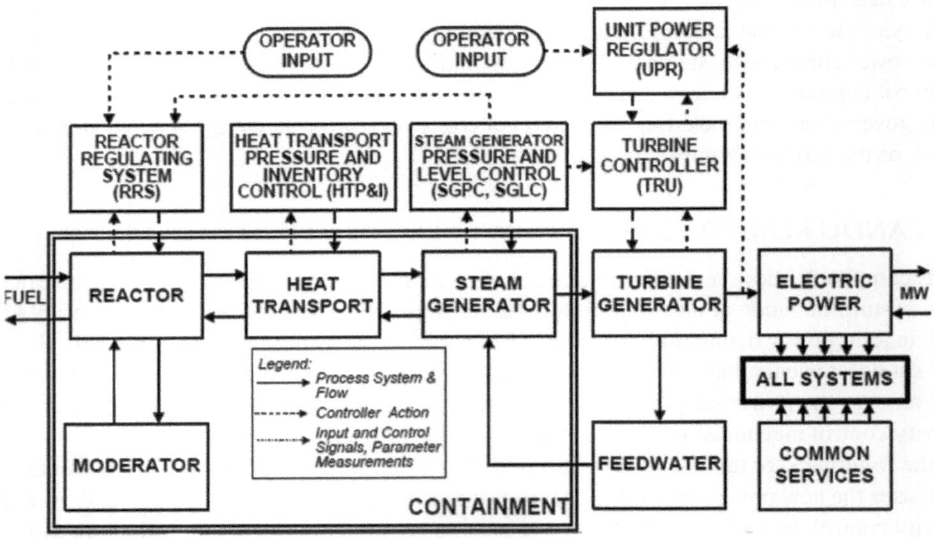

FIGURE 9.8 Block diagram of the main process and control system [9].

extended to create a set of coupled non-linear neutron kinetic equations. By characterizing the impact of neutrons in one zone on the fission processes in a neighboring zone, the coupling between the zones was achieved. Equations for the synthesis and decay of xenon and iodine are also used by the authors to incorporate fission product poisoning.

The multi-zone neutronic behavior and the utilization of diverse reactivity control characteristics of the CANDU reactor were described using non-linear ordinary differential equations (refer to Section 9.8). Adjustable rods, mechanical control absorbers, and liquid zone controllers were used to modulate core reactivity. External to the core, mechanical control rods are also utilized for reactor scram via gravity insertion. The 14 zones that comprise the CANDU reactor core are seen

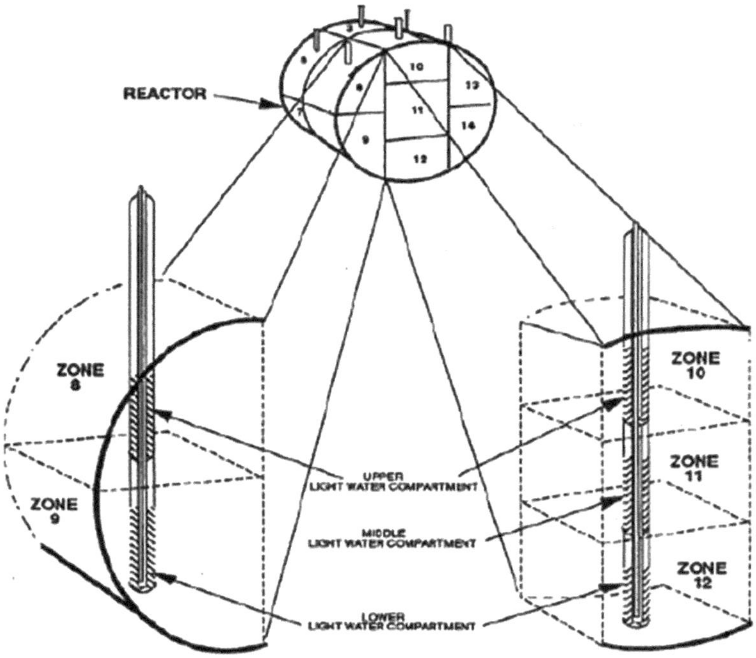

FIGURE 9.9 Division of the CANDU reactor core into 14 zones [12].

in Figure 9.9. Keep in mind that a set of zones shares the same physical dimensions and physics characteristics.

9.9.2 Reactor Power Response to Reactivity Insertion

Reference [12] detailed the reactor power response to a ramp insertion of reactivity in a zone and its impact on other zones. Over a period of thousands, a ramp reactivity was added from zero to 0.00015. The rate of ramp insertion is 1.5×10^{-7}/s, which is sluggish. The variance of fractional power change in zone 1 is seen in Figure 9.10. Fractional power units (FPUs) are used to represent power. Zones 2 through 7 reactor power transients are depicted in Figure 9.11. Figure 9.12 displays power transients in zones 8–14.

To reduce power tilts in the core and keep the overall fractional power equal to 1 (100%), control actions are used to modify the power levels in different zones. Please refer to reference [12] for further details. These simulations demonstrate that the CANDU reactor has special digital control actions to ensure appropriate power distribution in the core and to fulfill set point regulation.

9.10 REACTIVITY CONTROL DEVICES IN CANDU

In CANDU reactors, power is managed by regulating reactivity, much like in any other reactor. Reactivity devices, which can be added to or taken out of the core, are then used to regulate reactivity. Reactivity may be changed by adding or removing reactivity devices from the core, which also changes the absorption rate. This allows for changes in power, reactivity, and reactor shutdown. The Reactor Regulating System (RRS) or one of the two separate shutdown systems (SDS1 and SDS2) in CANDU reactors are in charge of the reactivity devices [14].

Heavy water is a very weak neutron absorber, which is advantageous with regard to fuel utilization. To compensate for reactivity changes in the fuel, the absorption properties of the moderator can be modified by adding a soluble neutron absorber in small concentrations. Gadolinium and

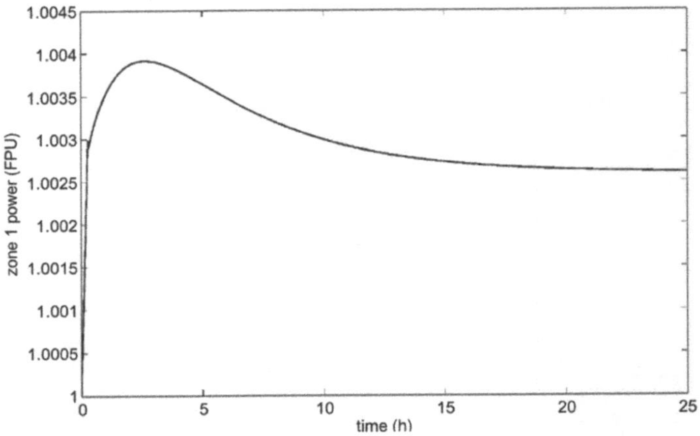

FIGURE 9.10 Power variation in Zone 1 due to ramp reactivity insertion in zone 1: FPU: Full Power Unit [12].

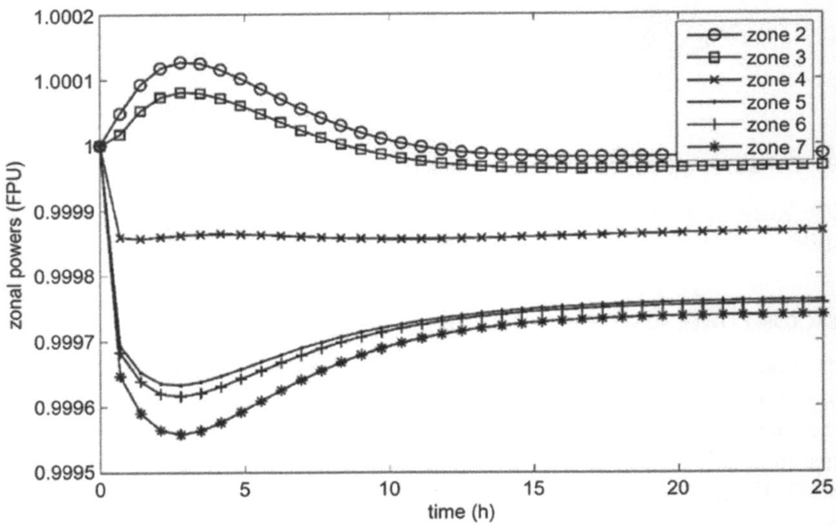

FIGURE 9.11 Power variation in Zones 2 to 7 due to ramp reactivity insertion in zone 1 for Full Power Unit (FPU) [12].

boron both have strong neutron absorbing isotopes, namely Gd-155, Gd-157, and B-10. These can be added through the liquid poison system and removed through the moderator purification system using ion exchangers. Light water is a stronger neutron absorber than heavy water, and therefore, the concentration of light water in the heavy water moderator must be limited to avoid excessive neutron absorption. The normal isotopic concentration of D_2O in the moderator is 99.5%. Light water, however, can be conveniently used as a control device within the reactor, where its volume in vertical tubes can be varied.

Following are the reactivity devices that the RRS uses to manage the typical CANDU-6 design.

- **Liquid Zone Control Compartments (H_2O-Filled):** The six liquid zone control absorbers, which are positioned vertically within the reactor core and are dotted with adjustment rods, shut-off rods, and mechanical control absorbers, are primarily responsible for maintaining the critical reactivity balance. These liquid zone absorbers have a total of 14

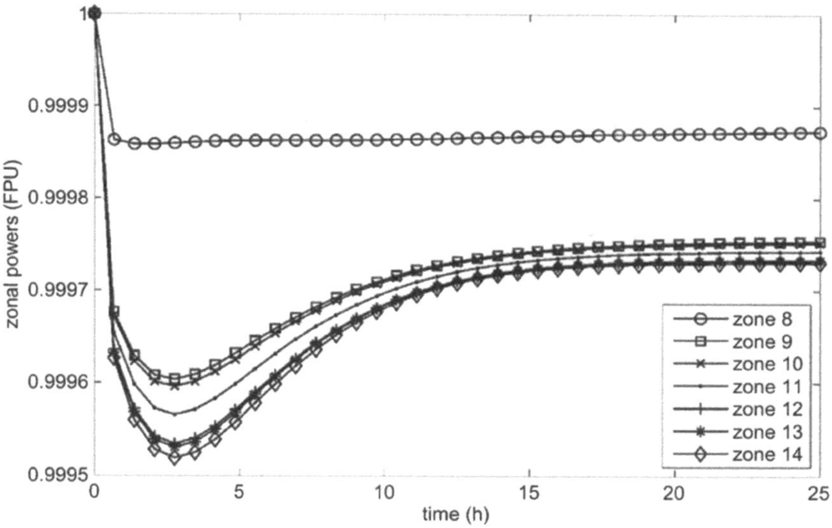

FIGURE 9.12 Power variation in Zones 8 to 14 due to ramp reactivity insertion in zone 1. FPU: Full Power Unit [12].

compartments, each of which contains light water in 2 or 3 distinct portions. The spatial control of neutron flow is made possible by these independent and distinct absorbers. In addition to acting as a control rod when light water is introduced or removed through inlet and exit pipes, where a variation in flow rate would affect the level, light water absorbs neutrons more strongly than heavy water. Helium gas is kept above the light water, and the level of light water is indicated by the pressure differential between the entrance and outlet, which are located at the bottom and top of the water column, respectively. Automated valves that receive signals from the flux detectors via the reactor regulating system's control computers regulate the amount of water in each compartment.

There are two functions here.

- Bulk control – exercised every 0.5 second

The water fills are manipulated all in the same direction

- To keep the reactor critical for steady operation, or
- To provide small positive or negative reactivity to increase or decrease power in a controlled manner
- Spatial control – exercised every 2 seconds

The water fills are manipulated differentially, to shape 3D power distribution towards desired reference shape.

- **Mechanical Rod Adjusters:** Adjuster rods are used in the reactor to optimize the neutron flux profile for reactor power, compensate for fuel burnup, and provide excess reactivity to overcome xenon-135 surges following a power reduction. CANDU reactors are designed to develop full power when these adjuster rods are fully inserted into the reactor core. There are 21 adjuster rods arranged in three rows vertically between the other control devices. The adjuster elements are raised or lowered individually by winching drives similar to those described for the shut-off rods, except that these adjusters do not fall into the reactor in the event of a shutdown signal from the flux detectors because they would normally be fully inserted during normal operation. When the adjusters are fully withdrawn, the reactivity gain is about 16 mk, which is sufficient to overcome minor xenon transients during power maneuvering.

Adjusters are moved in banks; maximum rate of change of reactivity per bank <0.1 mk/s and also help to flatten the power distribution, so that more total power can be produced without exceeding channel-power and bundle-power limits.

- **Mechanical Control Absorbers:** Four mechanical control absorbers, mostly composed of cadmium and enclosed in stainless steel, are included in addition to the liquid zone control absorbers. When a higher reactivity rate or depth is needed than what the liquid zone control absorbers can offer, these absorbers modify the flux level. They can cause a reactor to step back or other rapid reactor power reduction. They can be lowered into the core by releasing a clutch or pushed in or out of the core at varying speeds. The value of total reactivity is about −10 mk. The shut-off rods previously mentioned have a nearly identical design to these solid control absorbers.
- **Moderator "Poison":** Boron or gadolinium can be added as a poison to the moderator, to compensate for excess reactivity
- In the initial core, when all fuel in the core is fresh, and
- During and following reactor shutdown, when the ^{135}Xe concentration has decayed below normal levels.

Gadolinium is utilized after reactor shutdown, while Boron is used in the first core. The benefit of Gd is that its rate of burnout offsets the build-up of xenon. With the exception of the guaranteed shutdown state, typical toxic concentrations are just a few parts per million [15].

In addition, a complete reactor shutdown can be achieved by operating one or both of the following devices in addition to those listed above:

- **SDS-1: Mechanical Rod Shut-Off Absorbers:** A set of 28 vertical shut-off rods is controlled by the signals sent to Shutdown System 1 (SDS1) from the flux detectors or other plant indicators. These shut-off rods are used to turn off the reactor in both routine and emergency situations. They are made of cadmium and stainless steel absorber components. Perforated guide tubes constructed of zirconium alloy that are exposed to the moderator's pressure assist these rods as they are introduced into the core. The rod drive mechanism uses a stainless steel cable coiled on a drum to insert or remove the shut-off rods from the core. An electric motor with an electromagnetic friction clutch powers this drum. The clutch de-energizes in response to a shut-off signal, and the shut-off rods descend into the reactor core within the zirconium alloy guide tubes as a result of gravity. The electric drive motor may drive the shut-off rods into or out of the reactor core while the clutch is engaged.
- **SDS-2: Liquid Poison Injection System:** This system is a safety system rather than a control system, in contrast to the Moderator Poison Addition System. Instead of infusing a regulated amount to maintain reactivity at the necessary level, it injects a massive amount of soluble poison into the moderator to substantially lower reactivity.

Shutdown System 2 (SDS-2), as illustrated diagrammatically in Figure 9.13, is activated based on information received from the three horizontal in-core flux detectors or other plant indicators. The poison that is injected into the core is gadolinium nitrate dissolved in heavy water. This poison is contained in six tanks outside the reactor vault in an accessible area for routine maintenance. Each of the six poison tanks is connected to a nozzle pipe in the Calandria by stainless steel pipes. These

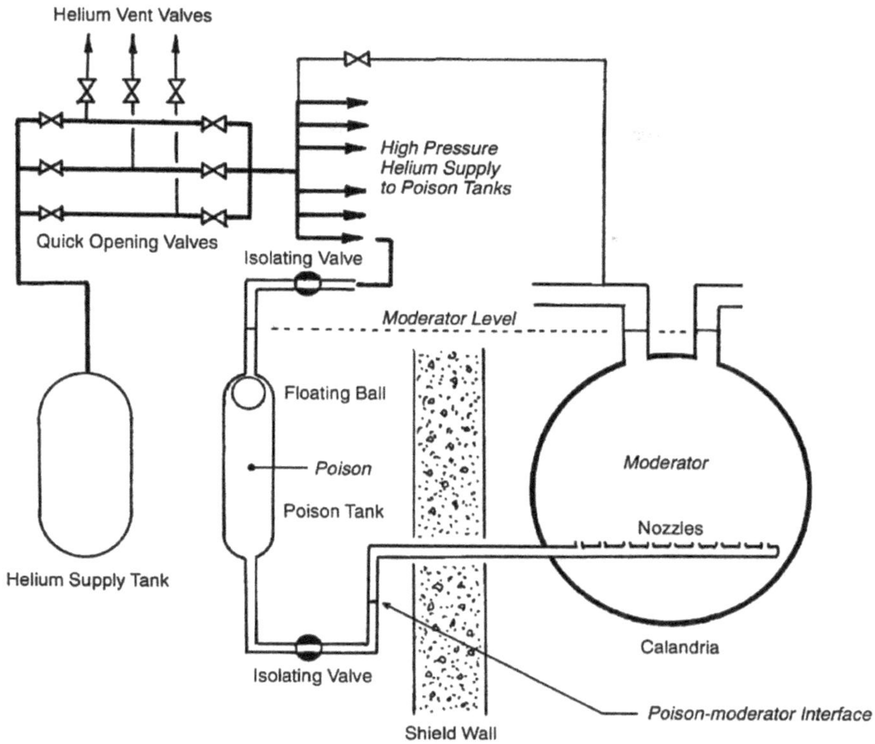

FIGURE 9.13 Shutdown System 2 (SDS2).

nozzle pipes are oriented horizontally in the reactor core, each with several orifice holes situated to provide maximum negative reactivity changes when the poison is injected. The nozzle pipes themselves are made of zirconium alloy.

Helium under pressure injects the gadolinium nitrate toxin into the core. Triple quick-acting valves separate the pressurized helium in this tank from the poison tanks. The poison is rapidly forced into the moderator by the pressure when these valves are opened. In the poison tanks, a ball keeps helium gas from blowing through and makes sure the calandria pressure stays below its intended value.

Generally speaking, SDS-2 can provide at least 50 mk of negative reactivity, however 30 mk is thought to be sufficient to cause a thorough and quick shutdown. In case SDS-1 fails or if the starting event is so severe or rapid that SDS-1 has not stopped the transient, this mechanism, which serves as a backup to SDS-1, should be triggered. It is impossible to recover quickly from such a trip since the moderator is pumped with liquid poison.

Two separate and independent shutdown systems that are set differently and run from various sources were determined to be part of the original CANDU design idea. In order to accomplish this goal, the SDS-1 and SDS-2 systems inject negative reactivity into the reactor in distinct ways, including in different orientations – one horizontally and one vertically.

Neutron flux detectors in the core and ion chambers adjacent to the core, combined with other sensors in the steam system and turbine generator, provide the right signals to run the control systems. Figure 9.14 displays the primary control devices as a component of the reactor assembly.

In general, CANDU 6 reactors can use the control devices listed below. Other CANDU reactors use the same basic concept, except if the reactor core is larger or smaller, there may be a difference in the number of devices.

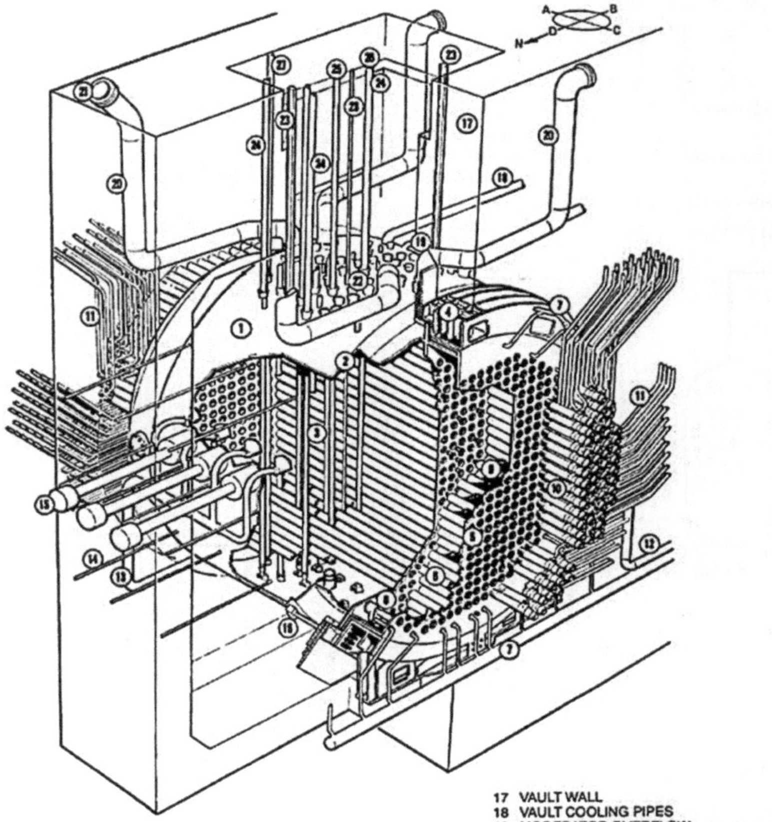

FIGURE 9.14 CANDU-6 reactor assembly.

TABLE 9.3
CANDU Reactivity Device Worth

Function	Device	Total Reactivity Worth (mk)	Maximum Reactivity Rate (mk/s)
Control	14 liquid zone controllers	7	±0.14
Control	21 adjusters	15	±0.10
Control	4 mechanical control absorbers	10	±0.075 (driving) −3.5 (dropping)
Control	Moderator poison	–	−0.01 (extracting)
Safety	28 shut-off units	80	−50
Safety	6 poison-injection nozzles	>300	−50

There are 7 in-core horizontal flux detectors and 26 in-core vertical flux detectors in a CANDU 6 reactor core. SDS-1 receives flux data from 12 detectors that are positioned vertically, while SDS-2 receives flux data from 3 horizontal flux detectors. The solid control rod absorbers and the liquid zone control absorbers get data from the remaining vertical flux detectors. Data from each of these in-core flux detectors may be used to map the reactor core's flux in three dimensions.

Table 9.3 summarizes the reactivity value and reactivity-insertion rates of the several CANDU reactivity devices. When all control devices are driven concurrently, the greatest positive reactivity-insertion rate that can be achieved is around 0.35 mk/s, which is well within the shutdown systems' design capabilities.

REFERENCES

1. E. E. Lewis, *Fundamentals of Nuclear Reactor Physics*, Academic Press, 2008; R. Isermann & P. Balle. Trends in the application of model-based fault detection and diagnosis of technical processes. *Control Engineering Practice*, 5(5), 709–719 (1997). https://doi.org/10.1016/S0967-0661(97)00053-1
2. https://www.iaea.org/ IAEA-TECDOC-1402, 2004; Nuclear Power Plants Information, 2010.
3. A. B. Van Rennes, J. C. Simons Jr., T. S. Gray, Dynamics and Control of Nuclear Power Plants, *Trans. Am. Inst. Electr. Eng.*, Part I: Commun. Electron. 76(3), 279–285 (1957); H. Y. Jeong, S. H. Choi, J. H. Park & , S. K. Kim. Transient identification of nuclear power plants using neural networks. *Annals of Nuclear Energy*. 23(9), 689–697 (1996). https://doi.org/10.1016/0306-4549(96)00004-5; H. Y. Jeong, S. H. Choi, J. H. Park & S. K. Kim, A study of pattern recognition for transient identification using neural networks. *Nuclear Technology*. 115(2), 179–186 (1996). https://doi.org/10.13182/NT96-A35371
4. J. Lewins, *Nuclear Reactor Kinetics and Control*, Pergamon Press, 1978.
5. T. W. Kerlin, B. R. Upadhyaya, *Dynamics and Control of Nuclear Reactors*, Academic Press, 2019.
6. General Electric Company, *BWR/6 Nuclear Steam Supply System: General Description*. GE Nuclear Energy Division, 1975.
7. R. T. Lahey Jr., F. J. Moody, *The Thermal-Hydraulics of a Boiling Water Nuclear Reactor*, Subsequent edition, Amer Nuclear Society, 1993.
8. B. Zohuri, S. K. Mousavi Balgehshiri, A. Zamani Paydar, Shielding of Ionizing Radiation in Nuclear Power Reactor Environment (A Technical Memorandum), *J. Mater. Sci. Manuf. Res., Sci. Res. Community* 3, no. 2, 1–7 (2022).
9. A. Zamani Paydar, Seyed Kamal Mousavi Balgehshiri, and Bahman Zohuri, *Advanced Reactor Concepts (ARC): A New Nuclear Power Plant Perspective for Energy*, 1st edition, Elsevier, 2023.
10. Dr. D. A. Meneley, Senior Advisor, *Dynamic Characteristics of CANDU Reactors*, Atomic Energy of Canada Ltd. and Adjunct Professor, Department of Engineering Physics McMaster University, 2000.
11. G. Bereznai & University of Ontario Institute of Technology. (2005). NUCLEAR POWER PLANT SYSTEMS and OPERATION. https://canteach.candu.org/Content%20Library/20071000.pdf?ID=1547&RootFolder=*
12. H. Javidnia, J. Jiang, M. Borairi, Modeling and Simulation of a CANDU Reactor for Control System Design and Analysis, *Nucl. Technol.* 165, no. 2, 174–189, (2009).
13. A. P. Tiwari, *Modeling and Control of a Large Pressurized Heavy Water Reactor*, PhD Dissertation Indian Institute of Technology, Bombay, 1999.
14. E. Nichita, *Reactor Dynamics*, UOIT and Benjamin Rouben, 12 & 1 Consulting, Adjunct Professor, McMaster & UOIT, 2014.
15. G. Harvel, CANDU Reactivity and Power Control, Associate Professor/Faculty of Energy Systems and Nuclear Science, UOIT, 2009.

Index

Q

R

S

T

U

V

Z

FSC
www.fsc.org
MIX
Papier | Fördert
gute Waldnutzung
FSC® C083411

Kolibri 360 GmbH
Ferdinand-Jühlke-Straße 7
99095 Erfurt, Deutschland
produktsicherheit@kolibri360.de